AF540000

Diagnostic Veterinary Microbiology

NIPA® GENX ELECTRONIC RESOURCES & SOLUTIONS P. LTD.
New Delhi-110 034

About the Authors

Dr. B.Puvarajan, Professor, TANUVAS-Regional Research and Educational Centre, Pudukottai Tamilnadu majoring Veterinary Microbiology specialized in Molecular Diagnosis of Animal Diseases has rich experience in Animal Virology, Bacteriology and Immunology. He has published national (145 nos) and International (48nos) research papers and fourteen books with ISBN(International and National) and 42 book chapters in leading Springers and Elsevier publications, 325 number of Popular articles in various regional dailies and newspapers and scientific forums conducted for farmers and farming community. He has guided 06 no. of M.V.Sc students and two no.of Ph.D students in Veterinary Science and Microbiology. The core area of research in molecular bacteriology, virology, gene sequencing, mapping of diseases, expertise in disease outbreak curtailment. Handled 5 major research projects under NADP, NABARD, State funded TAHDCO, Mahalir Thittam, GoI Schemes on Conservation of Ramnad White sheep and Self Financing Schemes on Desi chicken farming. Coordinated various World Bank Projects, National Livestock Mission etc., He has published booklets (14 numbers) on Goat farming and Healthy care for livestock and Ethnoveterinary practices for primary health care of livestock and poultry and pamplets published (115 nos- For the rural farmers to create disease awareness esp. on economically important diseases affecting livestock and poultry). **He has organized 09** no.of scientific workshops and 12 no. of Seminars and 04 no. of National and International webinars **He has received 20 no.of Awards and to mention a few are** Padma Sri B.V.Rao Award for Best Research on Poultry diseases for the year 2011, Appreciation from Hon.Vice Chancellor for conduct of Anti-Rabies Camp at Namakkal district Union 2011- One no., Best Thesis Award (Dr Nanjappa Gownder Award) on Poultry Diseases 2013, Scientific Oral Presentation Awards-19 in various Scientific conferences of national and International conducted all over India, Outstanding Award for Presentation of One Heath Concept at National Conference conducted at Kerala Veterinary University, Pookode, Kerala in 2015 November., Best Scientific Reviewer of International Journal of Livestock Research 2017-1 no., Best Popular article Award for Regional language (Tamil) by Agricultural Scientific society, New Delhi for the years continuously-2015, 2016, 2017 2018, 2019, 2020, 2021, 2022 Best Article award for farmers by Kalnadai Arviayal Tamil Iyakkam (KATI-2019) in the International Tamil Conference conducted by Veterinary Scientists at Chennai on 19.02.2019-01Most innovative Technology Award in Teaching by Pearl Foundation in 22.03.2020(March 2020)-1no.Best Article award for farmers by Kalnadai Arviayal Tamil Iyakkam (KATI-2021), Best Popular article Award -Agricultural Scientific society, New Delhi-2023, Best officer award 2023 received from Honorable District Collector, Pudukkottai and in 2025 received SPCA Award for prevention of rural and urbam rabies in Thanjavur District Honorable District Collector, Thanjavur for past 10 years.He has organized On campus and Off campus training for farmers on dairy farming, sheep and goat farming, disease prevention strategies for livelihood, piggery, dairy products and importance of clean milk production, Meat and meat hygiene , Backyard poultry and commercial poultry farming and Rabbit farming for farmers of Tamilnadu

Dr. J. Vijay Anand is an Assistant Professor in the Department of Animal Biotechnology at Madras Veterinary College, Chennai. With specialized expertise in animal biotechnology, his primary research encompasses mammary gland biology, proteomics, mammary epithelial cell culture, and reproductive biotechnology.Dr. Anand's contributions include pioneering work on buffalo mammary epithelial cells, chitinase-like proteins, and innovative ethnoveterinary approaches. His research articles have appeared in major journals, including*FASEB* Journal, Scientific Reports, Apoptosis, and Cell Biochemistry and Biophysics.He is currently involved in projects funded by ICAR-NASF and DBT, focusing on nano-micro matrices for the delivery of bioactives and phytosome-based alternatives to antibiotics in poultry.

Diagnostic Veterinary Microbiology An Update

B. Puvarajan, Ph.D.PGDEVP.,PGDAW
Professor and Head
TANUVAS Regional Research and Educational Centre
Pudukottai-622004 Tamilnadu

J. Vijay Anand
Assistant Professor
Department of Animal Biotechnology
Madras Veterinary College
Chennai-600007, Tamil Nadu

NIPA® GENX ELECTRONIC RESOURCES & SOLUTIONS P. LTD.
New Delhi-110 034

Disclaimer: The views expressed in the chapters including the contents are sole responsibility of the respective authors. The editors bear no responsibility with regard to source and authenticity of the contents.

NIPA® GENX ELECTRONIC
RESOURCES & SOLUTIONS P. LTD.

101,103, Vikas Surya Plaza, CU Block
L.S.C.Market, Pitam Pura, New Delhi-110 034
Ph : +91 11 4386 0225, 9717133558, 9540816132
E-mail: newindiapublishingagency@gmail.com
Website: www.niparesources.com

Print ISBN: 978-81-19235-75-9

ebook ISBN: 978-81-19235-77-3

Composed and Designed by NIPA®.

Preface

The book titled "**Diagnostic Veterinary Microbiology – An update**" has been completed and edited especially for the veterinary undergraduates, JRF, Veterinary Assistant Surgeon and aspirants for public service commission examinations.

The emergence and re- emergence of zoonotic diseases has redefined the role of veterinary researchers in infectious diseases. The animal models of diseases are of paramount importance to understand the origin, evolution, epidemiology and underlying molecular mechanisms involved in various diseases of animals and humans. No break through in science is possible in watertight close compartment. Therefore, in present scenario, climate and environmental changes share in emergence of various infectious diseases which lead to early diagnosis and an one health concept emphasizing a multi-disciplinary approach to devise efficacious diagnostics, therapeutics and vaccines. The use of Artificial intelligence and Bioinformatics for designing POCT and synthetic vaccines and mathematical model for disease forecasting are classical examples.

This book not only contains the most commonly address problems of paraclinical subject, the present book entitled " Diagnostic Veterinary Microbiology-An Update" would be helpful to scholars to design their research problems and would motivate them to cross – disciplinary boundaries as per the nature of their research problem. The book will also appraise field veterinarians of recent updates.

We express our sincere thanks with gratitude to our beloved teachers, friends and my dearest colleagues for their timely help, inspiration and guidance.

We place on record our sincere gratitude to the Prof (Dr).T.Sivakumar,Dean and Honorable Vice-Chancellor TANUVAS Chennai and Hon.Registrar TANUVAS Chennai for their constant engorgement in completing this book and kind support at every moment of this academic assignment We sincerely thank Prof. (Dr.) Murugan Professor Poultry Science VCRI Tirunelveli Tamilnadu for his continuous encouragement in preparation of this book. We are grateful to all the eminent resource persons from different institutions of India and abroad who sent their articles and enriched the content of this book.

Though the space and time do not allow us to name individually, we place on record the support provided by faculty and staff member of Department of Veterinary Microbiology, Veterinary College and Research Institute Orathanadu Tamilnadu.

This book is intended to provide an update on recent advances in the field of diagnostics to benefit the scholars, teachers and scientists working in the area of molecular biology. The book is a compilation invited papers on various, aspects of advanced diagnostic techniques in veterinary science. The contributors have been chosen based on their expertise in the field. Nevertheless, sincere effort has been made to include major topics of the recent concern. This book would serve as a reference for scholars, teachers and researchers of the veterinary science to provide a glimpse for the upcoming veterinary diagnostics. We are hopeful that the various chapters covered in this book will be immensely beneficial to the scientific community in general and students in particular.

Authors

Contents

1

Orientation to Clinical Microbiological Laboratory

Location and Layout

- As the laboratory will be handling materials hazardous to human and animal health, it must be isolated as much as possible from livestock yards, poultry farms and quarantine stations and should not function in or near establishments handling healthy animals, food products for human or animal consumption or near vaccine production laboratories.
- Animal houses, garages, workshops, stores, water tanks, residential accommodation, administrative blocks have to be constructed as per the need of ideal microbiological laboratory. Expansion possibilities should also be kept in mind. The building can be arranged in 5 main groups:
    - Administrative / laboratory rooms
    - Decontamination and disposal area
    - Animal house facilities
    - Workshops / stores/garages and
    - Residential accommodation for the staff, hospital, schools, playgrounds, parks, shops etc.
- The laboratory building may be rectangular in blocks, arranged in L, T or U shapes and may be of one or several stories. The temperature, wind velocity, rainfall, temperature fluctuations etc. should also be monitored in the area.

Work flow

- Efficient laboratories are designed as the specimens flow in one direction. Traditional microbiologic testing, follows a pathway of specimen receipt and plating, incubation and isolation of microorganisms, identification of microorganisms, antimicrobial susceptibility testing and result reporting. It is to minimizing the risk of amplicon carryover and specimen contamination.

Work Areas

- Work areas in laboratories are constructed on the basis of U-shaped modules or linear benches. Modules typically measure 10 x 10 ft. and can accommodate two or three persons. Advantages in the modular approach include minimized foot traffic in work areas, generous counter top and storage space, corner space, an increased sense of privacy for workers, and use of less floor space for aisles and corridors. Advantages in the linear bench approach include ease of cleaning, ease of moving about the laboratory, subjective sense of fewer clusters, easier location of large pieces of equipment such as incubators and refrigerators.
- The space surrounding work benches should be sufficient to accommodate ample waste containers both for paper and biohazardous waste. The laboratory should contain space for storing completed cultures, stock cultures, reference books and teaching materials. It is most efficient for completed cultures and reference materials to be located closer to the work benches.
- Laboratory case work can be built in (custom) or modular. The type of case work selected for a laboratory is determined by the size of the laboratory. Larger laboratory always prefer commercial modular furniture. It is more economical. Smaller laboratories, built in case work preferred to suit individuals convenient, it is less expensive, strong and durable.

Walls, Floors, Ceiling and Furniture

- Walls, ceilings and floors should be smooth, easily cleanable, impermeable to liquids and resistant to chemicals and disinfectants and should not be slippery. Acoustic tile ceiling are permissible for Bio safety level 1 and 2 lab. Sheetrock walls should be painted units epoxy so that they may be easily cleaned in the event of spill. Daily mopping with suitable detergents is mandatory. Flexible furniture systems that can be easily disassembled and moved are preferred. Bench tops should be sealed to the walls. Bench tops should be impervious to water and resistant to acids, alkalis, organic solvents and moderate heat. The tops made up of quarried stone, particle board or wood cores with acid-resistant plastic laminate surfaces and stainless steel are preferred.

Instrumentation

- CML should have ample space for large floor model incubators, refrigerators and centrifuges. Adequate electrical power supply necessary gas and water supplies to be incorporated into the area. As a general rule, a refrigerator and a centrifuge should be located in specimen processing area.
- Instruments commonly found in the microbiology laboratory includes
  - Bacterial detection devices
  - Bacterial identification and susceptibility testing devices
  - Gas liquid chromatography
  - Microscopes (light, fluorescent, phase dark field)
  - Centrifuges
  - Dry heat oven
  - Incubators (radiant heat, Co2)
  - Water baths
  - Heat baths
  - Vortex mixer
  - Anaerobe chamber
  - Refrigerators, freezers, Ultra-low freezers
  - Biological safety cabinet

Culture Room

- A microbiology laboratory should have essentially separate cubicles for bacteriology culture work and virology work because of strict aseptic precautions required for both kinds of works.
- Cubicle should contain all the required equipment so that frequent exit and entry can be avoided. Culture rooms must possess an anteroom for preparatory work before entering into a thoroughly sterilized cubicle. Anteroom equipped with isolation booths may be of an average size of 2.5m x 2.0m, with installations consisting of regular and ultraviolet lights, electrical points, gas, a bench for working in a sitting position, and air conditioning and heating, if required. Sterility in anteroom should also be maintained.

- The cabinet for instruments suitable for storage of sterile items must preferably be placed in the anteroom. Only essential items required at the time of culture work must be kept within the culture room. Protrusions, crevices, placement of machines etc, must be avoided to prevent accumulation of dust. Entry of visitors must be strictly prohibited, and cubicles may be fitted with glass partitions at the height of about 3 feet from the floor.

Electrical Power Supply

- Electrical outlets should be liberal in number and in excess power supply is required. Critical equipments should be wired into a central alarm system so that the appropriate person can be notified on any power failure.

IT and Telecommunication

- The modern clinical laboratory is highly dependent upon IT and telecommunications systems. Telecommunications systems need to support telefax units, sophisticated telephone systems, videoconferencing. The computer workstation with a variety of applications should be provided in offices and laboratories for immediate benefit to the laboratory staff.

Office and Administrative Support

- Office should be located close to but not within the laboratory. Office should be equipped with IT and telecommunications infrastructure, and other features that are necessary to make them efficient. As a thumb rule, individual offices should be no smaller than 100sqft.

Laboratory Storage

- Storage space should be adequate but not excessive. Insufficient storage space makes for a cluttered laboratory; unused storage space makes for a dusty one. Short term storage capacity should meet the daily needs. Long term storage of supplies should be in the main laboratory storage room. Storage space should be designed as, it can be cleaned easily.

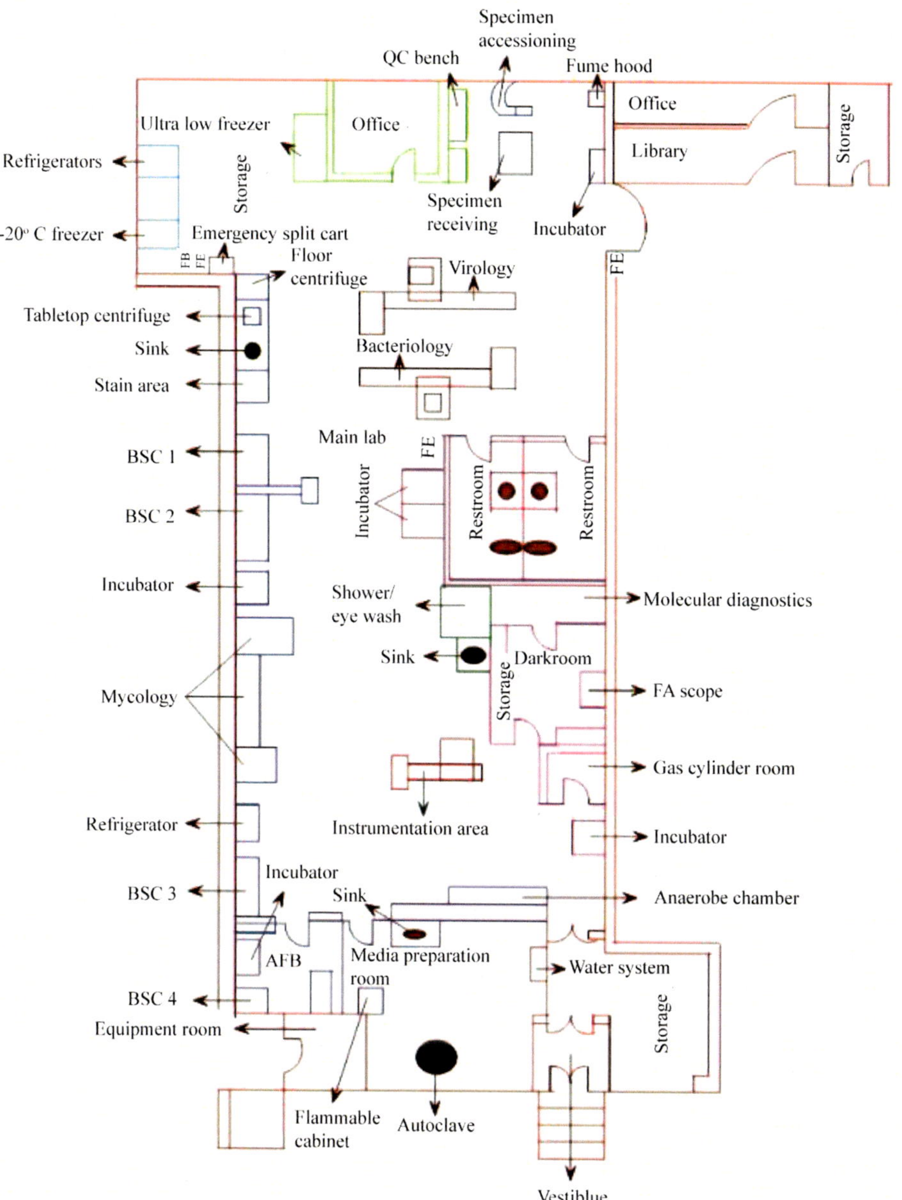

BSC - Bio-safety cabinet

2

Collection, Preservation and Dispatch of Clinical Specimens for Laboratory Diagnosis

On the face of an outbreak of a disease, collection of right clinical specimens in proper preservative, their transportation to the laboratory under ideal conditions and quick processing of specimens in the laboratory to identify the causative agent are crucial to successful control of the disease. This attains more importance when a highly contagious disease appears.

Collection of specimens from a virus-infected host is the first step for demonstration or isolation of a virus. Specimens for virus isolation should be collected at the time when the maximum amount of virus can be expected to be present. i.e. collect the specimens as early as possible after the onset of clinical symptoms. In many viral diseases maximum viral production occurs before the patient becomes ill so that it is ideal to collect the specimen at an even early time.

The nature of the specimen taken in any particular instance is determined by the patient's clinical syndrome and previous history. They are best considered under three main headings.

1. Specimen for virus isolation/antigen detection/ nucleic acid detection
2. Specimen for serological investigation
3. Specimen for direct examination

Most Clinical Materials Fall into the First Two Categories

Specimens for virus isolation are usually taken from affected area. Thus throat and nasal swabs or throat washings are taken for infections of the upper respiratory tract and sputum in lower respiratory tract infections. Scrapings are taken from the lesions on the skin and mucous membrane and if vesicles are present there collect the vesicle fluid also. When there is involvement of CNS collect cerebrospinal fluid. Stools/faecal samples or rectal swabs are also to be collected in such cases as most of the viruses causing CNS infection have their normal habitat in the alimentary tract. Stools and rectal swabs are also required in cases of respiratory disease agents such as Adenovirus and reo virus. In the case of generalized infections whole blood, faecal samples and urogenital swabs may be collected. Following autopsy, tissues from various organs including lymph nodes and liver may be collected.

During collection of materials, the following points should be remembered

1. Always use freshly sterilized equipments and materials to collect the specimens
2. Collect adequate amount of specimen to be tested and collect several specimens from the same premises if possible, especially in case of poultry and other small animals of low economic value
3. Select one or two animals of recent death and several animals in various states of illness if detailed examination is preferred.
4. Specimens from post-mortem animal should be collected as early as possible after death and opening of the body.

Methods of Collection

1. **Nasal swabs:** The posterior nasal passages are most likely to contain viruses and hence posterior nasal swabs should be collected. The swab should be moistened with transport medium immediately before use. Then it is introduced into the nostril and pushed along the floor in the nasal passages as far as possible. It is then gently rotated until it becomes generally coated with nasal secretion. The swab is then removed and it is placed into the bottle containing transport medium the extra length of stick is broken off and the vial capped.
2. **Throat swab:** The tongue should be pressed aside, the swab moistened and the posterior pharyngeal wall is rubbed around. The procedure is repeated on the other side, the swab is placed in a bottle, extra stick is broken off and the vial is capped.
3. **Rectal/cloacal swab:** The dry swab is inserted slowly through the anal orifice/cloaca into the rectum and rotated gently pressing against the rectal wall. Then it is withdrawn and placed in a bottle containing transport medium.
4. **Faecal samples:** Should be collected directly from the rectum of the animal and placed in a vial
5. **Cerebrospinal fluid:** Collect the fluid aseptically with the help of an 18 gauge needle and syringe and transfer to sterile tube.

6. **Oesopharyngeal fluid:** Some viruses localize in the pharyngeal region and it is advantageous to collect the pharyngeal scrapings. This is done using a probang cup. The probang cup is inserted into the mouth upto the pharyngeal region and moved up and down so as to collect scrapings of the pharyngeal wall and fluid. It should be withdrawn slowly and placed in the transport medium.
7. **Vesicular fluid:** With the help of a syringe fitted with a fine needle, the vesicular fluid is collected and placed in small volume (1ml) of transport medium.
8. **Lesion scrapings:** A sterile scalpel and a forceps are used to collect the skin scrapings. The epithelial covering of the vesicle is collected with the help of the forceps and placed in the transport medium
9. **Blood:** Blood is collected at the height of temp with a syringe containing heparin at the rate of 20 IU/ml of blood
10. **Post-mortem materials:** Specimens from dead animals or those that are necropsy should be collected within hours. Otherwise the virus may be destroyed by drying or bacterial decomposition. For viral isolation they are to be collected in transport medium. For histopathology tissues are to be immersed in 10% formal saline. The fixative should be 20 times the volume of the tissue. This can also be used for detection of the viral antigen.
11. **Serum samples for serological studies:** Paired serum samples are preferred. One collected during the acute phase of illness which will provide a baseline to compare antibody level on the second sample and the second collected during the convalescent phase- approximately 11-21 days later.

Draw blood aseptically from the jugular vein of large animals, wing vein or heart puncture of chicken or saphenous vein near the hock joint of ducks. Do not freeze. Allow it to clot at an angle and the clot is released from the wall by jarring it against the palm of the hand or with the help of a splinter. The tube is then incubated at 37℃ for 30 minutes and then transferred to 4℃ (refrigerator). The clot retraction usually expresses sufficient serum in 12 –18 hours. If not, centrifuge at 1000g for 10 minutes to separate the serum. Transfer the serum to sterile vials.

Preservation and Transport

In many instances there may be a time gap between collection of specimens and testing. During this period the specimen should be maintained in an environment favourable for the preservation of viral activities

The infectivity of the virus is unstable when exposed to extremes of pH, temperature above 37°C and desiccation. Therefore, the specimen must be immediately placed in suitable transport medium. When using the transport medium the following must be taken into account.

1. Fluid must be sufficient to withstand any likely evaporation but should not be too much to produce unnecessary dilution of the specimen.
2. It must be buffered to avoid extremes of pH, often found in clinical materials. This can be achieved by means of a phosphate bicarbonate buffer system.
3. Better to add some stabilizing agents like bovine serum albumin as it will keep the infectivity of the virus and also contribute to the buffering capacity of transport medium.
4. The transport medium for isolation should contain high concentration of antibiotics (Penicillin 1000 IU/ml, Streptomycin 1000µg/ml and Nystatin 250 IU/ml) as most of the clinical specimens are contaminated with bacteria and sometimes fungi.

Commonly used transport media are Hank's balanced salt solution, Phosphate buffered saline, viral transport media or even normal saline.

Specimens must be kept cold unless they can reach the laboratory within an hour of collection. In general they must be stored frozen (-20℃) but it is going to be tested during the first 24 hours; it can be stored at 4℃.

Blood with anticoagulant should not be frozen but shipped on ice. Similarly specimens for respiratory syncytial virus and urine for cytomegalo virus inclusion bodies also not be frozen but left at 4°C and processed without delay

For serum do not add any preservatives such as phenol or merthiolate if it is intended for serum neutralization test. However, preservatives can be added if it is for precipitation tests. Addition of 100 IU of penicillin and 100µg of streptomycin/ml of serum will prevent bacterial growth at the same time will not interfere with serological tests. It is to be inactivated by heating at 56℃ for 30 minutes to destroy any non-specific substances if present. Then they are to be stored at -20℃ after transferring to a screw capped vials.

Dispatch of Specimens

Specimens can be transported to the laboratory by the use of insulated containers. Plastic picnic bags can be used to hold dry ice but he vials must be sealed to prevent the entry of carbon dioxide, which is deleterious to many viruses. When mouthed containers are suitable are for holding specimens but stopper bottles are unsatisfactory and cotton plugged tubes are wholly unsatisfactory. All containers must be labelled.

The label should contain specimen number, date of collection and nature of specimen. The specimen must be accompanied by detailed history sheet. A history sheet should contain specimen number, date of collection, nature of specimen, particulars of the animals such as species, sex, age, details of preservatives added if any, particulars of illness such as temperature, duration, symptoms, lesions, name of the owner of the animal, general condition of the herd or flock, treatment if any, possibilities of contact with neighbouring farm animals and brief note on the post mortem findings.

It is also advantageous if the inner container is carefully wrapped in excess of absorbent paper or other materials to act as shock absorbent and to absorb any fluid that may be spilled. Where dry ice or vacuum flasks are neither suitable nor available, the materials can be transported in 50% glycerol saline, pH 7.4. It should be remembered that glycerol may be detrimental to viruses such as Rinderpest

The specimens must be neatly secured (without any leak or break) and packed before dispatch. Packages containing glassware must carry a caution on it namely "Glass handle with care". The specimen should reach the laboratory promptly. Specimens reaching the laboratory on holidays should be avoided. If unforeseen delay occurs in dispatching the specimens, the same must be preserved under refrigerated conditions.

3

Isolation and Identification of Bacteria by Gram's Staining and Cultural/ Biochemical Characteristics-I

The chief functions of Clinical Microbiological laboratory are to examine and culture specimens for microorganisms, to make accurate species identification of important isolates and to perform antibiotic susceptibility tests when indicated. These tasks will assist physicians in the diagnosis and treatment of infectious diseases.

Once the specimen is received in the laboratory, the specimens are examined visually depending on the physicians order and the nature of the specimen. Wet mounts and smears may be prepared and stained for microscopic examination. Observations may or may not be immediately reported to the physicians depending on the definitiveness of the results. Timely information may often be used to establish a presumptive diagnosis and institute a specific course of therapy. Specimens that require definitive identification of potentially pathogenic microbes are processed further. All agar plates are streaked for colony isolation; then plates and broths. All cultures are placed in an incubator with appropriate temperature and environmental conditions to maximize the growth and replication of microbes. Often a presumptive microbial identification can be made. A final report should be delayed while subcultures and additional test procedures are performed to identify the organisms definitively.

Preliminary Identification of Unknown Bacteria

A number of techniques may be used in the direct microscopic examination of clinical specimens to demonstrate the presence of microbes

The examination of wet mounts of unstained materials by phase contrast or dark field microscopy is useful for demonstrating motility, spirochaetes and endospores. Giemsa, Wright or acridine orange stains may be helpful in observing bacterial forms that stain poorly or that have little contrast from back ground material.

Gram stained reaction, when observed in conjunction with the types (cocci and bacilli) and arrangements of bacterial cells can be used to make presumptive identification.

Gram positive cocci in clusters suggest staphylococci; in chains, they suggest *Streptococci,* Gram positive, lancet – shaped diplococci particularly when seen in smears made from sputum samples, are characteristics of *Streptococci pneumoniae*; gram negative, kidney – shaped diplococci are characteristics of *Neisseria* species. Large gram positive bacilli suggest *Bacillus* or *Clostridium* species; small gram positive Bacilli suggest *Listeria* species or one of the coryneforms (diphtheriaoids) if "Chinese – letter" arrangements are observed. Curved, gram negative rods in diarrhoeal specimens suggest Vibrio species or if corkscrew forms are also seen *Camphylobacter* species.

Acid fast Staining Procedure used to Demonstrate *Mycobacterium* species

Although difficult to describe specifically, odours produced by the action of certain bacteria in plating media and in liquid media can be very helpful in the tentative identification of the microorganisms involved. Examples of micro organisms exhibiting distinctive odours include *Pseudomonas* species; grape juice; *Proteus* species, burned chocolate; *Alcaligenes faecalis*; freshly cut apples; *Corynebacterium* species; fruity; *Nocardia* and *Streptomyces* species; musty basement; *Clostridium* species; faecal, putrid; *Pasteurella multocida*; pungent (indole) like.

Bacterial species identification based on metabolic characteristics and selection of differential characteristics.

Observing colonial characteristics and gram – stained morphology, can make preliminary bacterial identification. However, the final characterization of an unknown bacterial isolate to identify it to the genus and species levels is usually accomplished by testing for certain enzyme systems that are unique to each species and serve as identification markers. In the laboratory, these enzyme systems are detected by inoculating a small portion of well-isolated bacterial colony into a series of cultures media containing specific substrates and chemical indicators that detect pH changes, or by the presence of specific by products. The clinical microbiologist must select appropriate sets or differential characteristics that will permit the identification of each group of bacteria.

Most tests used to assess the biochemical or metabolic activity of bacteria, by which, final species identification can be made.

For example, the lactose – utilizing properties of gram-negative bacilli can be directly evaluated from MacConkey agar by observing the red pigmentation of the colonies. H_2S production may be detected on Hektoen and XLD agars by observing colonies with black centers. The decarboxylation of lysine can also be suspected when observing colonies growing on XLD agar. A red halo around the colony, indicating an alkaline pH shift, indicates the decorboxylation of lysine.

Catalase test is frequently used to differentiate *Staphylococcus* (positive) from *Streptococcus* (negative). Hydrolysis of arginine, esculin and hippurate help in the identification of *Streptococci*. The bile solubility test is used to distinguish *Streptococcus pneumoniae.*

Oxidase test is useful for the initial categorization of many bacterial species that have distinctive colonial morphology. Oxidase positive colonies can be discounted as belonging to the *Enterobacteriaceae* and often help to identify certain bacterial species that produce cytochrome oxidase, such as *Aeromonas* species, *Plesiomonas* species, *Pseudomonas* species and *Pasteurella* speices

Indole production test is used to identify the *Escherichia coli, Pasteurella multocida* by growing the organism in tryptone broth.

The test used to measure the metabolic characteristics such as carbohydrate utilization, ONPG (O – Nitrophenyl – B-D-galactopyranoside) activity, indole production, methyl red test, production of acetyl methyl carbinol (acetoin), citrate utilization, urease production, decarboxylation of lysine, ornithine and arginine, phenylalanine deaminase production, hydrogen sulfide production and motility help to identify the organisms belong to the family *Enterobacteriaceae.*

In the clinical laboratory, commonly encountered Haemophilusspecies are identified on the basis of their hemolytic reactions on horse blood agar, their growth requirements for X and V factors and their satellitic property with V factor producing *Staphylococcus aureus*.

The catalase and oxidase test, growth on MacConkey medium, Haemolysis on blood agar, urease test and bipolar staining technique are helpful in identifying the organisms belong to the genus *Pasteurella.*

Brucella species are identified on the basis of CO_2 requirements for growth, biochemical tests such as catalase and oxidase tests, H_2S and urease production, growth in the presence of thionin and basic fuchsin dyes, growth on MacConkey medium and agglutination in antisera.

The characteristics such as relation to oxygen, colonial characteristics, haemolysis on blood agar, presence of spores, motility, production of lecithinase, lipase and catalase, growth in thioglycolate broth, reaction on milk medium, production of indole, hydrolysis of starch, esculin and gelatin and reduction of nitrate are used to identify the anaerobic organisms such as *Clostridium, Fusobacterium* and *Bacteriodes.*

Staining Methods-I

A. Loeffler's alkaline methylene blue

This stain is particularly useful to show McFadyean's reaction with Bacillus anthracis. This is also used to demonstrate metachromatic granules in case of Corynebacterium and as a counter stain with Ziehl-Neelsen's stain for acid-fast organisms.

Composition

Methylene blue saturated alcoholic soln.	30 ml
Pot. Hydroxide (1% aqueous soln.)	1 ml
Distilled water	99 ml

Add pot. Hydroxide soln. In water and combine with methylene blue solution before filtration

Materials

Staph. aureus and *E.coli* culture plates and staining solution.

Procedure

1. Prepare bacterial smears and fix on the flame
2. Pour stain on the smears and allow to act for 4 to 5 minutes.
3. Wash with water, blot and dry before examination.

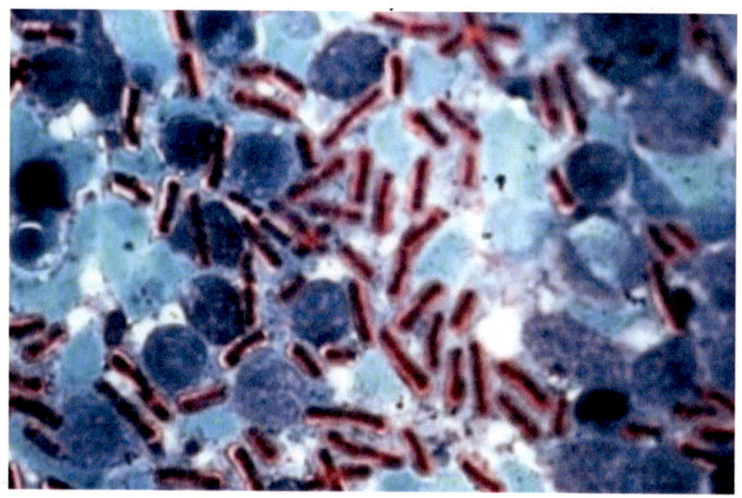

Differential Staining

A. Gram Staining

Composition

a. Ammonium oxalate – Crystal violet solution

Solution 1

Crystal violet	- 2 gm
Ethyl alcohol(95%)	- 20 ml

Solution 2

Ammonium oxalate	- 0.8 gm
Distilled water	- 80 ml

Mix solutions 1 and 2 and filter

b. Lugal's iodine solution

Iodine	- 1 gm
Potassium iodide	- 2 gm
Distilled water	- 300 ml

Dissolve the ingredients and filter

c. Ethyl alcohol (decolourizer)

d. Safranin (counter stain)

Safranin O (2.5% soln.) in 95% ethyl alcohol	- 10 ml
Distilled water	- 100 ml

Dissolve the stain and filter.

Materials

1. 24 hours culture smear fixed by flame

Procedure

1. Prepare bacterial smears and fix over flame.
2. Pour ammonium oxalate crystal violet stain and allow to act for 1 min.
3. Wash with tap water.
4. Add gram's iodine to act for 1 min.
5. Wash with tap water and blot dry.
6. Decolourize with 95% ethyl alcohol for about 30 seconds gently agitating the slide till no colour comes out from the smear. Blot dry.
7. Counter stain for about 10 to 20 seconds with safranin solution.
8. Wash with tap water, dry and examine.

Interpretation

Gram positive organisms will be deep violet and Gram negative red.

Gram Positive Rod/ Bacilli

Gram Negative Rod/ Bacilli

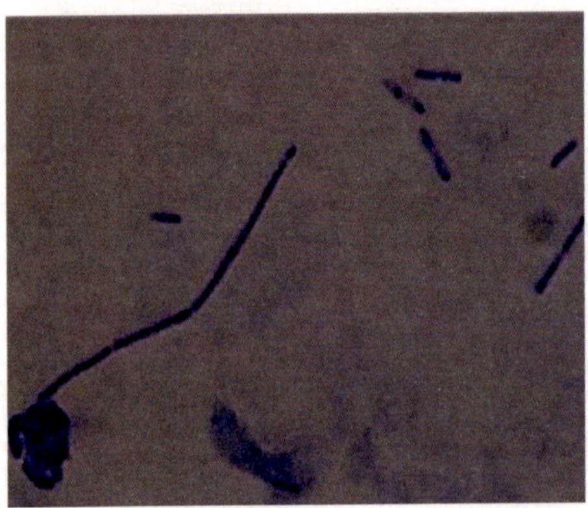

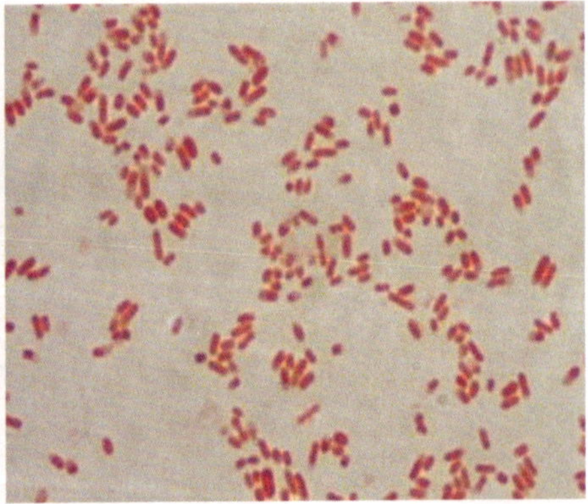

B. Ziehl - Nielsen's stain

Species belonging to the genus Mycobacterium do not stain readily by simple staining procedures. Their staining is facilitated by heat. Once stained they retain the colour of the dye even when treated with a suitable decolourizer. These organisms are designated as acid fast. The organisms, which are decolourized, take counter stain and are non-acid fast.

Composition

a. Concentrated carbol fuchsin

b. Acid alcohol (decolourizer)

Ethyl alcohol (95%)	-97 ml
Concentrated HCL	-1 ml

c. Loeffler's alkaline methylene blue (See Exercise 16).

Materials

A portion of the formalin preserved tuberculous lung, and staining solutions A, B and C.

Procedure

1. Cut a piece of the tubeculous lung, scratch the cut surface with a knife edge and spread evenly on the slide. Dry the smear and fix over the flame.
2. Floor the smear with carbol fuchsin and steam for 3 to 5 min. Do not boil or char the stain. Stain should not dry on the slide.
3. Wash with water.
4. Decolourize with acid alcohol until the preparation is faint pink or colourless (about 15 to 20 seconds).
5. Wash with water.
6. Counter stain with Loeffler's methylene blue for about 30 seconds.
7. Wash with water, blot carefully and dry before examination.

Interpretation

Acid fast bacteria will stain red and non-acid fast and tissue debris blue.

C. Leishmans staining method

Preparation

Dissolve 0.15 gm of Leishman's Powder in 100 ml acetone free methyl alcohol.

Materials

1. Blood smear from a case of anthrax
2. Blood smear from a case of haemorrhagic septicemia.
3. Muscle impression smear from a case of Black quarter.

Films

1. Apply 4 drops of stain
2. Rack gently for 12 – 15 seconds.
3. Add 8-12 drops of slightly, alkaline distilled water and mix by rocking.
4. Allow 15 minutes to stain.
5. Flood stain in stream of distilled water for 2-3 seconds.
6. Dry

Sections

1. Mix stain 1 part and water 2 parts and apply for 5 – 15 minutes.
2. Wash in Distilled water for 1 min.
3. Dehydrate, clear and mount in balsam and examine.

D. Wrights Staining

Preparation

Dissolve 0.15 gm of Wright's Powder in 100 ml acetone free methyl alcohol.

Materials

1. Blood smear from a case of anthrax
2. Blood smear from a case of haemorrhagic septicemia.
3. Muscle impression smear from a case of Black quarter.

Films

1. Apply 4 drops of stain.
2. Rack gently for 12 – 15 seconds.
3. Add 8-12 drops of slightly, alkaline distilled water and mix by rocking.
4. Allow 15 minutes to stain.
5. Flood stain in stream of distilled water for 2-3 seconds.
6. Dry

Sections

1. Mix stain 1 part and water 2 parts and apply for 5 – 15 minutes.
2. Wash in Distilled water for 1 min.
3. Dehydrate, clear and mount in balsam.

E. Giemsa Staining Method

Preparation

Azur II – eosin	- 3 g.
Azur II	- 0.8 g
Glycerin	- 250 g.
Methanol	- 250 g. (acetone free).

Procedure

1. Fix films in methanol for 3 minutes.
2. Mix 1 part stain to 2 parts slightly alkaline distilled water and stain for 15 minutes
3. Wash
4. Apply distilled water for 1 minute
5. Dry and examine

Distilled water is made slightly alkaline by adding 1% sodium carbonate. Some of the commonly used biochemical tests for the identification of bacteria

Catalase Test

Some bacteria and macrophages can reduce diatomic oxygen to hydrogen peroxide or superoxide. Both of these molecules are toxic to bacteria. Some bacteria, however, possess a defense mechanism, which can minimize the harm done by the two compounds. These resistant bacteria use two enzymes to catalyze the conversion of hydrogen peroxide and superoxide back into diatomic oxygen and water. One of these enzymes is catalase and its presence can be detected by a simple test. The catalase test involves adding hydrogen peroxide to a culture sample or agar slant. If the bacteria in question produce catalase, they will convert the hydrogen peroxide and oxygen gas will be evolved. The evolution of gas causes bubbles to form and is indicative of a positive test.

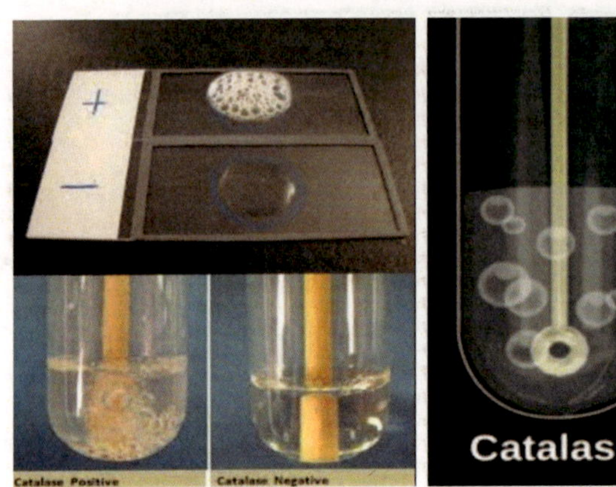

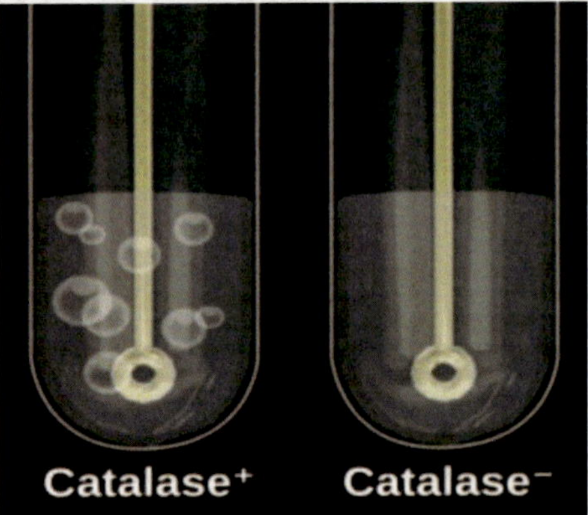

Positive – *Staphylococcus* **species** **Negative** – *Streptococcus* **species.**

Oxidase Test

Cytochrome oxidase is an enzyme found in some bacteria that transfers electrons to oxygen, the final electron acceptor in some electron transport chains. Thus, the enzyme oxidizes reduced cytochrome c to make this transfer of energy. Presence of cytochrome oxidase can be detected through the use of an Oxidase disk, which acts as an electron donator to cytochrome oxidase. If the bacteria oxidize the disk (remove electrons) the disk will turn purple, indicating a positive test. No color change indicates a negative test.

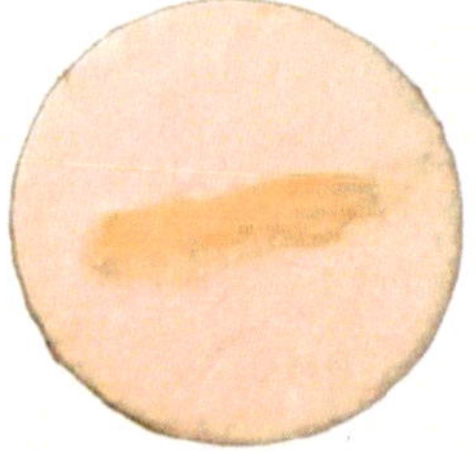

Positive – *Pseudomonas aeruginosa Oxidase Positive*

Negative - *Enterobacteriaceae Oxidase Negative*

Coagulase Test

Like the mannitol salts agar, the coagulase test is another method for differentiating between pathogenic and non-pathogenic strains of *Staphylococcus*. Bacteria that produce coagulase use it as a defense mechanism by clotting the areas of plasma around them, thereby enabling themselves to resist phagocytosis by the host's immune system. The sample in question is usually inoculated onto 0.5 ml of rabbit plasma and incubated at 37° C for one to four hours. A positive test is denoted by a clot formation in the test tube after the allotted time

Positive - Pathogenic *Staphylococci* **Negative - Non pathogenic** *staphylococci*

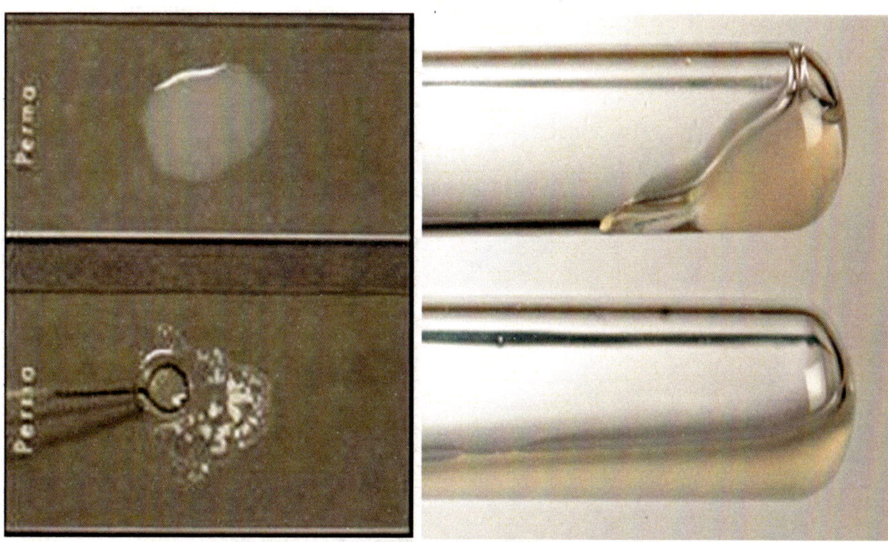

Slide Coagulase Test Tube Coagulase Test

Eosin Methylene Blue (Emb) Agar

EMB agar is a differential medium used in identification and isolation of Gram-negative enteric rods. EMB agar also inhibits the growth of Gram-positive organisms. The differential basis of this medium involves two indicator dyes, eosin and methylene blue, that distinguish between lactose fermenting and non-lactose fermenting organisms. Lactose fermenters form colonies with dark centers and clear borders while the non-lactose fermenters form completely colourless.

Positive – *E.coli*

Negative – *Enterobacter aerogenes*

Mannitol Salt Agar

A common medium used for the isolation of pathogenic staphylococci is the Mannitol Salts Agar. The high salt concentration of this medium inhibits the growth of most other organisms. Pathogenic staphylococci not only grow on the medium, but they also produce acid from it. This acid production turns the pH indicator from red to yellow. Non-pathogenic staphylococci can grow on the medium but produce no acid from it.

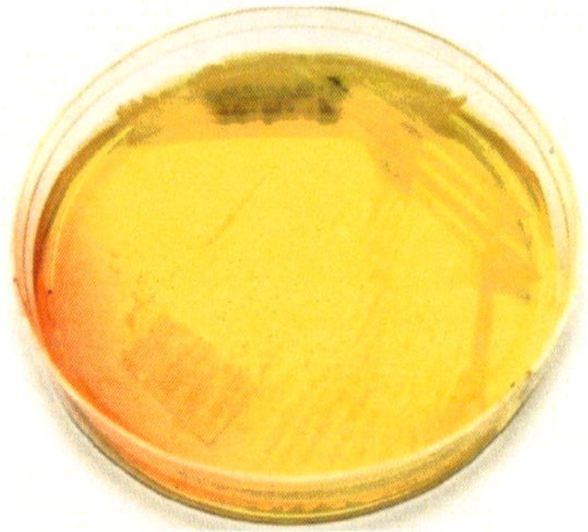

Streptococcus agalactiae

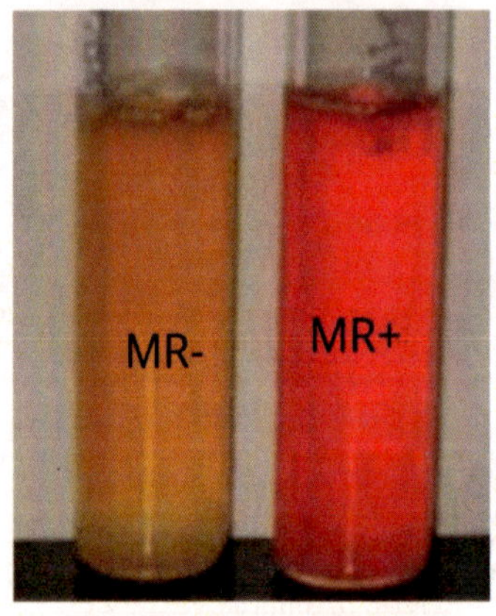

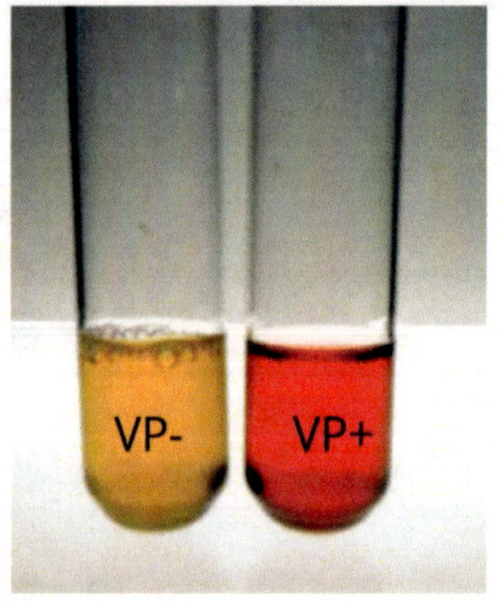

Positive – *Staphylococcus aureus* **Negative** - *Streptococcus agalactiae*

M.R. Test

Pyruvic acid, formed in the EMP pathway due to breakdown of glucose, is further metabolised through mixed acid fermentation pathway and results in the production of large quantities of acetic, lactic and formic acids with a marked drop in the pH of the test medium. This acidic condition is detected by methyl red, an indicator that is yellow at pH above 6.0 and red at pH below 6.0.

Positive – *E. coli* Negative - *Klebsiella pneumoniae*

V.P. Test

Pyruvic acid, the pivotal compound formed in the fermentative metabolism of glucose, is further metabolised through various metabolic pathways, depending on the enzyme systems possessed by different bacteria. One such pathway results in the production of acetoin (acetyl methyl carbinol), a neutral - reacting end product. Organisms such as members of the Klebsiella – *Enterobacter* – *Hafnia* – *Serratia* species produce acetoin as the chief end product of glucose metabolism. In the presence of atmospheric O2 and 40% KOH, acetoin is converted to diacetyl. Alpha- naphthol serves as a catalyst to bring out a red complex.

Positive – *K.. pneumoniae* Negative - *E.coli*

Indole Test

Indole is a componentof the amino acid tryptophan. Some bacteria have the ability to break down tryptophan for nutritional needs using the enzyme tryptophanase. When tryptophan is broken down, the presence of indole can be detected through the use of Kovacs' reagent. Kovac's reagent, which is yellow, reacts with indole and produces a red color on the surface of the test tube.

Positive – *E.coli* Negative - *K. .pneumoniae*

Citrate Test

The citrate test is used to determine the ability of a bacterium to utilize citrate as its only source of carbon. Bacteria can break the conjugate base salt of citrate into organic acids and carbon dioxide. The carbon dioxide can combine with the sodium from the conjugate base salt to form a basic compound, sodium carbonate. A pH indicator in the medium detects the presence of this compound by turning blue (a positive test).

Positive - *K..pneumoniae* Negative - *E.coli*

Urease Production Test

Microorganisms that possess the enzyme urease hydrolyse urea, thereby releasing ammonia, CO_2 and water. The ammonia reacts in solution to form ammonium carbonate resulting in alkalinisation and an increase in the pH of the medium. This alkaline reaction is detected by phenolphthalein indicator, which is colourless at pH below 8.1 and becomes pink-red at above pH 8.1.

Positive - *Proteus mirabilis* **Negative** - *E.coli*

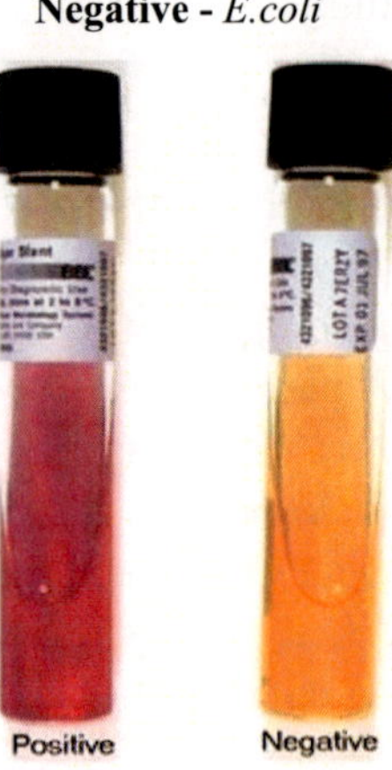

Indole Production Test Citrate Utilisation test Urease hydrolysis Test

Hydrogen Sulphide Production Test

The ability of certain bacterial species to liberate sulfur from sulfur containing amino acids or other compounds in the form of H_2S is an important characteristic for their identifications. The production of H_2S is detected by indicators such as ferrous sulfate, ferric citrate, lead acetate etc by forming an insoluble, heavy metal sulfide black precipitate.

Positive - *Salmonella enteritidis* Negative - *E.coli*

Nitrate Reduction Test

The ability of an organism to reduce nitrates to nitrites is an important characteristic used in the identification and species differentiation of many microorganisms. Organisms demonstrating nitrates reduction have the ability of extracting oxygen from nitrates to form nitrites and other reduction products. The presence of nitrites in the test medium is detected by the addition of α naphthalamine and sulfanilic acid with the formation of a red diazonium dye, P-sulfobenzene azo - α - naphthalamine. The development of a red colour within 30 seconds after adding the test reagents indicates the presence of nitrites and represents a positive reaction for nitrate reduction. If no colour develops after adding the test reagents, this may indicate either that nitrates, have not been reduced (a true negative reaction) or that they have been reduced to products other the nitrites such as ammonia, molecular nitrogen (N2), nitric Oxide (NO) or nitrous oxide (N_2O) and hydroxylamine. Because the test reagents detect only nitrites, the latter process would lead to a false negative reading. Thus it is necessary to add a small quantity of zinc dust to all negative reactions. Zinc ions reduce nitrates to nitrites and the development of a red colour after adding zinc dust indicates the presence of residual nitrates and confirms a true negative reaction.

Positive - *E.coli* Negative - *Mycobacterium avium.*

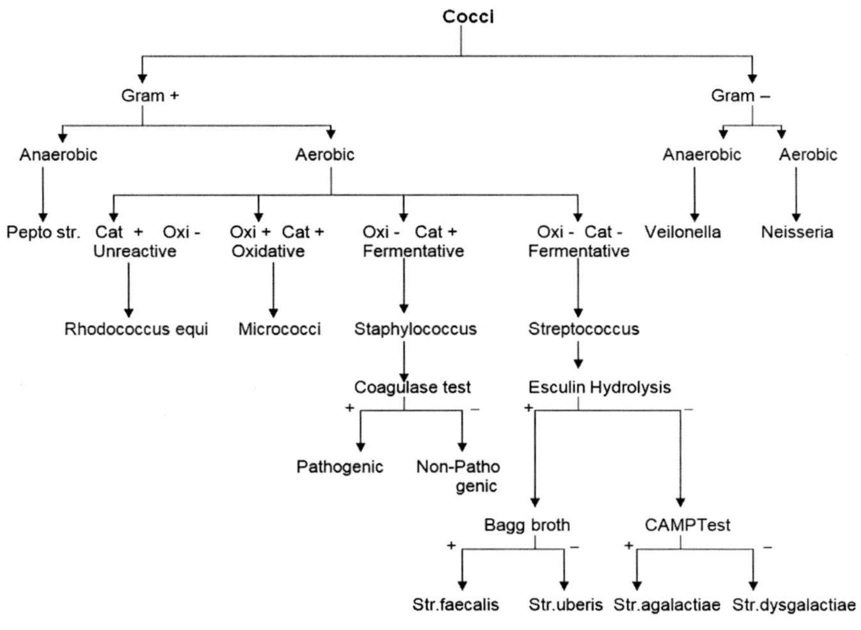

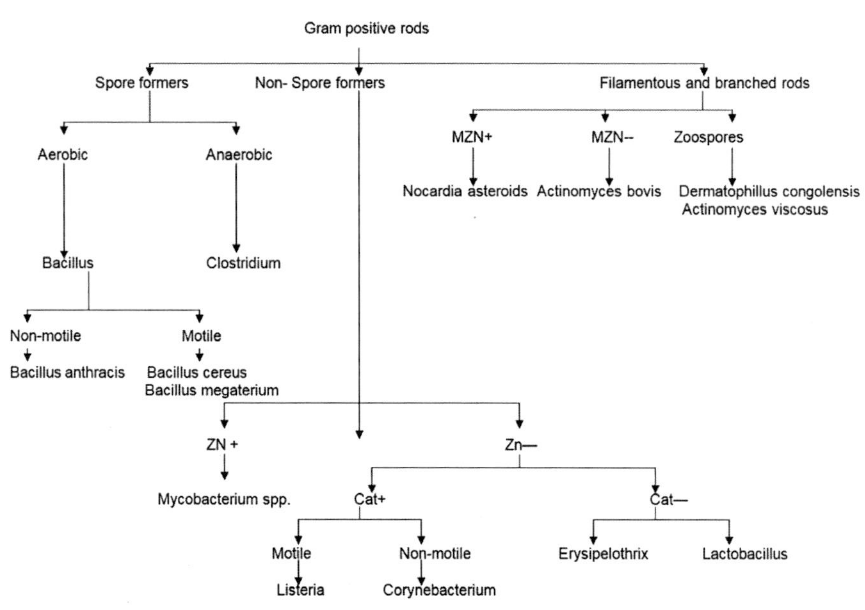

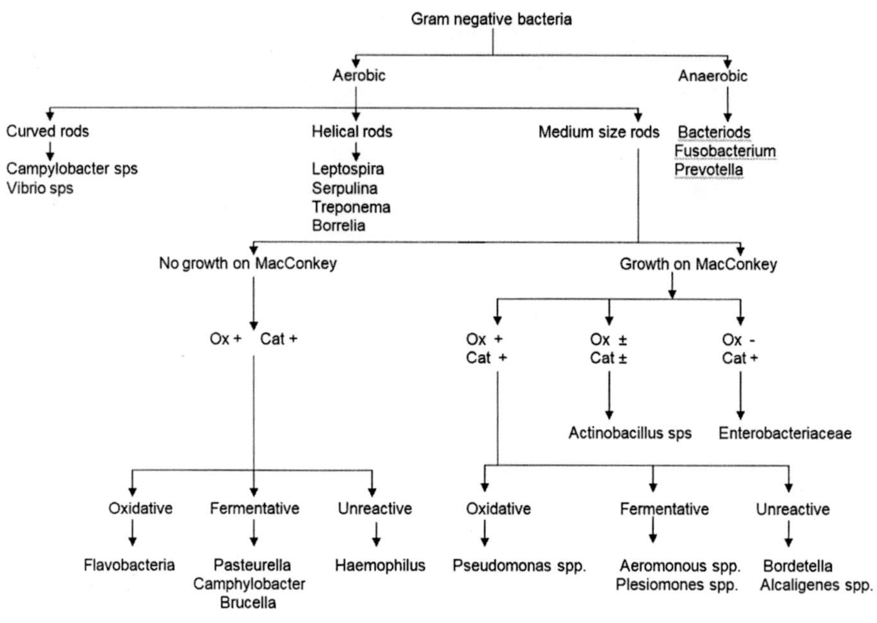
Gram negative bacteria
Aerobic
Anaerobic
Curved rods
Campylobacter sps
Vibrio sps
Helical rods
Leptospira
Serpulina
Treponema
Borrelia
Medium size rods
Bacteriods
Fusobacterium
Prevotella
No growth on MacConkey
Growth on MacConkey
Ox + Cat +
Ox +
Cat +
Ox ±
Cat ±
Ox -
Cat +
Actinobacillus sps
Enterobacteriaceae
Oxidative
Fermentative
Unreactive
Flavobacteria
Pasteurella
Camphylobacter
Brucella
Haemophilus
Oxidative
Fermentative
Unreactive
Pseudomonas spp.
Aeromonous spp.
Plesiomones spp.
Bordetella
Alcaligenes spp.

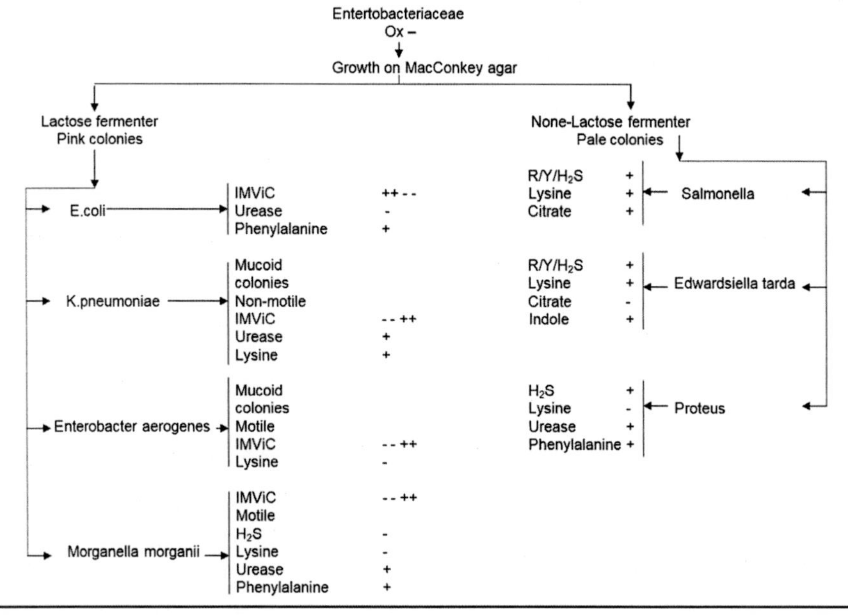
Entertobacteriaceae
Ox –
Growth on MacConkey agar
Lactose fermenter
Pink colonies
None-Lactose fermenter
Pale colonies
E.coli
IMViC ++ - -
Urease -
Phenylalanine +
K.pneumoniae
Mucoid
colonies
Non-motile
IMViC - - ++
Urease +
Lysine +
Enterobacter aerogenes
Mucoid
colonies
Motile
IMViC - - ++
Lysine -
Morganella morganii
IMViC - - ++
Motile
H2S -
Lysine -
Urease +
Phenylalanine +
R/Y/H2S +
Lysine +
Citrate +
Salmonella
R/Y/H2S +
Lysine +
Citrate -
Indole +
Edwardsiella tarda
H2S +
Lysine -
Urease +
Phenylalanine +
Proteus

4

Isolation and Identification of Bacteria by Gram's Staining and Cultural/ Biochemical Characteristics - II

Special Staining

A. Spore stain

Spores may be spherical or oval and may be central subterminal or terminal in position within the cell. Shape, size and position of the spore in the cell are useful for the identification of sporulated organisms. Spores are difficult to stain and when once stained, they are difficult to decolourize. With Gram's stain, mature spores remain unstained bodies. Modified Ziehl-Neelsen's method of staining will be used for the staining of spores.

Composition

a. Concentrated carbol fuchsin
b. Sulphuric acid 0.5 % solution.
c. Loeffler's alkaline methylene blue

Materials

Old cultures of *B. subtilis* and *Cl. Chauvoei* and staining solutions.

Procedure

1. Make smears and fix over flame.
2. Stain with carbol fuchsin and steam for 3 to 4 minutes.
3. Wash with tap water.
4. Decolourize with 0.5% sulphuric acid.
5. Wash with water.
6. Counter stain with Loeffler's methylene blue for 2 minutes.

Interpretation

The spores will stain red and vegetative cells blue

B. Capsule stain

Among pathogenic organisms, the presence of the capsule is considered an indication of their virulence, which may be absent in avirulent forms due to constant subculturing. Capsule is antigenic in nature and except in one case (B. anthracis) where it is polypeptide in nature, the chemical nature of capsule of most other bacteria is polysaccharide in nature. The composition of the medium in which the capsulated organisms are grown may influence the size of the capsule. In this exercise Hiss copper sulphate method of capsule staining will be used.

Composition

a. Saturated alcoholic soln. of gentian violet 5 ml

Distilled water 95 ml

Mix the ingredients and filter.

b. Copper sulphate 20 gm

Distilled water 100 ml

Dissolve copper sulphate crystals in the distilled water by gentle heating and filter.

c. Blood serum.

Materials

Freshly isolated virulent cultures of *P. multocida* and *Kleb. Pneumoniae,* and staining solutions.

Procedure

1. Make smears in serum, air dry and fix by gentle heating.
2. Apply gentian violet stain and heat gently until steam rises. Allow the dye to act for 15 to 20 seconds.
3. Wash the dye with 20% copper sulphate solution, applying several changes.
4. Do not wash with water. Blot dry and examine under oil immersion lens.

Interpretation

The capsule appears as a faint blue halo around the dark purple cell body against a faint purple background.

C. India ink method for capsule

This is the simplest and most generally applicable method for the demonstration of capsule.

Wipe a microscopic slide. Place a loopful of India ink on it. Emulsify a small portion of solid culture in the drop of ink or else mix a loopful of liquid culture with the ink. Place a clean coverslip on the ink drop and press the same down firmly through a sheet of blotting paper. Examine under oil immersion objective.

D. Flagella stain

Composition

A. Sodium chloride in distilled water, 1.5% soln.

B. Tannic acid in distilled water, 3% soln.

C. Basic fuchsin in ethyl alcohol, 1.2% soln.

D. Aqueous methylene blue, 1% soln.

Materials

Six to eight hour old broth culture of actively motile *Proteus vulgaris* and staining solutions.

Procedure

1. Gently wash the cells twice in centrifuge with distilled water at 3000 rpm for about 5 minutes, each time discarding the supernatant.
2. Gently resuspend the cells in distilled water to give a visible turbidity and incubate at 37°C for about 10 minutes. Check motility by hanging drop method.
3. Pick up a perfectly clean slide, flame for few seconds and before it cools, mark a heavily defined band around an area of about 1 X 1.5 sq. inch on the flamed slide with a grease pencil.
4. Incline the slide and place a loopful of the washed bacterial suspension at two points at one end of the enclosed area and allow them to run to the second end. Allow the slide to dry in air.

5. Place the slide on the staining rack. Thoroughly shake the staining bottle and slowly add about 1 ml basic fuchsin stain with a pipette into the marked area without soiling the grease pencil marks.
6. Allow the stain to act till there is formation of a fine precipitate in the stain. Precipitate formation is detected by the appearance of rust coloured cloud, which generally starts along one side of the stain and quickly spreads throughout the staining solution. This is the end of the staining period (about 10 to 30 minutes).
7. Flood the stain off the slide with a gentle stream of water.
8. Counter stain for 1 minute with 1% aqueous methylene blue.
9. Wash with water, air dry and examine under oil immersion lens.

Certain biochemical tests in identification of bacteria

Phenylalanine Deaminase Test

Phenylalanine is an amino acid that on deamination forms a ketoacid, phenyl pyruvic acid. Of the Enterobacteriaceae, only members of the proteus, morganella, and providencia genera possess the deaminase enzyme necessary for this conversion. This test depends on the detection of phenylpyruvic acid in the test medium. The test is positive, if a visible green colour develops on addition of a solution of 10% ferric chloride.

Positive - *Proteus* spp. Negative - *E. coli*

Esculin Hydrolysis Test

Esculin medium without bile is useful in differentiating several species of non-fermenting bacilli. Esculin is a substituted glucoside that can be hydrolysed by certain bacteria to yield glucose and esculetin. Esculin is a fluorescent compound that floureses, under long wave UV light at 360 nm. When esculin is hydrolysed, fluorescence is lost, and the medium turns black due to the reaction of esculin with the ferric ions in the medium. The development of black colour or the loss of fluorescence under UV light 360 nm) is interpreted as a positive test. Fluorescence or lack of black colour is interpreted as a negative test.

Positive - *Streptococcus faecalis* Negative - *Streptococcus agalactiae*

Decarboxylation Tests

Decarboxylases are a group of substrate-specific enzymes that are capable of reacting with the carboxyl (COOH) portion of amino acids, forming alkaline-reacting amines. This reaction known as decarboxylation, forms carbon dioxide as a second product. Each decarboxylase enzyme is specific for an amino acid. Lysine, ornithine, and arginine are the three amino acids routinely tested in the identification of the *Enterobacteriaceae.* The specific amine products are as follows:

Lysine → Cadaverine

Ornithine → Putrescine

Arginine → Citrulline

The conversion of arginine to citrulline is a dihydrolase, rather than a decarboxylase reaction, in which an NH_2 group is removed from arginine as a first step. Citrulline is next converted to ornithine, which then undergoes decarboxylation to form putrescine.

During the initial stages of incubation, both tubes turn yellow, owing to the fermentation of the small amount of glucose in the medium. If the amino acid is decarboxylated, alkaline amines are formed and the medium reverts to its original purple color.

Amino acid	Positive	Negative
Lysine	*E.aerogenes*	*E.cloacae*
Ornithine	*E.cloacae*	*K.pneumoniae*
Arginine	*E.cloacae*	*E.aerogenes.*

5

Drug Sensitivity and Rationale for Therapy

Antimicrobial Susceptibility Test

The routine use of the modified Kirby-Bauer method is recommended.

Material

- Sterile cotton swabs
- Soyabean – Casein digest broth or Tryptone soya broth
- Non selective agar such as nutrient agar.
- 0.5 Mc Farland turbidity standard:

  Solution A

  $Bacl_2$, $2H_2O$. 1.75G

  Make upto 100 ml with distilled water.

  Solution B

  H_2SO4 1.0 ml

  Make upto 100 ml with distilled water.

  Stock Solution

  0.5 ml of solution A + 99.5 ml of solution B. Dispense in 5 ml amounts. Seal and store in dark.

- Glass vials or tubes 10 ml capacity
- Muller – Hinton agar
- Antimicrobial discs
- Sterile forceps
- Ruler or calipers

Procedure

1. Atleast 4-5 well-isolated colonies, of the same morphological type are selected from a non-selective agar plate.
2. Touch the top of the colonies and transfer to a tube containing 4-5 ml of Soyabean-Casein digest broth or tryptone soyabroth.
3. Incubate the broth at 35-37°C upto 2-8 hours until a slight visible turbidity appears.
4. Adjust the turbidity of the broth by comparison with 0.5 McFarland turbidity standard.
5. Dip a sterile swab into the standardized suspension of bacteria and remove the excess fluid by pressing and rotating the swab firmly against the inside of the tube above the fluid level.
6. Streak the swab in three directions over the entire surface of the agar. Make a final sweep of the swab around the rim of the petridish.
7. Allow the inoculated plates to stand for 3-5 minutes but no longer than 15 minutes.
8. Place the antimicrobial discs onto the agar surface using sterile forceps or an antibiotic disc dispenser. The discs should be placed no closer than 24 mm (centre to centre). This is equivalent to 6 discs per standard 90 mm petridish.
9. Invert the plates and place in a 35-37°C incubator within 15 minutes of applying the discs. Incubate aerobically for 16-18 hours (24 hours for Staphylococci).
10. After incubation, measure the diameters of the zones of inhibition to the nearest mm using a ruler or calipers. Read the diameters from the back of the plates on Muller-Hinton agar but over the surface of the agar on blood agar plates. The diameter of the zones should be read across the center of the discs.

Analysis

An interpretation of the size of the zones of inhibition is made with reference to the standard chart.

The zone of inhibition is that area showing no obvious growth that can be detected by the unaided eye. The diameter of the zone is measured to each edge of this clear area. However, there are occasions when the exact extent of the zone is complicated.

a. Large colonies growing within an otherwise clear zone of inhibition should be subcultured, reidentified and retested. Faint growth of tiny colonies at the edge of the zone can be ignored.

b. *Proteus* species may swarm into areas of inhibited growth around certain antimicrobial discs. If the zones of inhibition are clearly outlined, then the veil of swarming can be disregarded.

c. With trimethoprim and the sulphonamides, slight growth in the zone can be ignored and the zone measured to the margin of heavy growth.

d. For bacteria that have to be grown on blood agar plates, the zone size for nafcillin, novobiocin, oxacillin and methicillin will be 2-3 mm smaller than the normal control limits.

e. Transmitted light should be used to examine for light growth within the zone in the case of methilillin-resistant strains of staphylococci.

The selection of the types of antimicrobial agents for use in the disc diffusion test will, to a large extent, depend on clinical considerations including the drugs that are available and in general use by the veterinarian. However, to make routine susceptibility testing relevant and practical the following guidelines may be helpful.

- A tetracycline disc will predict the result against all the other tetracyclines.
- Sulphisoxazole is a suitable representative for all the sulphonamides.
- Erythromycin will predict the result of all other macrolides.
- A Clindamycin disc will predict the result for lincomycin also.
- Aminoglycosides and quinolones should be tested separately. Individual members within each groups are not related closely enough to assume cross-resistance.
- Chloramphenicol, Vancomycin, Nitrofurantoin and trimethoprim / sulphamethoxazone are tested separately as required.
- Staphylococci should be tested against penicillin G and oxacillin (for resitance to methicillin, cephalosporins and other beta-lactams.)
- *Enterococcus faecalis* tested against penicillin G will predict the result against amplicillin, ampicillin – analogues, amoxicillin and the acylamino-penicillins.
- Streptococci should be tested against either penicillin G or ampicillin. Testing against both is not necessary.

- Staphylococci are usually susceptible to cephalosporins except for the methicillin-resistant strains; these should be reported as resistant even if the result of the in vitro test suggests otherwise.
- There is a lack of clinical correlation with *Enterococcus faecalis* against the cephalosporins and the results of the *in vitro* tests cannot be interpreted with accuracy.
- A cephalothin disc will indicate the result to cefaclor, cephapirin, cefazolin, cephaleridine, cephalexin and cefadroxin. Cefatozime represents ceftazidime, ceftizoxime and ceftriaxone.

Report the bacterium as susceptible, moderately susceptible, intermediate or resistant to each antimicrobial agent used in the test:

Susceptible: The infection may respond to the treatment at the normal dosage.

Moderately susceptible: The pathogen may be inhibited by attainable concentrations of the antimicrobial agent provided a higher dosage is used.

Intermediate: The result is equivocal and if the bacterium is not fully susceptible to an alternative drug, then the test should be repeated.

Resistant: The bacterium is not inhibited by the usually achievable systemic concentrations of the antimicrobial agent and efficacy has not been reliable in clinical studies.

6

Virus Detection and Cultivation of Animal Viruses

Viruses can be cultivated only in living cells. **Laboratory animals, embryonated avian eggs and cell cultures** are three systems available for this purpose. These three systems are used for the basic procedures of isolation and identification of viruses, maintenance of stock cultures and production of vaccines.

Cultivation of Viruses in Laboratory Animals

Animal inoculation helps to confirm the infectious nature of the disease. Laboratory animals used for many kinds of virological works such as clinical studies, pathogenicity of viruses, epizootiology, potency tests on vaccines and production of diagnostic and prophylactic antisera.

An ideal experimental animal is one that is

1. Uniformly and fully susceptible to the virus under study
2. Free from intercurrent infection and adventitious lesions and
3. Inexpensive purchase and maintain

If possible it is better to use **germ free, specific pathogen free or colostrums deprived animals** for experimental purpose. The same species of animals or atleast closely related species of the natural host is selected for inoculation, wherever possible. For example in case of swine fever, which shows a high degree of host specificity, the same species should be used for inoculation. in others such as in case of rabies with low host specificity, the causative virus will grow in a wide range of animals.

The virus material for animal inoculation should be made bacteriologically sterile and the material must be filtered or treated with antibiotics. Various methods may be employed for inoculating the virus material into the laboratory animals. The usual routes are intracerebral, intranasal, intramuscular, intravenous, intraperitoneal and subcutaneous. Other routes may be employed for specific purposes.

Different laboratory animals, routes of inoculation with respect to different diseases

Sl.No	Disease	Lab animals	Route of inoculation
1.	Rabies	Mouse	Intra-cerebral
2.	Malignant catarrhal fever	Calf	Intra-venous
3.	Teschen disease	Piglet	Intrea-muscular
4.	Foot & Mouth disease	Guinea pig	Intra-dermal
5.	Foot & Mouth disease	Unweaned mouse	Intra-peritoneal
6.	Swine vesicular disease	Piglet	Subcutaneous
7.	Rift valley fever	Ferret	Intra nasal
8.	Orf disease	Sheep	Scarification
9.	Vesicular stomatitis	Horse	Intra-lingual
10.	Poliomyelitis	Monkey	Oral
11.	Avian influenza	Chicken	Aerosal
12.	Infectious Laryngeotracheitis	Chicken	Intra-tracheal
13.	Pseudo-rabies	Mouse	Intra-conjunctival

Evidence for virus multiplication

1. Clinical symptoms
2. Rise in antibody titre
3. Reisolation of virus from secretions and excretions or from biopsy specimens
4. Examination of biopsy specimens by electron microscopy or by histological studies
5. Detection of viral antigens by FAT

At necropsy

1. Gross and histological examination of lesions
2. Virus isolation from tissues
3. Detection of viral antigens either by FAT, CFT or gel diffusion test in tissues

Cultivation of Viruses in Embryonated Eggs

Embryonated hen's egg provides the necessary conditions for the growth of a variety of viruses. Special advantages of the fertile eggs are

1. bacteriologically sterile,
2. immunologically incompetent,

3. relatively inexpensive and easy to handle.
4. the infected membranes and fluids provide large quantities of virus.

But it has the disadvantages of

1. Possessing maternal antibodies to poultry pathogens, which may be present in yolk
2. The egg may be contaminated with some infectious agents such as bacteria and virus that may pass through egg **(New castle disease virus, avian encephalomyelitis virus, avian leukosis virus, salmonella, mycoplasma and chlamydial agents etc.).**
3. they are also not susceptible for a wide range of viruses

The age of the embryo to be inoculated depends on the virus that is to be inoculated which will in turn decide the route of inoculation. The common routes of inoculation are:

Sl.No.	Route	Age of the embryo	Viruses inoculated
1.	Yolk sac	5-7 days	Rabies virus, Herpes simplex virus, Avian encephalomyelitis virus, Marek's Disease virus (Chlamydial and Rickettsial agents also)
2.	Allantoic cavity	9-10 days	Avian influenza, NDV, Avian adenovirus, IBV, Fowl plague, Inclusion body hepatitis virus
3.	Amniotic cavity	10-12 days	Influenza, type C, Mumps virus
4.	Chorio allantoic membrane	10-12 days	Pox virus, some herpes virus, IBD virus, duck plague virus
5.	Intra cerebral	13-15 days	Rabies virus
6.	Intra venous	15 days	Encephalomyelitis virus, blue tongue virus
7.	De-embryonated eggs	15 days	Influenza vaccine production

Evidence of virus multiplication in chick embryos

Sl.No	Indication	Virus
1.	Death of the embryo	NDV, fowl plague, paramyxovirus
2.	Specific lesions on the embryo	
	a. Pinpoint haemorrhages at the occipital region, under aspect of the abdomen, inter digital space	NDV
	b. Generalized congestion of the embryo	IBD, duck plague
	c. Necrosis of the liver, enlargement of spleen	IBD, duck plague, IBHV
	d. Curling and dwarfing of the embryo	Infectious bronchitis virus
	e. Odematous embryo	Duck-hepatitis virus
	f. Cherry red embryo	Blue-tongue virus

Sl.No	Indication	Virus
3.	Discoloration of the contents of egg a. Greenish allantoic fluid b. Urate precipitates	 Duck hepatitis-A virus IBV and IBD
4.	Change in quantity of allantoic / amniotic fluid a. Increase in quantity	Lentogenic strains of NDV, EDS-76 virus
5.	Changes in CAM a. Pock lesions on CAM Odematous membrane, congested membrane	Pox virus, some herpes virus, some reo virus and VSV Duck plague, IBD
6.	Haemagglutination of allantoic fluid	Orthomyxo and paramyxo viruses
7.	Haemagglutination of amniotic fluid	Orthomyxo and paramyxo viruses
8.	Direct EM examination	Orthomyxo and paramyxo viruses

Depending upon the virus that you want to propagate you may have to choose either chick embryos (for viruses mentioned earlier), **duck embryos** (for duck plague virus: EDS – '76 virus, etc.) **turkey eggs** (Haemorrhagic enteritis virus of turkeys) or even **pheasant's eggs** (Marble spleen disease virus).

Cultivation of Viruses in Cell Culture

The term tissue culture designates those cell systems established with tissue fragments or with masses of cells originating and migrating from the fragments. Cultures prepared with dispersed cells are designated as cell cultures.

Advantages of Cell Culture

1. Easier and more economical to prepare, maintain and manipulate
2. No antibody or other inhibitory factors to interfere with the growth of the virus
3. The effect of virus infection can be detected in a very short time. (hrs to days).
4. Metabolic studies and cytopathic changes may be studied conveniently
5. In some cases all cultures are more susceptible to the effects of virus than are animals and chick embryos
6. Different viruses produce specific metabolic and cytopathic changes
7. Cell cultures can be produced and preserved at –70° C or –196° C
8. Cell lines from different species and organs are available. Primary cell cultures can be prepared from different species or organs as per the needs of specific viruses
9. Cell cultures can be radio labeled to study the details of virus multiplication.

Disadvantages

1. It needs sophisticated, elaborate arrangements and required well trained persons
2. Problems of contaminations with bacteria and mycoplasma
3. Posssibilities of contamination with latent viruses and non-cytopathic effect (CPE) producing viruses.

Methods of Cell Culture

a) Organ Culture

Slices of organs carefully handled to maintain their original architecture and functions for days or weeks. Eg. Tracheal ring cultures. Not often used in virology these days, but can be used to study the histopathogenesis of the viruses and propagation of certain respiratory and enteric viruses (eg: Turkey rhino tracheitis virus, transmissible gastro enteritis virus in children)

b) Tissue culture/ Explant culture/ Maitland type culture

Tissue fragments taken from living or recently killed animals incubates in a nutrient medium in a well-aerated flasks, again not used for virus propagation these days.

c) Cell culture/ dispersion culture

Cells are disaggregated/ dispersed by trypsinisation and incubated in test tubes/ bottles/flasks/petridishes. The cells adhere to the inside of the container and grow to form a sheet of monolayer.

Types of Cell Cultures

1) Primary cell cultures

When cells are taken freshly from animals and placed in culture, the culture consists of a mixture of cell types most of which are capable of very limited growth invitro-usually **5-10 divisions**. Primary cell culture can be obtained either by

a) by allowing cells to migrate out from the tissue, which is adhering to a substrate or

b) by disaggregating the tissue mechanically or enzymatically (Trypsin treatment) to produce suspension of cells

Primary cell cultures retain most of the characteristics of the cells from which they originated. With regards to chromosomal number, primary cultures usually retain their diploid karyote. The most commonly employed tissues for preparing primary cultures are **kidney, lungs, liver, thyroid and testes**

Primary cultures are often preferred particularly for diagnostic purpose because of its greater virus susceptibility. The disadvantages of this type are the unpredictable batch-to-batch variation and the occasional presence of latent viruses, some of which are lethal to man.

2) Diploid Cell Strains/ Finite Cell Lines

These are cells of single type that are capable of undergoing upto **100 division's** invitro before dying. They retain their original chromosome throughout. They are widely used in **diagnostic virology and vaccine production** because of its satisfactory sensitivity and high yield of virus.

Eg: Fibroblasts established from human embryo lung: MRC 5, WI –38 (Wistar –38), IMR -90

3) Continuous Cell Lines

These are cells of a single type that are capable of **indefinite propagation** in vitro usually originated from cancer cells or transformation of diploid cell strains. They do not possess any close resemblance to their cells of origin.

The established cell lines have altered chromosome number, shorter doubling time and should be subcultured indefinitely. Established cell lines grow continuously. The cell lines show a great variation in karyotype. The cell lines retain their viability for years if stored at low temperatures.

Egs: 1. Cell lines derived from normal cells (transformation)

L929, and 3T3 derived from mice

BHK-21 from New born Syrian hamster kidney cells

PK 15 from normal pig kidney cells

RK 13 from rabbit kidney cells

CCL_4 from duck embryo fibroblast

MDBK from Madin Darbey Bovine Kidney cells

MDCK from Madin Darbey canine kidney cells

AGMK, Vero, BGM, BSC-1 from African Green Monkey Kidney cells

RhMK from Rhesus monkey kidney cells

2. Cell lines derived from cancer cells

HeLa, HEp2 and KB derived from human carcinomas

Application of Cell Cultures

1. Disease diagnosis: Primary cultures like calf kidney monolayer, chicken embryo fibroblast monolayer and cell lines like Vero, BHK-21 and MDBK are routinely used for virus isolation and identification
2. Specific identification of viruses by serum neutralization tests
3. Viral vaccine production: Cell culture system also plays a vital role in the development of viral vaccines. The best examples for cell culture vaccines are foot and mouth disease vaccine, rinderpest vaccine, sheep pox vaccine, rabies vaccine and Marek's disease vaccine
4. Production of

   a) Monoclonal antibodies and poly clonal antibodies against viral isolates

   b) Hormones

   eg: Glycoprotein harmones are produced in culture cells and have therapeutic use.

   c) Immunoregulators

   Animal cells are employed in the production of several non-antibody immunoregulators like interferons, interleukins etc.

   e) Tumour specific antigens

   The best example of this is the carcino-embryonic antigen (CEA). This tumour associated antigen produced by adenocarcinoma cells, which are found in sera or other body fluids indicating a state of malignancy. Production of CEA is achieved by its extraction from cultured adenocarcinoma cell lines. This tumour specific antigen is more useful in screening for anti-tumour agents and detecting malignancy in patients using immuno assays.
5. Reconstitution of skin
6. Gene transfer in genetic engineering
7. Toxicity testing of biologicals
8. For screening antitumour and antiviral drugs

The presence of viruses (Virus multiplication) can be detected by several methods as given below

1. Cytopathic effect (CPE)

Many animal viruses that replicate in susceptible cells in cultures are capable of providing **morphological changes** in cells such as size, shape, structural integrity, cells attachment to substrate surface, intracytoplasmic or intranuclear inclusion bodies, and vacuoles in the cytoplasm. These changes are visible either stained or unstained with inverted microscope or ordinary light microscope. These changes in the cells following virus infection are known as cytopathic effects (CPE).

The time at which those changes first become detectable depends on to some extent on the number of virions in the sample and on the growth rate of virus in question. This varies from **hours (entero virus, IBR virus) to several days (adeno virus, rubella virus).**

CPE due to virus activities are manifested in a number of different ways and they are mainly due to

a. virus proteins blocking the host cell protein synthesis

b. physical damage to the cell membranes

c. alteration of the plasma membrane by incorporating virus specific proteins

d. excess production of virus components inducing cytolysis

The CPE is characteristic for each type of virus. The CPE may be seen as,

1. Rounding, clumbing and complete destruction of the cell sheet. e.g. FMD

2. The formation of multinucleated cells, which are called giant cells and the dissociation of cell membranes with the fusion of cell cytoplasm to form syncytia or multinucleate cell (Polykaryons).

   e.g. Paramyxo viruses and Herpes viruses

3. The presence of intracytoplasmic or intranuclear inclusions. e.g. Pox viruses and adeno viruses

4. Transformation of the cells as shown by there altered morphology and piled up masses of cells due to loss of contact inhibition. e.g. Papova viruses.

5. Other changes in the appearance of the cytoplasm including increased granularity and distortion or fragmentation of the nucleus. e.g. Myxoma and some Poxvirus infections.

2. pH changes

Increase or decrease in pH of the cell culture fluid as in the case of Newcastle Disease virus and adeno virus respectively

3. Haemadsorption

Cultural cells infected with **orthomyxo, paramyxo and toga viruses,** which bud from the plasma membranes, acquire the ability to adsorb certain species of red cells (usually guinea pig red cells). This is due to incorporation into the plasma membrane of newly synthesized viral protein that has an affinity for red cells. This technique can be used to recognize infection with non-cytopathogenic viruses as well as the early stages of infection with cytocidal viruses.

4. Interference

The multiplication of one virus in a cell often inhibits the multiplication of another virus entering the cell subsequently. This phenomenon has proved useful for detecting latent or even hither to known infection in monolayer cultures by their reduced ability to support the growth of a challenge virus that is known to produce CPE in healthy cells of the same type.

e.g. Rubella virus interfere the multiplication of entero cytopathic orphan (ECHO) virus

5. Enhancement

As against interference the presence of one virus in a cell can enhance the multiplication of a challenge virus

e.g. Swine fever virus enhances the multiplication of New Castle disease virus.

6. Haemagglutination

Viruses that can agglutinate red blood cells may replicate in cell cultures with or without evidences of CPE. Such viruses can be detected by adding a small amount of the extra cellular cell culture fluid to a suspension of washed erythrocytes and observing the mixture for haemagglutination

7. Detection of viral antigens either in the infected cells or supernatant cell culture fluid employing various techniques such as FAT, CFT, immunoenzyme techniques etc.

8. Assay of cells or cell culture fluid in another host system.

7

Basic Molecular Biology

Molecular biology involves the study of **DNA, RNA and proteins** in cells. It emphasizes the biomolecules that carry out biological processes. One should understand the biomolecules participating in the process to understand the cellular function. In diagnostic microbiology, biomolecules are the analytes for testing and characterization. The analytes for molecular testing are the genomes (DNA), transcriptomes (RNA) and proteomes (Proteins) of the microorganism. Traditional methods of microorganism characterization depend on the culturing of the organisms, which is time-consuming and sometimes challenging to perform. Molecular techniques involve the characterization of DNA, RNA and protein to identify microorganisms. Nucleic acid sequence details are used to classify bacterial species, and mass spectrometry is used to identify microorganisms based on peptide profiles.

DNA (Deoxy ribonucleic acid)

DNA contains the genetic information required for the development and functioning of the living organism. It is the central informational molecule comprising a sequence of A, T, G and C nucelotides linked together by a phosphodiester bond. DNA is double helical in structure, composed of two complementary strands of nucleotides held together by hydrogen bonds between nitrogenous bases.

Structure of DNA: DNA is composed of two complementary strands of nucleotides that are twisted together to form a double helix structure. The nucleotides a recomposed of a sugar (deoxyribose), aphosphate group, and one of four nitrogenous bases: adenine (A), thymine (T), cytosine (C), and guanine (G). The two strands of DNA are held together by hydrogen bonds between the nitrogenous bases. Adenine pairs with thymine (A-T) and cytosine pairs with guanine (C-G). This base pairing is known as complementary base pairing. The two strands of DNA run in opposite directions, one strand runs in the 5’ to 3’ direction, the other runs in the 3’ to 5’ direction. This is known as the The sugar and phosphate groups form the backbone of the DNA molecule, and the nitrogenous bases stick out from the backbone

RNA (Ribonucleic acid)

RNA (ribonucleic acid) is a versatile biomolecule that plays a critical role in storing and transferring genetic information. The structure of RNA is essential for its various functions within cells. RNA is typically a single-stranded molecule, unlike DNA which is double-stranded. The single-stranded nature of RNA allows for greater flexibility in its function. RNA is composed of nucleotides, which consist of a sugar(ribose), a phosphate group, and one of fournitro genous bases: adenine (A), uracil (U),cytosine(C), and guanine(G). RNAcan form base pairs with other nucleic acids, just like DNA, but unlike DNA which pairs A-T and C-G, RNA pairs A-U and C-G. RNA molecules can form specific secondary structures, such as **stem-loop** structures and **hairpin loops**, through base pairing between nucleotides within the molecule. These structures play an important role in the function of RNA. RNA molecules can also form specific three-dimensional structures, such as cloverleaf and L-shaped structures, through base pairing and non-base pairing interactions. These structures are important for the specific functions of some types of RNAs. There are different types of RNA, each with specific functions. These include messenger RNA (mRNA), transfer RNA (tRNA), ribosomal RNA (rRNA) and micro-RNA (miRNA), among others.

Proteins

Proteins are complex biomolecules that perform awide range of functions in living organisms. Proteins are composed of amino acids that are linked together by peptide bonds. The sequence of amino acids, or primary structure, determines the unique properties of each protein. The secondary structure refers to the arrangement of the amino acids in regular repeating patterns, such as alpha helices and beta sheets. The tertiary structure refers to the overall three-dimensional shape of the protein. The three dimensional structure is determined by the interactions between the amino acid side chains. The quaternary structure refers to the organization of multiple protein sub units into a functional protein complex. Environmental factors such as changes in temperature, pH, and the presence of denaturants affect the proteins structure. The stability of a protein is related to its ability to maintain its three-dimensional structure under these conditions. Hydrogen bonds, ionic bonds, disulfide bonds, and vander Waals interactions are important in maintaining the stability of the protein structure. Proteins interact with other molecules, such as other proteins, nucleic acids lipids, and small molecules, in order to perform their functions. These interactions can be of non-covalent interactions, such as hydrogen bonding ionic interactions, and hydrophobic interactions, and covalent interactions such as disulfide bonds and post-translational modifications. Proteins that catalyze chemical reactions are called enzymes. Enzymes speed up chemical reactions by lowering the activation energy needed for the reaction to proceed. Enzymes are specific to the substrates they act on, and the active site of the enzyme is the region where the substrate binds, and there action takes place. Proteins often interact with small molecules called ligands, such as hormones, vitamins, and neurotransmitters. Ligand binding can cause a change in the protein's conformation, leading to a change in its activity. Proteins can also interact with other proteins to form complexes. These interactions can be transient or stable, and specific domains on the surface of the proteins can mediate them. Understanding protein structure, stability, and interactions is crucial for the design of drugs and other biomolecules that can bind to specific proteins and modulate their activity. It is also important for the study of enzymes, protein-ligand interactions, and protein-protein interactions in order to understand the molecular basis of diseases and to develop new therapeutic strategies

The Central Dogma of Molecular Biology

The central dogma of molecular biology is a concept that describes the flow of genetic information within a cell. It states that genetic information flows from DNA to RNA to proteins, and that this flow is unidirectional. The first step in the central dogma is the transcription of DNA into RNA. The enzyme RNA polymerase catalyse the formation of a single-stranded RNA molecule which is complementary to the template DNA strand. The second step in the central dogma is the translation of RNA into proteins. This process occurs in the ribosomes, which are the site of protein synthesis. During the process translation, the sequence of nucleotides in the RNA molecule is read in groups of three, called codons, and each codon corresponds to a specific amino acid. Proteins are the end products of the central dogma and are responsible for living organisms' structural and functional diversity. They have various functions like enzymes, transport proteins, structural proteins, etc

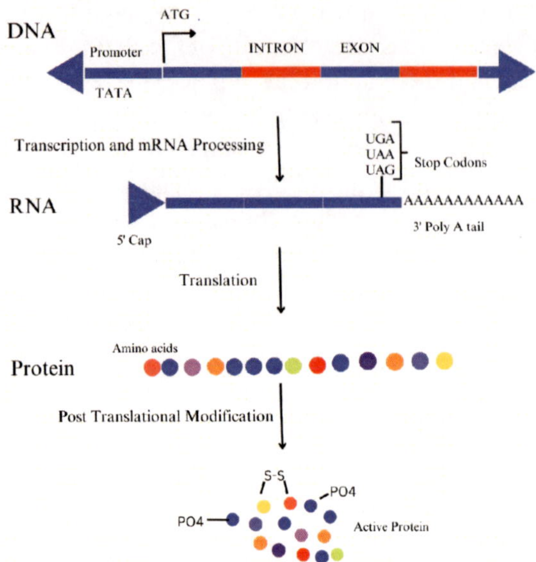

Central Dogma of Molecular Biology: Eukaryotic Mode

Exceptions to the central dogma

Reverse transcription: This is the process by which RNA is converted back into DNA. This process occurs in certain viruses such as retro viruses, where the viral RNA is reverse transcribed into DNA, integrating into the host genome.

RNA Editing: This is the process by which the sequence of nucleotides in an RNA molecule is altered after transcription. This can occur by the addition, deletion or substitution of nucleotides

RNA interference: This is a mechanism by which small RNA molecules can regulate the expression of specific genes. These molecules bind to specific mRNA molecules and prevent the translation of the corresponding protein.

Prion: Prions are protein-based infectious agents that can cause a range of neurodegenerative diseases. They can convert normal proteins into their infectious forms by changing the conformation of the protein

The central dogma of molecular biology is a fundamental concept in biochemistry, and understanding the exceptions to the central dogma is essential for understanding the diversity of molecular interactions in living organisms

Strategies in Veterinary Molecular Diagnosis

Veterinary Molecular diagnosis generally involve samples from farm and pet animals collected by veterinarians. In farm animals, herd is the diagnostic unit, whereas the pet animal diagnosis is individual centric. Molecular techniques are widely used for vaccine development, study on host-pathogens interactions, microbio me and resis to me analyses. In the veterinary micro biology diagnostic setting, molecular diagnostics primarily focuses on the detection, identification and genotyping of viruses, bacteria and fungi. Molecular microbiology has emerged as the leading area in clinical microbiology laboratory and created new opportunities for laboratory diagnosis to affect patient care. It also helps in disease prognosis and monitoring the response to treatment. However, in most cases the new molecular methods supplement rather than replace the conventional laboratory tests for Culture has long been the 'gold standard' for infectious disease diagnosis but for several diseases molecular methods have replaced the culture as the 'gold standard'. Molecular methods have been found to be advantageous in situations in which conventional methods are slow, insensitive, expensive or not available. High loads of pathogens (bacterial/virus) remains a major concern in veterinary molecular diagnosis. Higher concentration of DNA /RNA in the samples , especially in diseases such as FMD /Classical swine fever, will often lead to cross contamination during sampling, handling and processing of samples. Hence proper care and strict implementation of laboratory safety measures is required to avoid this problems in molecular diagnosis laboratories.

Veterinary molecular diagnosis in herd diagnostic units can be used to for diagnostic and monitoring the specific disease status of the herd. In monitoring programmes, testing of representative samples are done at different time points using serological methods. However, vaccination interferes with the serological tests, in such cases, molecular tests are the method of choice. The outcome of Molecular diagnostics can be a **detection** of pathogen, **typing** of the pathogen and **complete genomic sequence determination** of the pathogen. The methods such Polymerase chain reaction (PCR) detects the presence or absence of DNA/RNA of specific pathogen. For further typing of the pathogen, genotyping techniques single nucleotide polymorphism (SNP), multi-locus sequence typing (MLST), multi- locus variable number tandem repeat analys is or 16 rDNA sequencing can be used. With genotyping additional information about the pathogen can be in corporated into the test. In some conditions there will be a need for whole genome sequence information of the pathogen; this Can be done with whole genome sequencing (WGS).WGS iseasie to perform in smaller genomes like virus than that of bacteria. Genotyping techniques should have better discriminatory power; how well the technique can distinguish the unrelated strains of the pathogen. **Hunter-Gaston Diversity index (HGDI)** gives the discriminatory power of the genotyping technique. TheHGDIindexisbasedontheprobabilitythattwounrelatedstrainssampled from a test population will be placed in different typing groups.

Molecular methods are mainly of two major types, amplified methods and non-amplified methods as follows:

Amplified methods

1. Polymerase Chain Reaction (PCR)
2. Transcription Mediated Amplification (TMA)
3. Nucleic acid sequence based amplification (NASBA)
4. Ligase chain reaction (LCR)

Non-amplified method

Hybridisation based methods/Nucleicacidprobes

Amplified Methods

Amplification of nucleic acid is the basis of all the amplified methods. Amplified nucleic acid may be either DNA or RNA.

1. Polymerase Chain Reaction (PCR)

Polymerase chain reaction (PCR) is the target amplification system. PCR permits the synthesis of essentially limitless quantities of at ar get nucleic acid sequence. All the target amplification techniques are sensitive to contamination and can lead to false positive reactions. This drawback must be addressed before any of these techniques are used in the clinical laboratory. However, false positive reactions can be reduced by special laboratory design practices. Other target amplification techniques include TMA and NASBA. Besides originally described PCR, other types of PCR include reverse-transcriptase PCR (RT-PCR), nested PCR, multiplex PCR and real-time PCR.

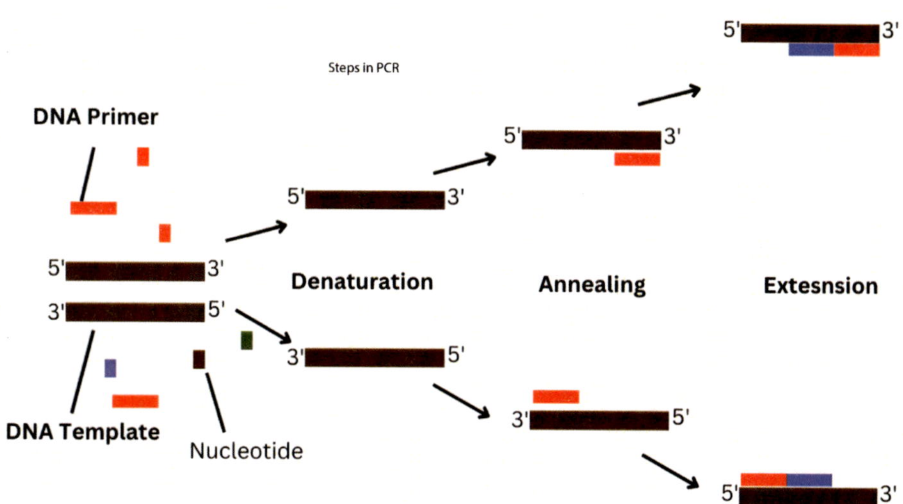

(i) RT-PCR: Reverse-transcriptase PCR (RT-PCR) was developed to amplify RNA. In this technique target is RNA instead of DNA. In this process complementary DNA(cDNA) is first produced from RNA with the help of the enzyme reverse transcriptase and then cDNA is amplified by PCR.

(ii) Nested PCR: Nested PCR was developed to increase the sensitivity and specificity of PCR technique. It uses two pairs of amplification primers. One primer pair is used in the first round of PCR to amplify the desired sequence. The amplified product of the first round is then subjected to second round of PCR with the second set of primers which anneal to the sequences found only in the first round products.

(iii) Multiplex PCR: This technique includes two or more primer sets designed for amplifying different targets. This will help in amplification of more than one target sequence in a clinical specimen. Multiplex PCRs are usually less sensitive than PCRs with single set of primers

(iv) Real Time PCR: Real-time PCR is a major breakthrough for the detection of PCR products. Other names for real-time PCR include kinetic PCR and homogeneous PCR. In real-time PCR, amplified products (amplicons) are detected as they accumulate after each cycle, in contrast to standard PCR where amplicons are seen at the end of the procedure. Thus, a positive result can be obtained quickly, often while the assay units. This technique does not use an agarose gel for detecting amplicons. The imaging system for detection of amplicons is a part of the real time instrument. Another major benefit is that the reactions occur in closed tubes which do not have to be opened for detection. Hence there is very small chance that amplicon from a real- time PCR will contaminate equipment, reagents, and workspaces. Real-time PCR can also be used to quantitate nucleic acids, which is useful for monitoring the certain diseases, such as HIV and hepatitis. Real-time PCR uses a fluorescent reporter dye. Real-time PCR instrument can measure the increase in reporter fluorescence as PCR product accumulates. The computer system records fluorescent peaks as fluorescence intensity versus PCR cycle number

2. Transcription Mediated Amplification (TMA)

Transcription-mediated amplification (TMA) is an isothermal RNA amplification method. RNA target is reverse transcribed in toc DNA and then RNA copies are synthesised with the help of RNA polymerase. A 10” fold amplification of the target RNA can be achieved in about 2 hours. The advantages of TMA include no requirement for a thermal cycler, and contamination risk is minimised.

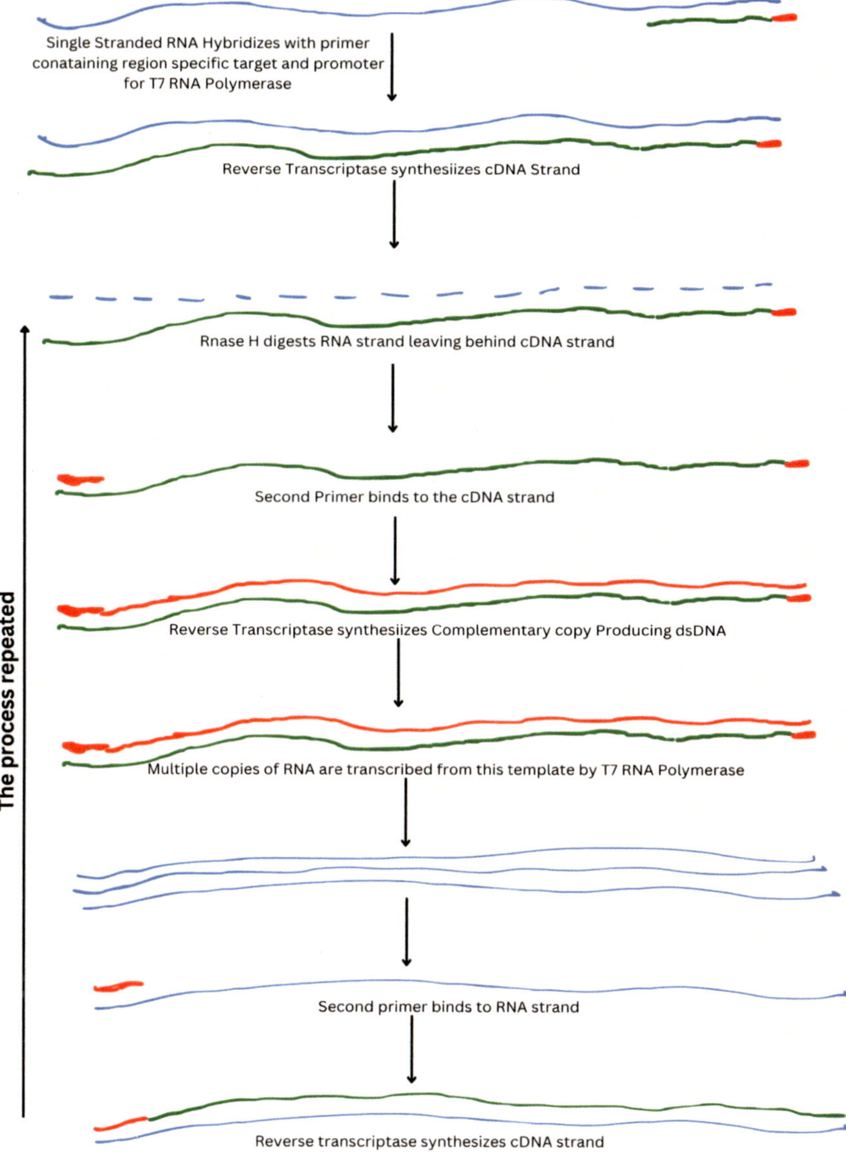

3. Nucleic Acid Sequence-Based Amplification(NASBA)

Like TMA, it is also aniso thermal RNA amplification method. The method is Similar to TMA, RNA target is reverse transcribed into cDNA and then RNA Copies are synthesised with the help of RNA polymerase. It also does not require thermal cycler. AMVRT, RNaseH and T7 bacteriophage RNA polymerase are used in NASBA, while TMA uses an RT enzyme with endogenous RNaseH activity and T7 RNA polymerase. NASBA-based kits for detecting and quantitating HIV-1 RNA and CMV RNA are available.

4. Ligase Chain Reaction (LCR)

Ligase chain reaction (LCR) is a probe amplification method in contrast to above mentioned methods which are target amplification methods. The probe amplification method differs from target amplification method in that amplified products contain only a sequence present in the initial probes. In LCR, two oligonucleotide probes hybridise adjacent to one another on each of the denatured target DNA strands so that a 'nick' is formed. Enzyme DNA ligase then seals then nick by joining thee 3' endofoneprobe and the 5'' endofoneprobeandthe5' end of the other. Each ligated product and the original target serve as a template in subsequent rounds of denaturation, annealing and ligation. It results in an exponential increase of ligated products. Like PCR, LCR also requires a thermal cycler

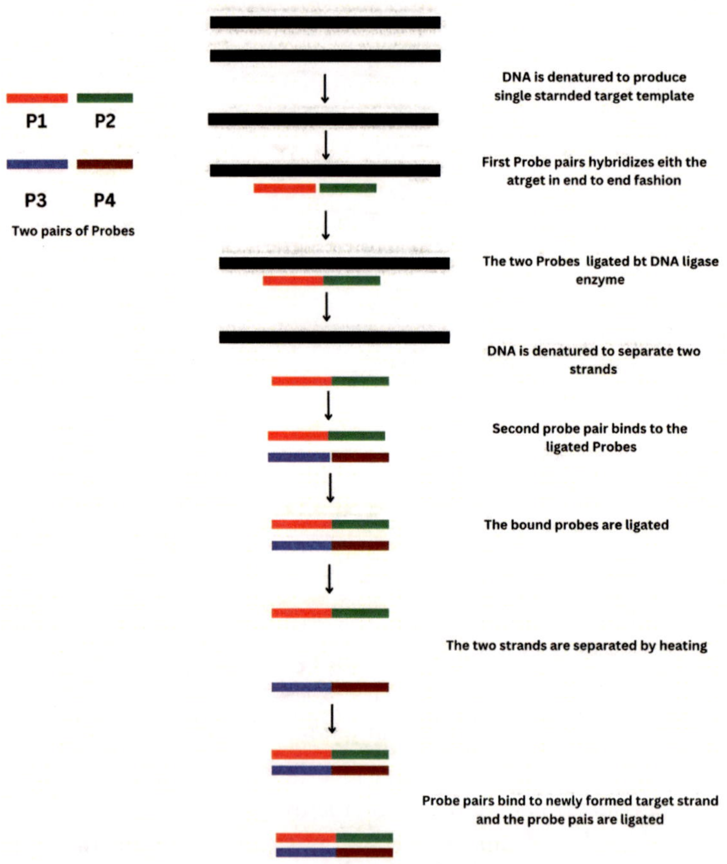

Hybridisation based methods/Nucleic acid probes

Nucleic acid hybridisation can be used to identify DNA or RNA that match a specific sequence. It uses single-strand DNA as a 'probe' to identify specific DNA fragments or clones in a collection that have a sequence complementary to the probe DNA. This in volves the separation of two strands of DNA by exposing to high temperature, low salt conditions or chemicals such as formamide etc. The denaturation process is reversed by lowering temperature, high salt conditions and removing the chemical agents; the process is called renaturation or annealing.

Hybridisation based methods are classified based on its application as

a) Solutionphase

b) Solidphase

c) In -situ

Solution phase hybridization

In this method the mixture of 'target' and 'probe' interact in an aqueous environment. The target DNA is denatured and the single stranded target is mixed with probe. The unhybridized single stranded DNAi a removed using Enzymatic digestion (S1nuclease). The hybridized double stranded DNA is separated using Trichloroacetic acid precipitation (TCA).

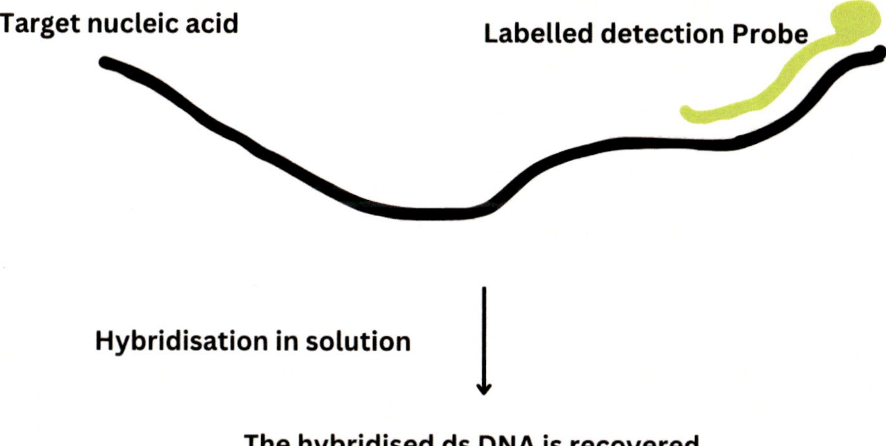

Solid Phase Hybridization

In this method hybridization reaction occurs on a solid support. The following methods are based on solid phase hybridization

a) Dot/slot hybridization

b) Sandwich hybridiation

c) Southern blot hybridization

d) Northern blot hybridization.

In situ Hybridization

The most common and widely used hybridization based method is in situ hybridization which could use fluorescent (FISH) or Chromogenic (CISH) molecules.

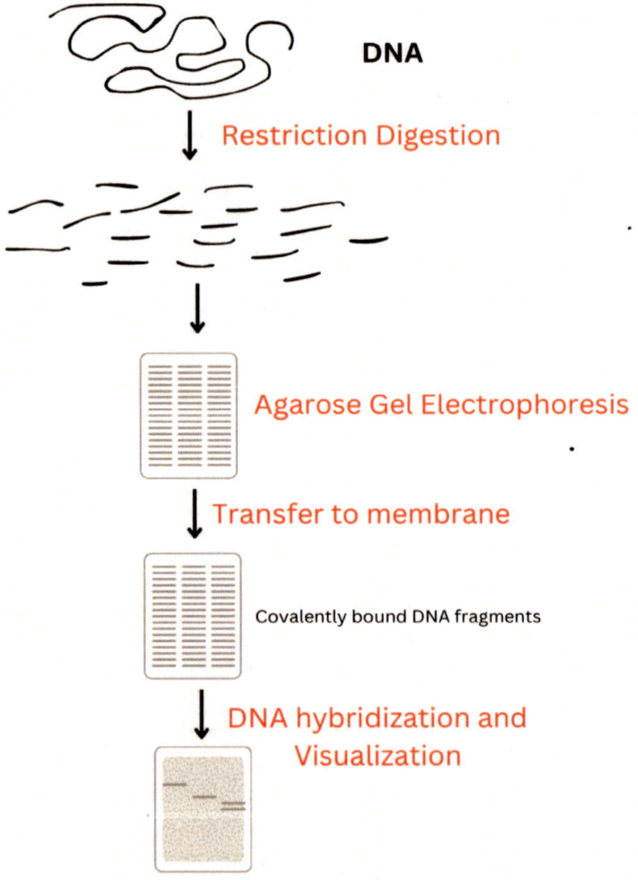

Use of hybridization based methods in Veterinary Microbiology

The chromogenic insitu hybridization assays in used in rapid characterization of mycobacterium, dimorphic fungi in positive culture samples. Fluorescent in situ hybridization based assay for identification and differentiation of *Mycobacterium tuberculosis*.

Use of amplification based methods in Veterinary Microbiology

Polymerase chain reaction is the most common toll used worldwide for the detection of microbial pathogens. Nucleic acid amplification methods are amongst the best in detecting pathogens with high sensitivity and specificity in the clinical samples. Realtime PCR, Multiplex PCR and Reverse transcriptase PCR are the common variants of the amplification methods. Taqman real time PCR RT-PCR assay has been developed for detection of the Japanese encephalitis virus, Realtime quantification is devised for the detection of avian orthro virus and *Salmonella* species.

Use of Genome sequencing in Veterinary Microbiology

The Next Generation Sequencing helped sequence the complete viral genome such as avian avian influenza virus, classical swine fever virus. Nanopore sequencing has proved a revolutionary diagnostic tool in detecting the Ebola virus, influenza viruses and porcine viral enteric disease complexes.

Molecular Epidemiology

Anepidemicis a disease condition that affects manyun related individuals simultaneously. A pandemicisa disease condition hat sweeps across wider geographical area. Epidemiology study involves collecting and analysing environmental, microbiological and clinical data. Molecular epidemiology studies causative genetic and ecological factors at the molecular level to find the origin, distribution and strategies to prevent the disease. In microbial diseases, it involves the determination of genetic similarities and differences between microbial isolates.

Epidemiological Genotyping methods

Plasmidanalysis

Restriction endonuclease mapping

Pulsed field gel electrophoresis

Ribotyping

RAPD PCR

Meltcurve Analyis

Massparallel sequencing

The future of molecular diagnostics

Geno typing methods have significantly increased epidemiological studies and helped implement control programmes efficiently. However, several issues still need to be resolved in the field of molecular typing, such as the lack of a 'gold standard' by which to judge the typing method and issues related to the standardization of laboratory procedures. Despite these problems, molecular techniques will provide for rapid diagnosis with increased specificity and sensitivity and will become the standard practice in microbiology. The future of molecular techniques relies on the innovations on simplification of the methods without the need for skilled personnel and sophisticated laboratories.

8

Agar Gel Precipitation Test

Precipitation and agglutination reactions are similar except in that for precipitation reaction antigen must be in soluble form. The precipitation reaction is highly specific and very sensitive for the detection of antigen being capable of demonstrating antigens in dilutions of 1:100,000 to 1: 1,000,000 or more. It is important to bear in mind that antigens used in precipitation reactions are of molecular size not cellular. A soluble antigen reacted with its homologous antibody at the optimum proportions produce an insoluble precipitate.

The essential ingredients for precipitation test are soluble antigen (precipitinogen), antibody (precipitin) and an electrolyte (normal saline).

a) Precipitation ring test

The antigen is carefully layered over the antiserum, without mixing so that an interface is formed. Diffusion of each reagent will then occur into the other. If the system is homologous, precipitation will occur at the point in the tube, where the proper ratio of antigen to antibody is reached.

Eg: Ascoli's test for anthrax diagnosis.

In advanced decomposition of animals when cultures would be almost impossible, the precipitation method of Ascoli is used for diagnosis of Anthrax. This method is a mixing of immunized rabbit serum with an extract of organs under examination for the production of a precipitate.

Procedure

1. Grind up organs or blood of suspected animals.
2. Suspend in saline.
3. Boil for five minutes
4. Filter through filter paper and allow to cool
5. Carefully stratify 0.5 ml of this extract on 0.5 ml immune serum in small test tubes.

6. Allow to stand for 15 min at room temperature.
7. Definite ring like precipitate forms at junction of two fluids in positive cases.
8. Always run control with known anthrax extract and with normal saline.

b) Precipitation in Gels-Gel Diffusion Test:

Agar gel precipitation test (AGPT) or Ouchterlony test.

Principle

Antigen and antibody placed in wells cut in agar gels diffuse and form an opaque band of precipitate of the antigen-antibody complex in the zone of optimal proportions. Since each antigen combines with its own specific antibody, an antigenic mixture will give multiple bands. Setting up the reaction in adjacent wells assesses the immunological relationship between two antigens.

a) When the lines are confluent it indicates immunological identity in terms of antiserum used.
b) A spur develops with partially related antigens.
c) The lines cross indicate unrelated antigens.

Materials

- Antigen (IBDV)
- Antisera (IBDV antiserum)
- Slide or petri-dish.
- 1% agarose in normal saline. In case of poultry pathogens the composition is Agarose 1g; Sodium chloride 8g, Distilled water –100ml (To avoid contamination during storage and incubation add sodium azide 0.05g or merthiolate 1 in 10,000 concentration).
- Template
- Gel cutter
- Needle
- 50µl micropipette & tips
- Normal saline
- Staining solution.

Coomassie brilliant blue (R 250)	0.5 g.
Ethanol (96%)	45.0 ml
Glacial acetic acid	10.0 ml
Distilled water	45.0 ml

- Destaining solution

  Staining solution without Coomassie brilliant blue.

Procedure

1. Select a clean, scratch free, flat bottom Petridish and pour the melted agarose.
2. After the agarose has solidified, with the help of the template and gel cutter, pattern of wells are cut. The inter-well distance can be varied from 4 to 6mm. The agarose from the wells can be removed with the help of needle. Seal the bottom of wells with the help of 0.5% melted agarose gel.
3. Add IBDV in the central well with the help of separate syringe. Add IBDV antiserum in the peripheral wells
4. Cover the plate with the lid and incubate at 37°C in a box under humid atmosphere (to prevent drying of the gel) for 24 to 48 h. Results can be read after 24-48h. A single antigen will give rise to a single precipitation line in presence of homologous antibody. Antigenic preparation from a bacterium may possess several somatic, flagellar or capsular antigens, each of which will form a separate precipitation line, if the corresponding antibodies are present in the antiserum. Also observe for line of identity, line of partial identity and line of non identity.

Staining of Precipitation Bands

After optimal development of precipitation patterns, the plates can be stained for considerable improvement in sensitivity.

The plates are submerged in normal saline and change saline in hourly intervals several times to remove non-precipitated proteins. Cracking of gel because of air bubbles can be avoided by filling the wells with a drop of agarose solution. Salt is removed by keeping the plates in distilled water for one hour. Gels are finally dried at 37°C overnight by keeping a wet filter paper over them. After drying, filter paper may be removed by slight wetting. The gels are then stained for 15-30 minutes by using the Coomassie brilliant blue stain followed by destaining.

9

Enzyme Immunoassay

Among various immunodiagnostic tools enzyme immunoassay becomes increasingly popular for the detection of humoral immune response (seroconversion) or microbial antigens for the purpose of diagnosis.

Enzyme Linked Immunosorbent Assay (ELISA) is widely used in studies of infectious diseases.Despite several merits of ELISA, it has some limitations such as requirement of ELISA reader to obtain accurate measurements. To overcome this, the dot immunobinding assay was developed .In Dot Immunoassay nitrocellulose membrane is used as an adsorbent matrix instead of plastic surface.The antigen is dotted onto the nitrocellulose membrane and the membrane then incubated with test antibody, and secondly with enzyme conjugated antibody directed against the first antibody.In case of positve reaction, (specific antigen –antibody reaction) color dot is developed against the white background at the site of antigen deposition.In case of negative reaction, no color development takes place at the site of antigen deposition .

Principle

- In direct ELISA antibody is directly attached to solid phase and targeted by added antigen (detecting antigen).
- These added antigens are targeted by antispecies antibodies linked to enzyme termed as conjugates.
- The bound enzyme is developed by the addition of substrate /chromogen, then stopped and finally read using a spectrophotometer

Materials

- ELISA plates
- Micro pipettes ,both single and multiple channel
- Antigen
- Antibody
- Phosphate buffered saline (PBS) pH 7.4

- Carbonate bicarbonate buffer pH 9.6
- Substrate solution
- 1% Bovine serum albumin
- Antispecies IgG – HRP conjugate
- Stopping reagent – 1M H_2So_4
- ELISA reader

Procedure

- Calculate the amount of antibody to be used for coating the plates by checker board titration.
- Add 100 µl of antibody to all the wells and incubate the plates at humid chamber at 4° C over night.
- Empty the plates and wash thrice (5min each) in PBST washing solution. Tap the plate against any soft paper or towel after each washing to get rid of any unabsorbed material in the well.
- For blocking the unsaturated sites of polystyrene plate well, add 100 µl of 1 % BSA to all the wells and incubate at 37°C for one hour.
- Wash the plates 3 times with washing solution and tape as mentioned above.
- Add 100 µl of appropriate diluted antigen to the first well and serially dilute antigen as required
- Keep negative control.
- Keep positive controls.
- Incubate the plate at 37°C for 1 hr .
- Wash and tape the plate as mentioned above.
- Add appropriate dilution of 100 µl of anti species IgG HRP conjugate to all the wells and incubate the plate at 37° C for 1 hr
- Wash and tape the plate as mentioned above.
- Add 100 µl of substrate to all the wells. Keeping a substrate controls, the plates are incubated at room temperature for 30 minutes for development of colour.
- At the end of incubation period stop the reaction by adding 100 µ of appropriate stopping reagent.
- Read the absorbance value on a ELISA reader using 492 nm wavelength filter.

10

Agglutination Test

Principle

The principle of the test is that antibodies possess two antigen-binding sites that will attach specifically with the antigenic determinants on the surfaces of the bacteria, cells or particles. Under suitable conditions one antibody molecule will combine with the determinant group on the surfaces of two bacteria and in this way a lattice is formed. These lattices are generally visible with the naked eye as clumps and they sediment readily due to the large size of the clump.

Pre-requisites for Agglutination Reaction

a) The antigen (agglutinogen) should be in particulate form, which will remain in suspension for sufficiently long time.

b) The antibody (agglutinin) should be directed to the target surface epitope.

c) Antigen – antibody should be present in optimal proportions.

d) An electrolyte is necessary for the reaction to take place and usually the reaction is carried out in physiological saline solutions (0.85% NaCl). Because of the overall negative charge on red cells or bacteria at neutral pH it is difficult for them to come close enough, so that for molecule such as IgG antibodies are able to form a bridge between epitope of two different cells or bacteria. The presence of salt tends to neutralize the charge effects to allow agglutination.

Plate Agglutination Test

This is employed to screen sera within a short period for preliminary testing.

Materials

- *Brucella abortus* coloured antigen
- suspected serum
- glass plate.

Procedure

1. Place one loopful of the suspected serum and one loopful of *B. abortus* antigen adjacent to each other on a glass plate.
2. Mix the reagents with an inoculation loop and rotate the plate while tilting the sides slowly. Watch for the agglutination reaction.
3. Check the antigen for auto agglutination by replacing the serum with a loopful of saline.

Abortus Bang ring Test (ABRT)

This test is also known as Milk Ring Test (MRT). This test is employed to test milk from brucella infected cow/buffalo herd.

Materials

- Milk from brucella infected and non-infected herd
- Tetrazolium stained *B. abortus* antigen
- Test tubes for conducting the test.

Procedure

1. Pipette 2 ml of milk in 2 different tubes from each sample. Add 1 drop of brucella ring test antigen and mix thoroughly.
2. Incubate the milk for 1 hour at room temperature or for 30 min at 37°C in a water bath.
3. Observe the result and record.

Interpretation

The clumps of agglutinated organisms are carried to the surface of the milk along with fat globules. A positive test is indicated by a column of decoloured milk with a red cream layer on the top. A negative test will show red colour of the milk throughout with an uncoloured cream layer on top.

Tube Agglutination Test for Brucella

The tube agglutination tests are more accurate than the plate test, but require relatively long incubation time.

Materials

- Dryer's agglutination tube.
- Test serum and positive serum

- Brucella abortus plain antigen
- Normal saline

Procedure

1. Set up six agglutination tubes in a row and add 0.8 ml of saline to tube No.1 and 0.5 ml to the other five tubes.
2. Add 0.2 ml of test serum to tube No. 1. Mix well and then transfer 0.5 ml to tube No.
3. Repeat two fold dilutions upto tube No.6 and then discard 0.5 ml (dilution 1:5, 1:10 and so on in each tube).
4. Add 0.5 ml of plain antigen to each tube and mix well (final dilution of serum resulted will be 1:10, 1:20 and so on).
5. Control tubes are also set in the test.
   a. Positive control with 0.5 ml of known positive serum and 0.5 ml plain antigen.
   b. Negative control –with 0.5 ml saline and 0.5 ml plain antigen.
   6. Because of the special significance of the 50% end point, a control tube should be set up with 0.75 ml saline and 0.25 ml plain antigen.
7. All tubes are then incubated at 37°C for 18-20 h before the results are read.

Result Recording

The degree of agglutination is determined by reading the degree of clearing before shaking the tubes. The highest serum dilution showing 50% agglutination (ie 50% clearing) or more is taken as the end point or titre in the serum.

A titre of 1:40 or above indicates an infected animal.

11

Haemagglutination Test

Principle

Some viruses like Orthomyxo, Paramyxoviruses etc. have the property to clump the erythrocytes of certain species of birds or animals. This is not an antigen-antibody reaction but a network of erythrocytes formed by the bridging virions. These virions on their envelop have peplomers that binds with the receptors on the erythrocytes.

If large quantities of virus are used for HA and allowed to act on red blood cells for a long period of time, the agglutinated red cells shed the virus particles which then float freely in the suspending fluid and neither attach to the same red cells nor agglutinate them. This process of dissociation of virus from the red cells is known as elution, and is due to the destruction of receptors on red cells by the virion associated receptor destroying enzyme - neuraminidase. The red cells freed from the virus particles are known as stabilized red cells. The elution will occur most rapidly if the temperature is raised to 37°C as the enzyme neuraminidase will be more active at this temperature.

Practical Application of Haemagglutination

a) Spot test: For rapid detection of the presence of a haemagglutinating virus. This can be done by mixing two or three drops of suspected sample (infected chick embryo fluids or cell culture fluids) with equal quantity of 5% chicken red cells in normal saline. The appearance of the clumps of agglutinated red cells, indicate the presence of a haemagglutinating virus in the sample.

b) Titration of the virus using the haemagglutinating property.

Micro Haemagglutination Test

Materials

- Normal saline

  0.85 g sodium chloride in 100 ml of distilled water.

- Alsevers solution - anticogulant used during collection of blood for preparation of erythrocyte suspension.

  Sodium chloride – 4.2 g.

  Sodium citrate - 8.0 g.

  Dextrose - 20.5 g.

  Citric acid - 0.55 g.

  Distilled Water - 1000 ml

  Sterilize at 10 lbs for 10 min and store at 4°C

- 1% chicken erythrocyte suspension:- Collect blood from four numbers of 6-week-old chickens in Alsevers solution in the proportion of 1.1 ml of Alsevers solution for each ml of blood collected. Wash the erythrocytes thrice by gentle centrifugation in sterile normal saline at 1500 rpm for 5 min. Prepare 1% suspension by adding 99 ml of sterile normal saline to 1 ml of the packed erythrocytes and store at 4°C.
- Microtitre plates – 'V' bottom.
- Micropipette – 25 μl and tips.
- Antigen: Allantoic fluid collected from embryonated eggs infected with Newcastle disease virus.

Procedure

1. Add 25 μl of normal saline in all the wells in the first row.
2. Add 25μl of the virus containing allantoic fluid in the first well and make serial two-fold dilution. Discard 25μl of the suspension from the eleventh well.
3. Add 25 μl of 1% chicken RBCs in all the 12 wells. The last well is considered as RBC control.
4. Shake the plate and leave it for 20 min.
5. Look for lattice formation ie. haemagglutination. The end point is the reciprocal of the highest dilution showing 100% haemagglutination. The control well should show button formation.
6. Calculation of 4HA

$$4\text{ HA} = \frac{\text{HA titre}}{4}$$

(e.g). HA titre is 1 in 128.

$$4\ \text{HA} = \frac{128}{4} = 32$$

1 ml of the virus infected allantoic fluid + 31 ml of the sterile normal saline will give a suspension that has 4 HA in every 25 μl. This will be used to conduct HI test.

Haemagglutination – Inhibition Test

Principle

The ability to cause agglutination of erythrocyte of certain species of animals and birds is found in many virus families. This property of haemagglutination is inhibited by the specific antiserum of the virus. If specific antibody and virus are mixed prior to the addition of red blood cells, haemagglutination is inhibited. The HI test is used either to type antigen or to measure antibody.

The procedure for the haemagglutination inhibition (HI) test consists of adding either a standard amount of test serum to dilutions of a virus (Alpha method) or a standard amount of virus to dilutions of the test serum (Beta method). The Beta method is the more usual procedure and is also the more accurate whereas alpha method enables large numbers of sera to be tested in the shortest time possible during an out break of disease in domestic animals or birds.

Micro Haemagglutination Inhibition Test

Materials

- No.1 to 5 as given in Haemagglutination test
- 4 HA antigen

Procedure

1. Add 25 μl of normal saline in all the wells in the first row.
2. Add 25 μl of the serum in the first well and make serial two-fold dilution. Discard 25 μl of the suspension from the last well.
3. Add 25 μl of virus suspension containing 4 HA units to all the wells. Leave for 20 min.
4. Add 25 μl of 1% chicken RBC to all the wells and leave for 45 min.
5. RBC control:- Add 25 μl of saline to a well to which add 25μl of 1% chicken RBC.

6. Check titration:- Add 25µl of saline to 4 wells in a row. Add 25µl of the 4 HA virus suspension to the first well and make serial two –fold dilution. Discard 25µl of the suspension from the fourth well. Add 25µl of 1% chicken RBC to all the four wells. Leave for 20 min. The first two wells will show haemagglutination if the viral suspension prepared is having 4HA units of virus in every 25 µl.
7. The end point or HI titre of the serum is the reciprocal of the highest dilution showing 100% inhibition of haemagglutination.

12

Introduction and History of Microbiology

Microbiology is the study of microorganisms or microbes. They are very small in size and hence cannot be seen with naked eyes. **Microorganisms** are excellent models for understanding cell function in higher organisms, including humans. Hence the science of microbiology is the foundation of all the biological sciences and microorganisms consist of bacteria, fungi, algae, protozoa and viruses. The fields according to the organisms studied are henceforth termed as:

Bacteriology	- Bacteria (singular: *bacterium*)
Phycology (*phyco*, seaweed)	- Algae (singular: *alga*)
Mycology (*myco*, a fungus)	- Fungi (singular: *fungus*)
Protozoology (*proto*, first; *zoo*, animal)	- Protozoa (singular: *protozoan*)
Virology	- Viruses

Microbiology began when people learned to make lenses from glass and combine them to produce magnifications great enough to see microbes. The importance of bacteria in the lives of people, animals and the environment has been unfolded in the past 100 years. It was also believed that the incidence of the disease was due to divine reasons since man had no idea of microbes at this point of time. In the earlier days, leprosy was considered as a dangerous infectious disease and spread of the disease was attributed to direct contact with the affected persons. Hence, certain procedures were specified for segregation of the affected persons. At this point of time various theories were also attributed to the genesis of disease. Some of the important theories are as follows:

1. **Theurgical theory** – Disease was attributed to the wrath of divine spirits as a punishment of individual sins.
2. **Miasmatic theory** – This theory was proposed by Hippocrates and further developed by his student Galen. It states that the disease was due to the emanations from the earth, the influence of the stars, moon, wind, water and the seasons.

3. **Pore theory** was supported by Asclepiades (124 B.C.), Themison (143 – 23B.C.) and Thessalus (60 A.D.). This theory was based on the fact that a symmetry of proportions of the pores resulted in health and a disproportion of pores caused disease.
4. Galen (120 – 200 A.D.) revived the Miasmatic theory and elaborated them with his own observations and imaginations.
5. At this same point of time some of the naturally occurring events also aroused man's curiosity like appearance of maggots on carcass and a new theory called **theory of spontaneous generation** was developed. This theory states that living beings are produced spontaneously from non living matter. The formation of life from non living matter is called **abiogenesis**. It was believed that maggots would emerge from dead flesh without eggs being laid and snakes and toads could be born in soil without prior contact with snakes or toads.

Various workers had conducted experiments to disprove this theory. **Francesco Redi (1688)** an Italian physician disproved this idea of spontaneous generation by showing that rotting meat carefully kept from flies will not spontaneously produce maggots. He filled six jars with decaying meat and covered three of them with lids. After several days meat in the open jars had maggots while the meat in the sealed jars had no maggots. **Spallanzani** (1769), a monk, boiled and sealed broths. When he was careful no microbes developed. **Needham** said that other factors, excluded by Spallanzani, were needed for Spontaneous Generation, notably air. He conducted the same experiments but sloppily done, in which microbes grew from contamination. The theory's brief success was due to the lack of knowledge about sterilisation and microbes.

Louis Pasteur (1861) took up the challenge and utilized broths allowing air but disallowing microbes. He used broths closed with cotton and showed that the germs accumulated on cotton. He performed the **Swan neck flask experiment** and concluded the following observations:

- The bended neck allowed air to enter the bottle and the liquid but trapped any particulates including microorganisms.
- No microbial growth was present as long as the liquid broth did not come in contact with the microbes.
- Hence air alone was not sufficient to generate life.

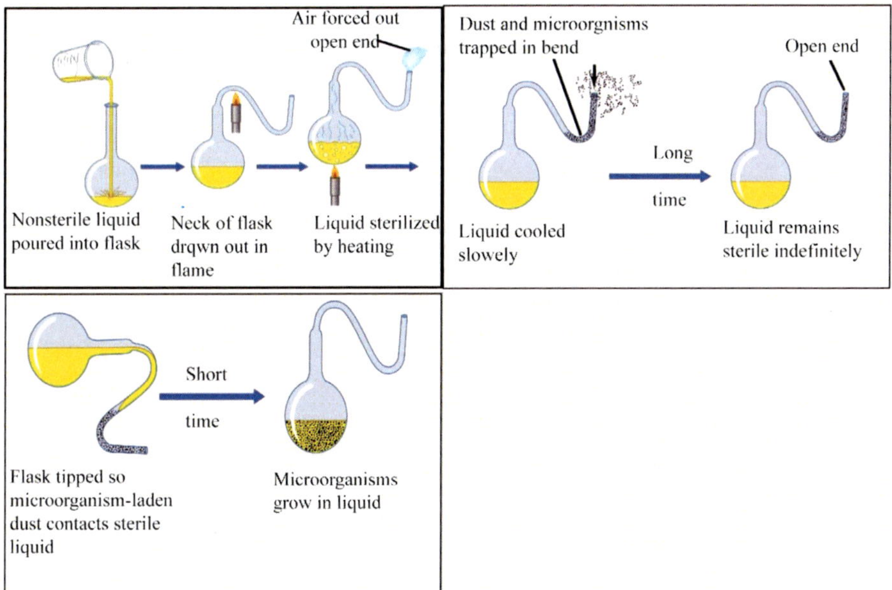

6. Theory of biogenesis - Rudolph Virchow (1858) claimed that living cells can only arise from pre-existing living cells. This was proved by Louis Pasteur and other scientists also proved this concept called **biogenesis.**

7. Germ theory of disease was laid by **Jacob Henle** in 1840. The first absolute proof of the germ theory came as a result of the classical work of **Robert Koch** on anthrax in 1876. **Joseph Lister** proved that pathogenic microbes could be transmitted from patient to patient in surgical wards if physician's hands were not washed, instruments were not sterilized and surgical wounds were not disinfected.

The early developments in the field of microbiology are listed below in chronological order

- **Fracastorius of Verna** (1546) said that a *Contagium vivum* is the cause of infectious disease. It is based on microbial origin of disease. Later, **von Plenciz** (1762) substantiated this theory.
- Around 1665, the English scientist, **Robert Hooke** built a **compound microscope** and used it to observe thin slices of cork. He coined the term "***cell***" to describe the orderly arrangement of small boxes that he saw because they reminded him of the cells (small, bare rooms) of monks.
- **Antony Van Leeuwenhoek (1632- 1723)** of Holland got the credit for the discovery of the microbial world. He ground lenses and

made microscopes as a hobby. He observed cells with an improved microscope. The resolution was enough to allow about 300 to 400 X useful magnification.

Leeuwenhoek saw bacteria, protozoa, yeasts and described all the microbial forms we now know, excepting viruses. He described the microbes he saw under the microscope as "**animalcules**" meaning **little animals.**

- **Carl Linnaeus** (1735) developed the system of binomial naming of living organisms.
- **Muller** (1786) - First classification of bacteria
- **Schwann** (1837) - Identified alcoholic fermentation, (father of fermentation)
- **Tyndall** – Established a sterilisation procedure called ***Tyndallisation*** in which the fluids were heated at periodical time intervals. This method was proved effective in killing vegetative organisms and spores.
- **Joseph Lister** (1827-1912) – Developed anti-septic wound dressing procedure, which saved the lives of persons with, suspected wound contamination.
- **Ignaz Sommelweis** (1850) advocated washing of hands to stop the spread of disease
- **Hansen** (1874) – Identified the bacteria causing leprosy in man.
- **Hess** (1882) developed agar for the cultivation of bacteria.
- **Paul Ehrlich** (1882) - developed acid fast stain
- **Hans Christian Gram** (1884) – Grouped bacteria into two groups based on a staining procedure called Grams staining.
- **Loeffler** (1884) – Identified the bacteria causing diphtheria.
- **Escherich** (1885) - Identification of *Escherichia coli* as natural inhabitant of human gut
- **Fraenkel** (1886) – Identification of Pneumococcus
- **Petri** (1887) developed petri dish that is still widely used in the cultivation of bacteria.
- **Pfeiffer** (1890) - Identification of Pfeiffer bacillus, *Haemophilus influenza*

- **von Behring and Kitasato** (1890) - Immunization of animals with diphtheria toxin
- **Dmitri Iwanowsky (1892)** – Discovery of virus (Tobacco mosaic virus)
- **Bordet** (1895) - Discovery of alexin (complement) and hemolysis
- **Widal and Grunbaum** (1896) - Development of diagnostic test based on agglutination of typhoid bacilli by immune serum
- **Loeffler and Frosch** (1898) - Discovery of Foot and mouth disease virus
- **Beijerinck** (1899) - Discovery of intracellular reproduction of tobacco mosaic virus
- **Twort** (1915) - Discovery of viruses that infect bacteria (bacteriophages)
- **d'Herelle** (1917) - Independent rediscovery of viruses that infect bacteria
- **Alexander Fleming** (1928) – Discovery of Penicillin
- **Ernst Ruska** (1932) - First Electron microscope

Contribution of Eminent Scientists in Microbiology

- **Louis Pasteur** (1822-95) – No other person has contributed so significantly than Louis Pasteur for the subject in earlier days.
    - He has studied the fermentation process in wine manufacture and identified certain substances which are formed only by the living organisms and laid down certain guidelines for clean production of wine which saved the French wine industry in those days. Hence he is also called ***saviour of French wine industry.***
    - He also developed technique for laboratory cultivation of bacteria.
    - His accidental finding on loss of virulence by *Pasteurella multocida,* causing fowl cholera and systematic work on *Bacillus anthracis* causing anthrax and rabies virus paved the way for control of infectious diseases by a process called vaccination.
    - He evolved the practice of heating to 120°C under pressure for sterilisation, which even killed spores (a forerunner for sterilisation by moist heat (autoclaving) and dry heat (hot air oven - 170 °C).
    - In1885, he performed the first rabies vaccination

For these significant contributions, Louis Pasteur is called *Father of Microbiology*.

- **Robert Koch** (1843-1910) – A German physician who moved from the field of medicine to microbiology and has contributed significantly.
  - His notable achievements were the two important concepts called *Koch's Postulates* in 1882(which deal with the establishment of aetiology of infectious diseases) and *Koch's Phenomenon* (which gave firm basis for cell mediated immune response).
  - He also established the procedure for preparation, fixation and staining of smears for bacterial identification by aniline dyes.
  - He also developed nutrient gelatin and nutrient agar (solid media) for artificial cultivation of bacteria (1881).
  - He identified bacteria causing tuberculosis (*Mycobacteium*-1882) and cholera (*Vibrio- 1883*)and disproved the idea that tuberculosis was inherited. He made several visits to Africa, at least two visits to Asia, and one to the United States. He conducted research on malaria, typhoid fever, sleeping sickness, and several other diseases. His studies of tuberculosis won him the Nobel Prize for physiology and medicine in 1905.
  - His visit to IVRI at Mukteshwar, India is noteworthy to remember.

Koch's Postulates

1. The specific causative agent must be found in every case of the disease.
2. The disease organism must be isolated in pure culture.
3. Inoculation of a sample of the culture into a healthy, susceptible animal must produce the same disease.
4. The disease organism must be recovered from the inoculated animal. Edward Jenner (1796): Small pox vaccination

In his time milkmaids and villagers often said that "if you want to marry a woman who will never be scarred by the pox, marry a milkmaid." Hence Jenner thought that previous infection with cowpox could offer protection against smallpox. He collected scrapings of cowpox lesions from the fingers of **Sarah Nelmes**, a young milkmaid and injected it into **James Phipps**, an 8-year-old boy. James got mild fever and typical cow pox lesions. A few weeks after recovery, Jenner injected James with the live smallpox virus and found that the boy was protected from the disease. This method was very efficient that by 1840 the British government had banned alternative preventive measures against smallpox. Jenner invented the word **"Vaccination"** for his treatment (from Latin ***vacca***, a **cow**). Pasteur adopted this word for immunization against any disease.

Discovery of important causative organisms

Year	Disease	Organism	Scientist
1876	Anthrax	*Bacillus anthracis*	Robert Koch
1880	Typhoid	*Salmonella typhi*	Eberth
1882	Glanders	*Burkholderia mallei*	Loeffler and Schultz
1884	Diphtheria	*Corynebacterium diphtheria*	Loeffler
1885	Tetanus	*Clostridium tetani*	Nicolaier
1894	Plague	*Yersinia pestis*	Kitasato and Yersin
1895	Fowl typhoid	*Salmonella gallinarum*	Moore
1897	Brucellosis	*Brucella abortus*	Bang

Golden Age of Microbiology

The period between 1857 and 1914 is called as Golden age of Microbiology since during this period numbers of important discoveries were made. The two noted scientists during this period were Louis Pasteur and Robert Koch.

13

Classification and Nomenclature of Bacteria

Classification of Organisms

Over the years, scientists have developed several systems for the classification of organisms. Greek philosopher Aristotle (384-322 BC) grouped life forms as either plants or animals. Microscopic organisms were unknown at that time. Fungi were included in plants. Aristotle's system distinguished only between plants and animals on the basis of movement, feeding mechanism, and growth patterns. In 1735, Carolus Linnaeus formalized the use of two Latin names to identify each organism, a system called **Binomial nomenclature**. He grouped closely related organisms and introduced the modern classification groups: kingdom, phylum, class, order, family, genus, and species. At that time single-celled organisms were observed but not classified. This classification had two kingdoms.

Plantae: containing Plants and Fungi

Animalia: containing Animals

In 1866, Ernst Haeckel proposed a third kingdom- **Protista,** to include all single-celled organisms like amoebas and diatoms. Some taxonomists also placed simple multicellular organisms, such as seaweeds, in Kingdom Protista.

In 1938, Herbert Copeland proposed a fourth kingdom, **Monera,** to include only bacteria. This was the first classification proposal to separate prokaryotes from eukaryotes, at the kingdom level.

In 1957, Robert H. Whittaker proposed a fifth kingdom –**Fungi.**Fungi do not ingest food as animals do, nor do they make their own food, as plants do. They secrete digestive enzymes around their food, breaking it down before absorbing it into their cells.

	Prokaryote	Eukaryote			
Kingdom:	Monera	Protista	Fungi	Plantae	Animalia
Organisms:	Bacteria	Amoebas, diatoms, and other single-celled eukaryotes, and sometimes simple multi-cellular organism, such as seaweeds.	Multicellular, filamentous organisms that absorb food	Multicellular organisms that obtain food through photosynthesis	Multicellular organisms that ingest food

In 1990, **Carl Woese** proposed a new category called a **Domain,** to reflect evidence from nucleic acid studies that more precisely reveal evolutionary, or family, relationships. He found that a group of organisms previously classified under Bacteria belong to a separate taxon. They are the **Archea.** They have unique molecular structures and physiological characteristics. They live in extremely hot, salty or acidic anaerobic environments. He suggested three domains, **Archaea**, **Bacteria** and **Eucarya**, based largely on the type of ribonucleic acid (RNA) in cells.

The Archaea and Bacteria domains contain prokaryotic organisms. The Eucarya domain contains eukaryotes. This domain is further subdivided into the kingdoms - Protista, Fungi, Plantae, and Animalia.

	Prokaryotes			Eukaryotes			
Domain	Archaea		Bacteria	Eucarya			
Kingdom	Crenarchaeota	Euryachaeota		Protista	Fungi	Plantae	Animalia
Organisms	Ancient bacteria that produce methane	Ancient bacteria that grow in high temperatures					

What are the differences among microbes?

Characters	Protozoa	Algae	Fungi	Bacteria	Virus
Chlorophyll	-	+	-	+ or -	-
Cell wall	-	+	+	+	NA

How are prokaryotes and eukaryotes differentiated?

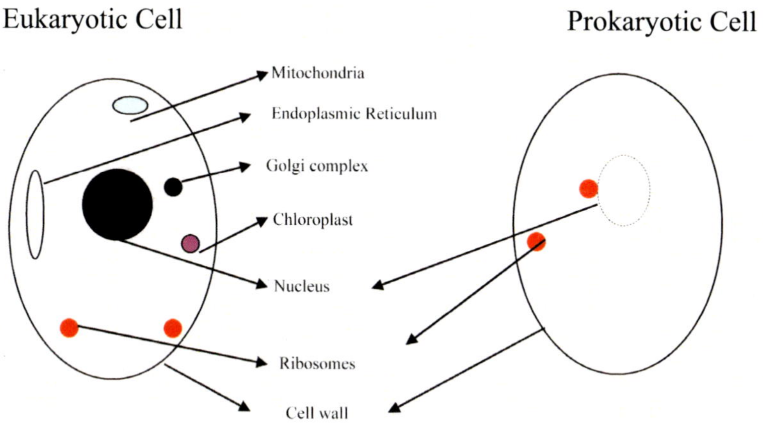

Eukaryotic Cell

Prokaryotic Cell

Mitochondria
Endoplasmic Reticulum
Golgi complex
Chloroplast
Nucleus
Ribosomes
Cell wall

Characteristics	Prokaryotic cells	Eukaryotic cells
1. Major groups	Bacteria, blue green algae	Algae, fungi, protozoa, plants and animals
2. Cell wall	Contains peptidoglycan, lipids and proteins	Absent. If present contains chitin or cellulose (Green plants)
3. Nuclear structure i) Nuclear membrane ii) Chromosome iii) Ploidy iv) Transcription/ Translation v) Histones	 - Absent - Single, closed, circular, ds DNA --- Haploid - Continuous, with short lived mRNA and polysome formation - Absent	 - Present - Multiple, linear chromosomes - Diploid, haploid (Fungi) - Discontinuous, long lived mRNA. Transcribed in nucleus and translated in cytoplasm. - Present
4. Cytoplasm Ribosomes Mitochondria Golgi complex Endoplasmic- -reticulum Cytoplasmic- -membrane	Present – 70 S Absent Present, made of Phospholipids and no sterols (except Mycoplasma)	Present – 80 S Present Present, made of Phospholipids and sterols.
5. Motility	Flagella (Simple)	Flagella (Complex)
6. Energy generation	Cytoplasmic membrane associated	Mitochondria
7. Reproduction	Asexual	Sexual
8. Recombination/Gene exchange	Chromosomal or plasmid gene exchange via transformation, transduction or conjugation	Meiosis results in genetic recombination.

Classification And Nomenclature of Pathogenic Bacteria

Classification is the arrangement of bacteria into taxonomic groups on basis of their similarities or relationships.

I. **Taxonomic groups**

Bacterial taxonomy ranks from kingdom to subspecies and each of them are described as follows;

1. Kingdom

All prokaryotic organisms including bacterial pathogens of man and animals and eukaryotes including yeast, fungi and protozoa as well as higher plants, domestic animals & man are classified under the kingdom.

2. Division

Bacterial pathogens of domestic animals and man are classified into three divisions based on cell wall type:

Division	Type of cell wall
Firmicutes	Gram positive
Gracilicutes	Gram negative
Tenericutes	No rigid cell wall

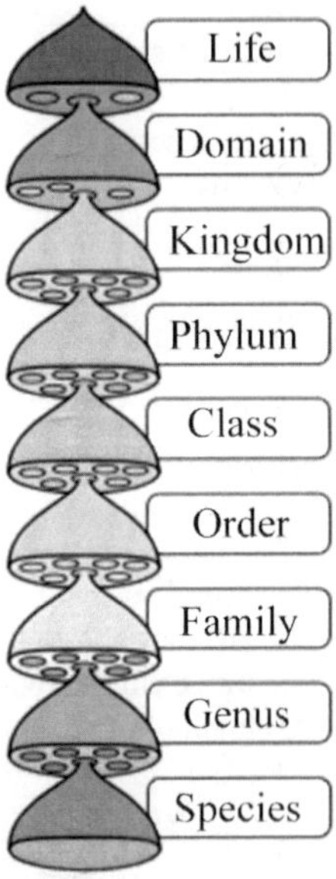

Life
Domain
Kingdom
Phylum
Class
Order
Family
Genus
Species

3. Classes

- are seldom used.

4. Order

Orders are seldom used in systematic classification with exceptions like chlamydiales and Rickettsiales, which contain the obligate intracellular pathogen.

5. Families

Families represent groups of related genera or groups of genera that are based on practical convenience and have limited applicability in understanding their characteristics.

6. Genera

The bacterial genus is usually a well defined group that is clearly separated from other genera. The genus name often indicates the morphology of the species or the organism's discoverer.

7. Species

The species is a collection of strains that share many properties. The species is the basic taxon of bacteria except in leptospira and salmonella in which serovar is the basic taxon.

8. Subspecies

Subspecies is the lowest taxonomic rank in nomenclature. A species may be divided into two or more subspecies. It is often used to facilitate the description of organism within a species with notable differences in host range or other important characteristics.

Taxonomic Classification and Terminology within Species

i) **Strains:** A strain is made up of descendants derived from an initial colony isolated from an exogenous source, such as pathologic lesions. A type strain serves as the permanent example of the species. A reference strain is often used for comparative infectious disease studies.

ii) **Biovars:** Biovars have special biochemical or physiological properties.

iii) **Serovars:** Serovars (serotypes) have distinctive antigenic properties. For serotyping bacteria, capsular antigen, somatic antigen and flagellar antigen are used.

iv) **Pathovar:** Bacteria having pathogenic properties for certain hosts.

v) **Phagovar:** Bacteria with the ability to be lysed by certain bacteriophages.

vi) **Morphovar:** Bacteria having special morphological features.

The most important properties considered for the bacterial classification include

- Colony morphology
- Cell shape & arrangement
- Cell wall structure (Gram staining)
- Special cellular structures
- Biochemical characteristics

Other Types of Classification of Bacteria Include

a) **Phylogenetic classification**

- Based on evolutionary evidence obtained from fossil records

b) **Official classification**

c) **Nucleic acid based classification**

- based on G+C content or nucleic acid hybridisation or sequencing

d) **Classification based on Serological tests**

Bergey's Manual of Systematic Bacteriology

- In 1927, David Bergey & colleagues published *Bergey's Manual of Determinative Bacteriology*, a manual that grouped bacteria into phenetic groups, used in identification of unknowns. It is now in its 9th edition.
- In 1984, a more detailed work entitled *Bergey's Manual of Systematic Bacteriology* was published, still primarily phenetic in its classification.
- Publication of the second edition of *Bergey's Manual of Systematic Bacteriology* was begun in 2001. The 2nd edition gives the most up-to-date phylogenic classification of prokaryotic organisms, including both eubacteria and archaea.
- The classification in *Bergey's Manual* is accepted by most microbiologists as the best consensus for prokaryotic taxonomy.

Phylogeny of Domain *Bacteria*

The current grouping of bacteria as per Bergey's Manual of Systematic Bacteriology is as follows:

- Volume 1 (2001): The Archaea and the deeply branching and phototrophic Bacteria
- Volume 2 (2005): The Proteobacteria—divided into three books:
- 2A: Introductory essays
- 2B: The Gammaproteobacteria
- 2C: Other classes of Proteobacteria
- Volume 3 (2009): The Firmicutes
- Volume 4 (2011): The Bacteroidetes, Spirochaetes, Tenericutes (Mollicutes), Acidobacteria, Fibrobacteres, Fusobacteria, Dictyoglomi, Gemmatimonadetes, Lentisphaerae, Verrucomicrobia, Chlamydiae, and Planctomycetes
- Volume 5 (in two parts) (2012): The Actinobacteria

The 2nd edition of *Bergey's Manual of Systematic Bacteriology* divides domain *Bacteria* into 23 phyla. Nine of the more notable phyla are described here.

- **Phylum *Aquiflexa***
  - The earliest "deepest" branch of the *Bacteria*
  - Contains genera *Aquiflex* and *Hydrogenobacter* that can obtain energy from hydrogen via chemolithotrophic pathways
- **Phylum *Cyanobacteria***
  - Oxygenic photosynthetic bacteria
- **Phylum Chlorobi**
  - The "green sulfur bacteria"
  - Anoxygenic photosynthesis
  - Includes genus *Chlorobium*
- **Phylum *Proteobacteria***
  - The largest group of gram-negative bacteria
  - Extremely complex group, with over 400 genera and 1300 named species
  - All major nutritional types are represented: phototrophy, heterotrophy, and several types of chemolithotrophy
  - Sometimes called the "purple bacteria," although very few are purple; the term refers to a hypothetical purple photosynthetic bacterium from which the group is believed to have evolved
  - Divided into 5 classes: *Alphaproteobacteria*, *Betaproteobacteria*, *Gammaproteobacteria*, *Deltaproteobacteria*, *Epsilonproteobacteria*

Significant Groups and Genera Include

- Photosynthetic genera such as *Rhodospirillum* (a purple non-sulfur bacterium) and *Chromatium* (a purple sulfur bacterium)
- Sulfur chemolithotrophs, genera *Thiobacillus* and *Beggiatoa*
- Nitrogen chemolithotrophs (nitrifying bacteria), genera *Nitrobacter* and *Nitrosomonas*

- Other chemolithotrophs, genera *Alcaligenes, Methylobacilllus, Burkholderia*
- The family *Enterobacteriaceae*, the "gram-negative enteric bacteria," which includes genera *Escherichia, Proteus, Enterobacter, Klebsiella, Salmonella, Shigella, Serratia,* and others
- The family *Pseudomonadaceae*, which includes genus *Pseudomonas* and related genera
- Other medically important *Proteobacteria* include genera *Haemophilus, Vibrio, Camphylobacter, Helicobacter, Ricketssia, Brucella*

- **Phylum *Firmicutes***
    - "Low G + C gram-positive" bacteria
    - Divided into 3 classes
    - Class I – Clostridia; includes genera *Clostridium* and *Desulfotomaculatum,* and others
    - Class II – *Mollicutes*; bacteria in this class cannot make peptidoglycan and lack cell walls; includes genera *Mycoplasma, Ureaplasma,* and others
    - Class III – Bacilli; includes genera *Bacillus, Lactobacillus, Streptococcus, Lactococcus, Geobacillus, Enterococcus, Listeria, Staphylococcus*, and others

- **Phylum *Actinobacteria***
- "High G + C gram-positive" bacteria

Includes genera *Actinomyces, Streptomyces, Corynebacterium, Micrococcus, Mycobacterium, Propionibacterium*

- **Phylum *Chlamydiae***
    - Small phylum containing the genus *Chlamydia*
- **Phylum *Spirochaetes***
    - The spirochaetes
    - Characterized by flexible, helical cells with a modified outer membrane (the outer sheath) and modified flagella (axial filaments) located within the outer sheath

- Important pathogenic genera include *Treponema, Borrelia,* and *Leptospira*

- **Phylum *Bacteroidetes***

    - Includes genera *Bacteroides,Flavobacterium, Flexibacter,* and *Cytophyga*; *Flexibacter* and *Cytophyga* are motile by means of "gliding motility"

Prokaryotes are divided into four divisions – Gracilicutes, Firmicutes, Tenericutes and Mendosicutes. All these terms are derived from Latin words *gracilis* (thin), *firmus* (thick), *tener* (soft) and *mendosus* (faults) respectively.

PROKARYOTES - BACTERIA

Gracilicutes — *Gram-negative cell wall*

Firmicutes — *Gram positive cell wall*

Tenericutes — *Lack a cell wall and called Mycoplasmas*

Mendosicutes — ***Archaeobacteria**- live in extreme conditions, Halophilic, methanogens and thermo-acidophils very primitive*

*Gracilicutes, firmicutes and tenericutes are collectively referred at **Eubacteria***

Eubacteria are also grouped into **Cyanobacteria** and **Otherbacteria**. The bacteria that cause disease in animals belong to Eubacteria.

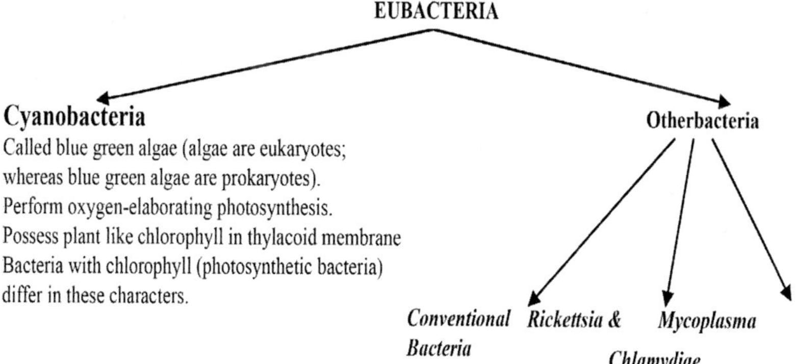

The classification of bacteria is described below in detail;

Kingdom – **Prokaryotae**

Division: Frimicutes	Gracilicutes	Tenericutes
(1)Gram positive cocci Family: Micrococcaceae Genus: Micrococcus Genus: Staphylococcus **Other genera** Genus: Streptococcus Peptococcus Peptostreptococcus	1) **The Spirochetes** Order: Spirachaetales Family: Spirochaetaceae Genus: Borrelia Genus: Treponema Family: Leptospiraceae Genus: Leptospira	Class: Mollicutes Order: Mycoplasmatales Family: Mycoplasmataceae Genus: Mycoplasma Genus: Ureaplasma Family: Acholeplasmataceae Genus: Acholeplasma
(2) Endospore forming gram positive rod Genus: Bacillus Genus: Clostridium	2) **Aerobic helical / Vibrioid bacteria**. Genus: Campylobacter	
(3) **Regular non sporing Grampositive rod** Genus: Lactobacillus Genus: Erysipelothrix	3) **Gram negative aerobic rods and cocci** Family: Pseudomonadaceae Genus: Pseudomonas Family: Legionellaceae Genus: Legionella Genus: Listeria Family: Nisseriaceae Genus: Nisseria	
	Other genera: Genus: Alcaligenes Genus: Bordetella Genus: Brucella Genus: Francisella &Moraxella	
(4) **Irregular non sporing Gram Positive rod** Genus: Actinomyces Genus: Corynebacterium Genus: Bifidobacterium	4) **Facultatively anaerobic Gram negative rods** Family: Enterobacteriaceae Genus: Citrobacter, Edwardsiella, Enterobacter, Escherichia, Klebsiella, Morganella, Proteus, Providencia, Salmonella, Serratia, Shigella, Yersenia	

<table>
<tr><td>Division: Frimicutes</td><td>Gracilicutes</td><td>Tenericutes</td></tr>
<tr><td>(5) The Mycobacteria
Family: Mycobacteriaceae
Genus: Mycobacterium</td><td>5) Anaerobic Gram negative rods
Family: Bacteriodaceae
Genus: Bacteriodes, Fusobacterium</td><td></td></tr>
<tr><td>(6) The Nocardia forms

Genus: Nocardia
Genus: Rhodococcus</td><td>6) Anaerobic Gram negative cocci
Family: Veillonellaceae
Genus: Veillonella</td><td></td></tr>
<tr><td colspan="3">The Rickettsias and Chlamydias
Order: Rickettsiales
Family: Rickettsiaceae
Genus: Cowdria, Coxiella, Ehrlichia, Neorickettisia, Rickettsia
Family: Anaplasmataceae
Genus: Anaplasma, Eperythrozoan, Haemobartonella
Order: Chlamydiales
Family: Chlamydiaceae
Genus: Chlamydia</td></tr>
</table>

<table>
<tr><th colspan="2">Scientific classification</th></tr>
<tr><td>Domain:</td><td>Bacteria</td></tr>
<tr><td>Kingdom:</td><td>Eubacteria</td></tr>
<tr><td>Phylum:</td><td>Firmicutes</td></tr>
<tr><td>Class:</td><td>Bacilli</td></tr>
<tr><td>Order:</td><td>Bacillales</td></tr>
<tr><td>Family:</td><td>Staphylococcaceae</td></tr>
<tr><td>Genus:</td><td>Staphylococcus</td></tr>
<tr><td>Species:</td><td>S. aureus</td></tr>
<tr><td colspan="2">Binomial name</td></tr>
<tr><td colspan="2">Staphylococcus aureus Rosenbach 1884</td></tr>
</table>

II. Nomenclature

Nomenclature is the scientific naming of organisms. The nomenclature of the different kinds of living creatures falls into two parts;

a) Informal or vernacular names

b) Scientific names

The scientific names are regulated by the International code of Nomenclature of bacteria and these scientific names are in italicized form. The Binomial nomenclature is based on the genus and species of bacteria.

(eg) *Streptococcusequi*
genus species.

Rules of nomenclature

- There is only one correct name
- Names causing confusion and error should be rejected.
- All names are latinized regardless of origin. (The first word **genus** is always capitalised, the second word **species** is not capitalized. Both genus and species names together are referred to as the species and are underlined or italicized if printed).
- Any new or revised species designations are valid only after publication in the International Journal of Systematic Bacteriology. The type strain should be deposited in reference type culture collections ATCC (American type culture collection) and NTCC (National type culture collection)

14

Morphology and Structure of Bacteria

1. Morphology of Bacteria

Structure of Bacteria

Bacteria are very small with a diameter of 0.5 – 1 μm in size. Due to this small size the surface area / volume ratio is very high compared to that of large organisms of similar shape. The large surface area compared to a small volume facilitates easy nourishment of all areas of the cell. So there is no circulatory mechanism to distribute the nutrients and there is little or no cytoplasmic movement within the cell.

i) Shape

Bacteria exhibit three types of morphology;

- Spherical - cocci (coccus) - *Greek word kokkos, meaning a berry*
- Straight rods - bacilli (bacillus) - *Latin word bacillus, meaning a stick*
- Helically curved rods -spirilla (spirillum)

Variations in these basic forms are also seen. Eg: cocobacillary, ovoid and filamentous.

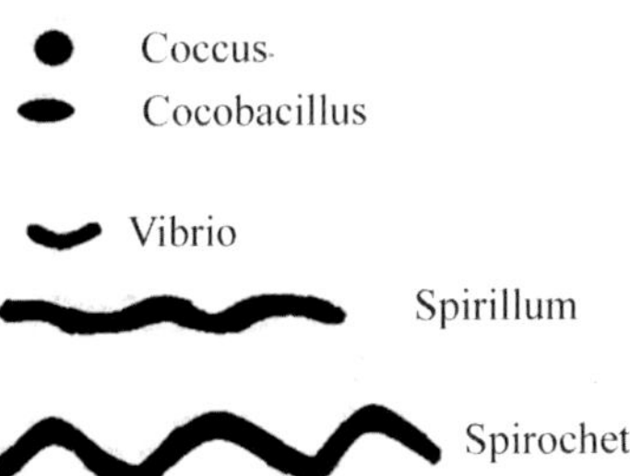

Bacilli range from 2 - 5 μm in length and 0.5 - 1 μm in width. Length of spirochetes is upto 20 μm. But their width is 0.1 – 0.2 μm. Cocci are approximately 1 μm in diameter. Based on size, bacteria can be grouped as below:

Large bacteria – Spirochetes, *Bacillus,Clostridium*

Medium – *Escherichia coli, Proteus*

Small – *Brucella, Pasteurella, Haemophilus*

Very small – *Rickettsia, Chlamydia, Mycoplasma*

ii) Arrangement

Bacteria are usually arranged in a manner characteristic of their species.

a. Cocci

It occurs in several characteristic arrangements depending on the plane of cellular division and whether daughter cells stay together following division. They are about 1 μm in diameter occurring in the following characteristic cell grouping:

Diplococci: Cells divide in one plane and remain attached predominantly in pairs.

Streptococci: Cells divide in one plane and remain attached to form chains.

Tetracocci: Cells divide in two planes and characteristically form groups of four cells.

Staphylococci: Cells divide in three planes, in an irregular pattern, producing “bunches” of cocci. (Greek: staphule, grapes)

Sarcinae: Cells divide in three planes, in a regular pattern, producing cuboidal arrangement of cells.

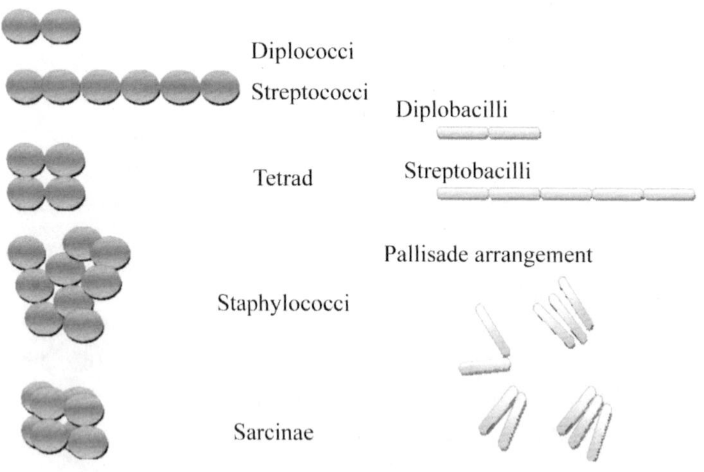

b. Bacilli

Rod shaped (i.e. cylindrical) organisms. Usually straight, varying in length from 2-10 **µm** and breadth from 1-2 **µm** according to species.

Diplobacilli: These occur mostly singly or in pairs.

Streptobacilli: formation of chains (eg: *Bacillus subtilis*).

Palisade arrangement: cells are lined up side by side like match sticks and at angles to one another (Corynebacterium).

Trichomes are similar to chains but have a much higher area of contact between adjacent cells (Eg; *Saprospira* sp.)

Streptomyces species form long, branched, multinucleate filaments called **hyphae**. The hyphae collectively form a mycelium.

Curved rods are usually curved with a twist or turn. Bacteria with less than one complete twist or turn are called **vibrioid** (eg: Vibrio). Those with one or more complete turns have a helical shape.

Spirilla: rigid helical bacteria. Spiral shaped, non flexous filaments

Spirochetes: highly flexible. They can twist and contort their shape.(e.g. Leptospira)

Mycelia: Branching filaments which may tend to fragment (Actinomyces) or to remain filamentous (Streptomyces).

Structure of Bacteria

The principal structures of a bacterial cell are broadly classified into;

a) Structures external to cell wall

Glycocalyx Flagella Fimbriae Pili
Axial filament

b) The Cell wall

c) Structures inside the cell wall

Plasma membrane Ribosomes Inclusions Nuclear Plasmids
material

The bacterial structure can be summarised as an outer most cell envelope, which is composed of three structures namely capsule or slime, cell wall and cytoplasmic membrane. Inside the cytoplasmic membrane is the fluid like cytoplasm. The cytoplasm consists of organelles like ribosomes and mesosomes. It also contains the nuclear material that is both DNA and RNA. Apart from these two basic structures, bacterial cell also possesses certain appendages like flagella that are organ of locomotion; fimbriae and pili that are useful in adhesion and sexual conjugation respectively. Certain special structures like plasmids (extra chromosomal DNA) are characteristic of bacterial cells.

The following is a brief description of the structure and properties of the various constituents of the bacterial cells.

A. Structures External to Cell Wall

1. Glycocalyx

The term **glycocalyx** is used to indicate the material that surrounds the cell. The bacterial glycocalyx is a viscous gelatinous polymer lying outside the cell wall. The capsules of most bacteria are simple polysaccharides composed of one or few different sugars.

Some bacteria contain hyaluronic acid in their capsule (streptococci in Lancefield groups A and C). A few capsules are made of polypeptides, as for example, capsule of *Bacillus anthracis* is made of a polymer of glutamic acid. *Yersinia pestis* has a protein capsule.

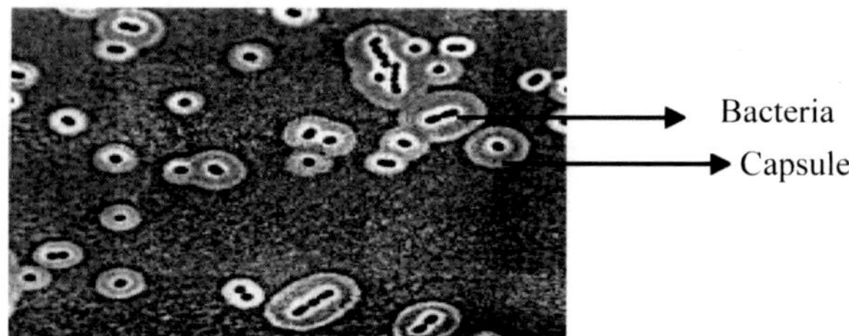

When the glycocalyx substance is firmly attached to cell wall it is referred as **capsule** and if it is arranged loosely attached to cell wall it is referred as **slime layer.** If it is too thin to be seen by light microscopy then it is called as microcapsule.

The important **functions** of bacterial capsule are:

- anti-phagocytic property
- an important virulence factor
- adhesion of bacteria to cells or surfaces,
- supply of nutrition during adverse conditions,
- protection of bacteria from dehydration and
- prevention of migration of nutrients from bacteria.

2. Flagella

It is an external appendage that helps the bacteria to propel. Flagellum is the organ of locomotion for bacteria. Flagella are thin, filamentous, whip like structures of uniform thickness with a diameter of 12-19nm. The position, number and arrangement of flagella in a bacterial cell are of taxonomic importance. The flagella that originate from the end of the cell are called *polar flagella.*

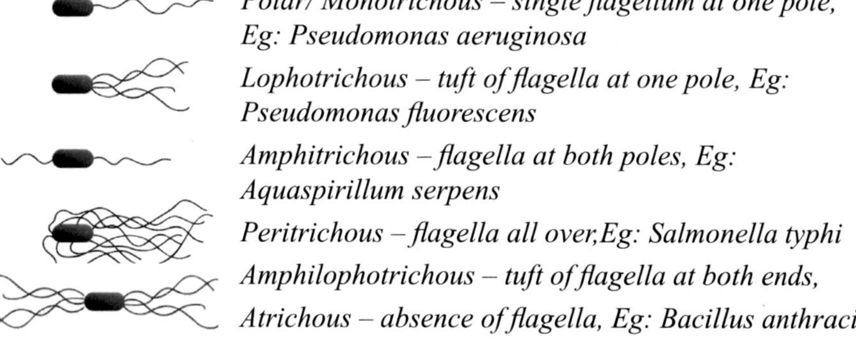

The flagellum is composed of three parts - filament, hook and basal body. The filament is made up of number of units of a protein called ***flagellin.*** It is arranged as chains over a hollow. The protein of hook is different from filament protein ie. Flagellin.

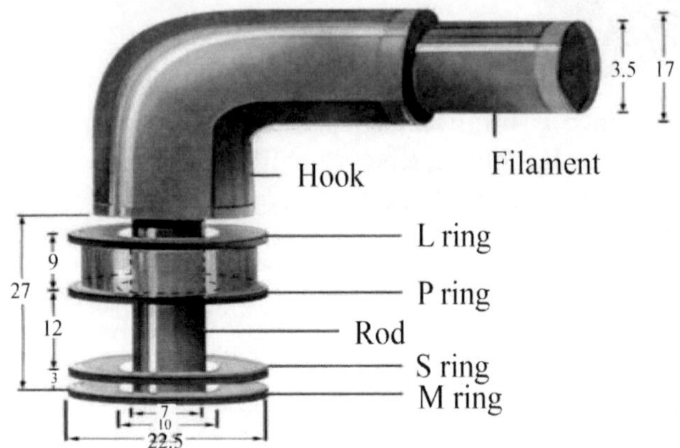

The basal body is composed of set of rings and is embedded in the plasma membrane that gives the flagella rotary motion (the way the shaft in electrical motor rotates) and propels the organisms. The number of basal plates varies between Gram positive and Gram negative organisms. The energy required for such motion is provided by the proton gradient of cytoplasmic membrane.

Approximately 256 protons must cross the cytoplasmic membrane to power a single flagellum. The flagellum can rotate at speeds of 1200 rpm that enables the bacteria to move at speeds up to 100mm/second or 0.0002 mile/hour. The bacterial motion is referred as tumble, run or swim. The bacterial flagellum provides the bacterium with a mechanism of swimming towards or away from chemical stimuli that is known as **chemotaxis.** The chemicals that attract bacterium are known as ***chemo*attractants** and that repel bacterium are known as ***chemorepellents.*** Some bacteria move in response to a magnetic field and such movement is known as **magnetotaxis**. Magnetotaxis is mediated by a special inclusion structures known as **magnetosomes** that are magnetic iron oxide (Fe_3O_4). Some bacteria detect and move in response to variations in light intensity and such movement is called **phototaxis**.

The important **functions** of bacterial flagella are:

- It is an apparatus for locomotion
- It offers greater absorptive surface for the bacterium

- It imparts type specificity to the bacterium
- It helps in identification and classification of bacteria.

3. Axial Filament

The equivalent of flagella in certain bacteria (Spirochetes) is referred as **axial filaments**. They are bundles of fibrils that arise at the ends of the cell wall beneath the outer sheath (periplasmic space) and spiral around the cell. Hence they are called as periplasmic flagella or axial filament or endoflagella. They are anchored at one end and the rotation of filaments propels the bacteria like the way the cork screw moves through a cork.

Spiroplasmas also exhibit motility but they lack any organelle for motility even periplasmic flagella. Mechanism of this motility is not known. Gliding motility is seen in some bacteria, Eg: *Cytophaga* species. They are motile only when they are in contact with solid surface. They exhibit a sinuous, flexing motion. It is a slow motion covering only a few μm/second. The mechanism of this motility is not known.

4. Pili or Fimbriae

These structures refer to short, thin, straight, hair like projections that rise from surface of some bacteria and are involved in attachment processes. Chemically they are phosphate-carbohydrate-protein complexes containing a single type of peptide unit called pilin. The main role of pili is attachment and not motility. They play a major role in attachment of bacteria to epithelial cells lining the respiratory, intestinal and genitourinary tracts. This helps bacteria from being washed away by the flow of mucous or body fluids and permits infection to be established. Fimbriae are located either at the poles or throughout the entire surface.

The type of attachment is used to distinguish between pili and fimbriae. Pili refer to attachment between the bacterial cells during conjugation and fimbriae refer to all other types of attachment. The pilus involved in the reproductive process is also referred as **F pilus** and is responsible for the formation of conjugation bridge for the transfer of DNA from donor to recipient cells. Pili also act as receptors for bacteriophages. In some bacteria pili are considered as one of the virulence factors. A bacterium may have either one or two pili per cell. The pili are comparatively bigger and longer than fimbriae.

5. Spore

When some species of bacteria are in an environment unfavorable for rapid multiplication they will develop into spores, which are dormant forms capable of survival for long periods of time under adverse conditions. They are resistant stages in the life cycle of the Gram positive genera Bacillus and Clostridium. Each bacterium develops into one spore.

Endospore: Spore produced within the cell ; Exospore: External to the cell

Endospores are unique to bacteria. They are thick walled highly retractile bodies produced by Bacillus, Clostridium, Sporosarcina and few other genera. Shape and location of the spore within the cell vary depending on the species – central, sub terminal and terminal. Terminal endospores (Eg; Clostridium tetani) are seen at the poles of cells, whereas central endospores (Eg; Bacillus cereus) are more or less in the middle. Subterminal endospores (Eg; Bacillus subtilis) are those between these two extremes, usually seen far enough towards the poles but close enough to the center so as not to be considered either terminal or central. Lateral endospores are seen occasionally.

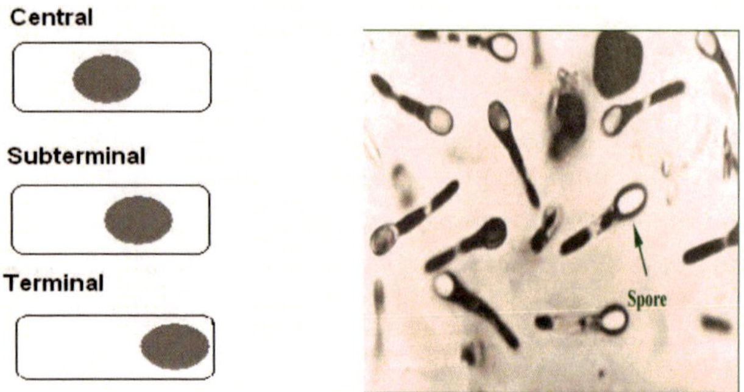

Sometimes the endospore can be so large the cell can be distended around the endospore, as in *Clostridium tetani*.

Endospores cannot be stained by regular staining due to the impermeability of the endospore wall to dyes and stains. The rest of a bacterial cell may stain but the endospore is left colourless. Endospores can be stained by the Schaffer-Fulton stain, which stains endospores green and bacterial bodies red.

Endospores are resistant to dessication, heat, radiation, disinfecting agents and staining. The degree of heat resistance of endospores varies with bacterial species. Most spores can resist 90 C for atleast 10 minutes. Dehydrated state and dipicolinic acid (DPA) may be responsible for heat resistance. DPA is unique to endospores. DPA constitutes 10-15% of the spore's dry weight. DPA occurs in combination with large amount of calcium. Calcium dipicolonate may play a role in heat resistance. DPA synthesis and uptake of calcium occur during advanced stages of sporulation. Spores are usually produced by cells growing in rich media which are approaching the end of active growth.

Sporulation

When environmental conditions are become unfavourable sporulation is started. It takes about eight hours. The DNA is replicated and a membrane wall called spore septum begins to form between it and the rest of the cell. The plasma membrane of the cell surrounds this wall and pinches off to leave a double membrane around the DNA. The developing structure is now known as a **forespore**. Calcium dipicolinate is incorporated into the forespore at this stage. Next the peptidoglycan cortex develops between the two layers and the bacterium adds a **spore coat** to the outside of the forespore. Sporulation is now complete, and the mature endospore will be released when the surrounding vegetative cell is degraded.

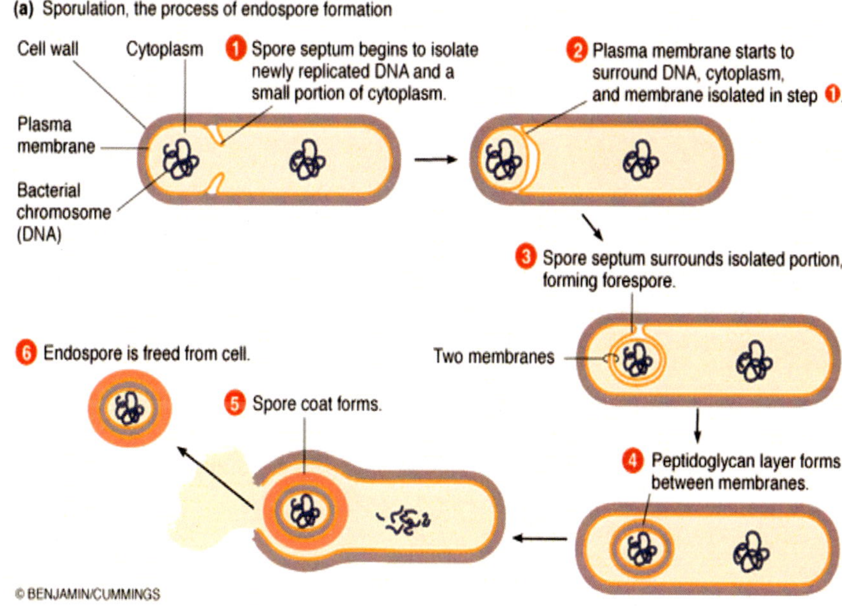

Structure of Endospore

- Endospores are made up of a central spore cell, which is surrounded by various layers. The outermost layer is the **exosporium** - a thin protein covering.

- Below this is the **spore coat** which is composed of highly cross-linked keratin and layers of spore-specific proteins. The **cortex** consists of loosely cross-linked peptidoglycan and contains dipicolinic acid (DPA), which is particular to all bacterial endospores.
- **Core -** The core contains the cell wall and, cytoplasmic membrane, nucleoid, and cytoplasm. The core only has 10-30% of the water content of vegetative cells; therefore the core cytoplasm is in a gel state. The low water content contributes to the endospores survival in dry environments. However, the low water content and gel cytoplasm contributes to the inactivity of the enzymes in cytoplasm. **Small acid soluble spore proteins (SASPs),** are formed during sporulation and bind to DNA in the core. SASPs protect the DNA from UV light, desiccation, and dry heat. SASPs also serve as a carbon energy source during germination.

Structure of endospore

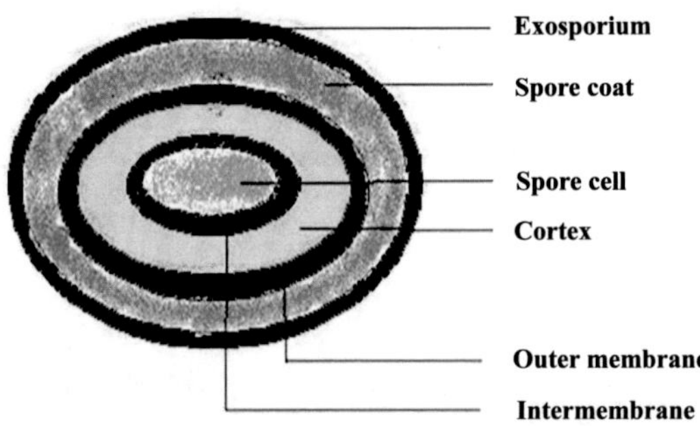

Germination

Formation of vegetative cells from spores is called germination. During germination endospores lose their resistance to heat and staining. Outgrowth occurs, characterized by synthesis of new cell material and development of the organism into a new cell.

Exospores

Methylosinus genus forms exospores by budding at one end of the cell. They are heat and dessication resistant but they do not contain DPA.

B. Cell wall

All bacterial pathogens of man and domestic animals possess complex rigid cell walls except the Mycoplasma and Ureaplasmas, which lack a cell wall and are surrounded by a cytoplasmic membrane. Cell wall constitutes as much as 10-40% of the dry weight of the cell. The cell wall consists of **peptidoglycan (Murein)** which gives rigidity to the cell wall and is located adjacent to the cytoplasmic membrane. It is composed of a disaccharide polymer with alternating monomers of N-acetylglucosamine (**NAG**) and N-acetyl muramic acid (**NAM**) that are linked by β1-4 glycoside bonds and peptides. The peptides consist of 4-5 amino acids, including D–alanine, D–glutamic acid, L-Lysine, L-alanine and Diaminopimelic acid. The peptidoglycan often makes up 80-90% of the Gram+ve cell wall but only upto 10% of the Gram-ve cell wall.

1. Gram Positive Bacteria

The cell wall has a single thick layer of peptidoglycan which makes up 50 – 90 % of the cell wall. Thickness of the cell wall in the Gram positive bacteria range from 150 to 800Å. Teichoic acids, polysaccharides, proteins and sometimes lipids are other cell wall components. Teichoic acids make up 20 – 40 % of the cell wall of the Gram positive bacteria. Teichoic acids are polymeric chains of glycerol or ribitose molecules linked to each other by phosphodiester bridges (Eg: *Staphylococcus aureus* and *Streptococcus faecalis)*. They bind magnesium ions which help to protect the bacteria from thermal injury by stabilization of the cytoplasmic membrane. Very little lipid is present in the cell walls of most Gram positive bacteria. Cell wall of *Streptococcus pyogenes* has polysaccharides covalently linked to the peptidoglycan.

Teichoic acid is responsible for regulating the movement of cations into and out of the cells and also to prevent the cell from lysis at low pH. It also provides antigenic character to bacteria.

An acid fast bacterium contains high levels of lipids in their cell walls which make them difficult to stain. Mycolic acid is present in the cell wall of Mycobacteria and Corynebacterium which gives them the acid fast character. A mycolic acid derivative called cord factor (trehalose dimycolate) is toxic and plays a role in the disease caused by these organisms.

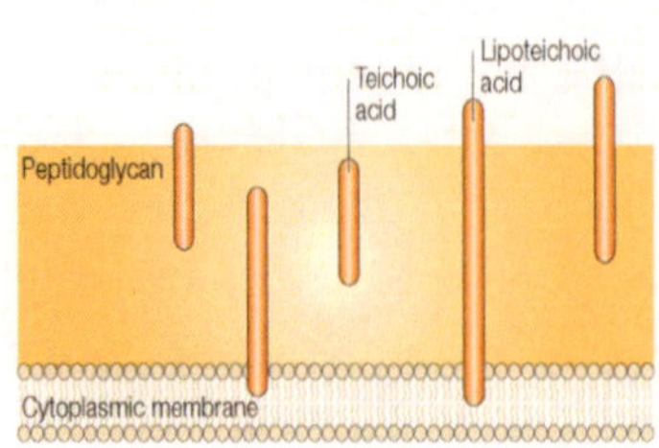

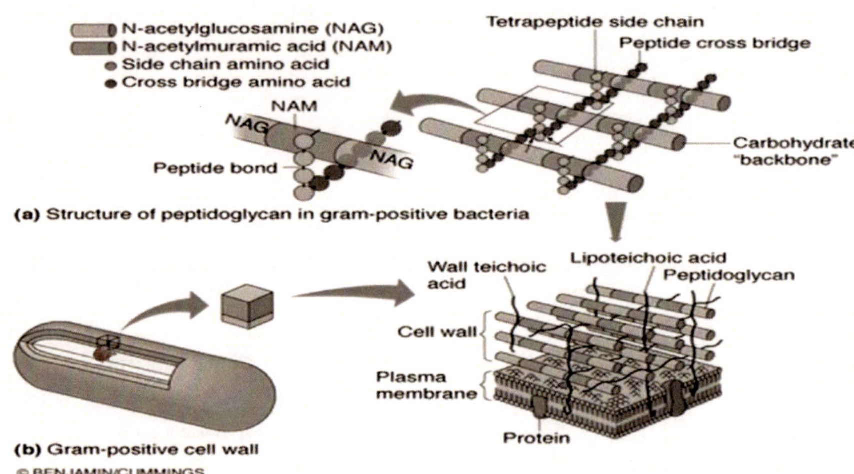

(a) Structure of peptidoglycan in gram-positive bacteria

(b) Gram-positive cell wall

2. Gram Negative Bacteria

The cell wall has a very thin layer of peptidoglycan layer which makes up 10 % of the cell wall. Gram negative bacterial is approximately 100 Å in thickness. The outer membrane is composed of **phospholipids, lipopolysaccharides (LPS) and proteins**.

The **LPS** are a class of macromolecules unique to Gram negative bacteria. It is composed of the core polysaccharide, Lipid A and O antigens. Lipid A is the principal endotoxic component of the LPS. Lipid A contains glucosamine-4-phosohate, ethanolamine and a variety of long chain fatty acids. The endotoxicity is due to Beta –OH myristic acid.

A trisaccharide of 2- Keto 3-Deoxyoctonic acid links the Lipid A to the core polysaccharide via a ketoacetic linkage. The inner portion of the core polysaccharide is composed of ethanolamine, heptoses and phosphate. The outer portion consists of glucose, galactose and N- acetylglucosamine. The O antigens are responsible for the serologic specificity. It is the extension of the core polysaccharide made up of repeating oligosaccharide unit each composed of 3-4 different hexoses. The porin proteins form pores or channels in the outer membrane of the bacterium which allows nutrients to diffuse through the cell wall. Other proteins in the outer membrane provide specific transport systems for nutrients whose passage through the pores might be too slow.

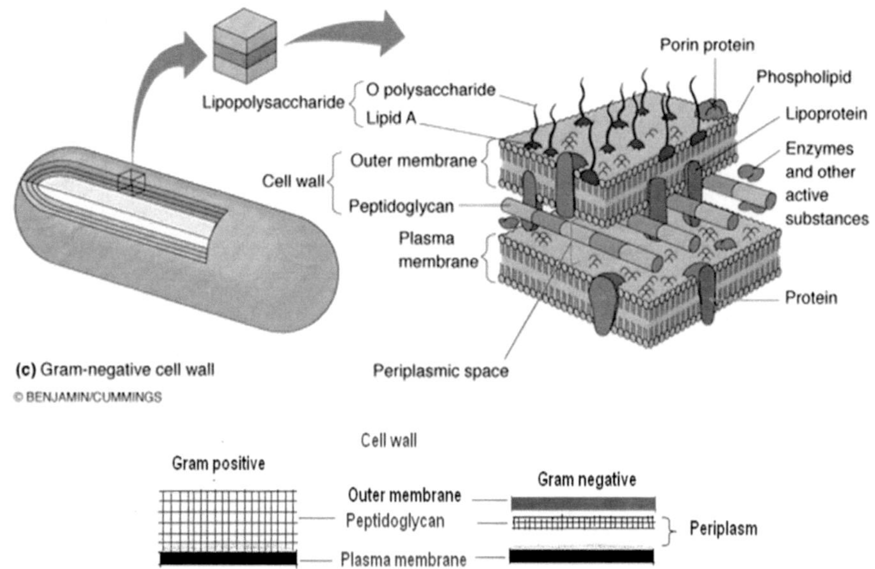

3. Atypical Cell Wall

Mycoplasma in the prokaryotes group do not have cell walls whereas the cell walls of archaeobacteria are composed of pseudomurin.

Protoplasts

As a result of damage to cell wall by lysozymes, the gram-positive bacteria lose the cell wall, but still are surrounded by plasma membrane and can carry out metabolism. Such structures are referred as **protoplasts**.

Spheroplasts

Damage to cell wall of Gram-negative bacteria caused by lysozyme is not as extensive as in Gram-positive bacteria. The Gram-negative cells retain certain layers of outer membrane and plasma membrane and such structures are referred as **spheroplasts**.

Periplasm

The periplasm is the space **located between the cell wall and cytoplasmic membrane**. It serves as a buffer component between the internal environment of bacterial cells and the external environment.

The physiological action in Gram positive bacteria is not known. In Gram negative bacteria it contains a variety of hydrolytic enzymes and proteins that aid in the transport of various compounds.

C. Cell Membrane or Cytoplasmic Membrane

It lies between the cell wall and the bacterial protoplasm. It is 7.5 nm thick and composed of a lipid bilayer in which proteins are embedded. The lipids maintain the structural integrity of the membrane and proteins (60-70%) are responsible for the specialized functions like biosynthetic enzymes of cell wall, transport enzymes and cytochrome enzymes. The lipid layer is composed of phospholipids (20-30%). The polar head groups are exposed to the aqueous phase of the membrane surface and the tail is embedded in the interior.

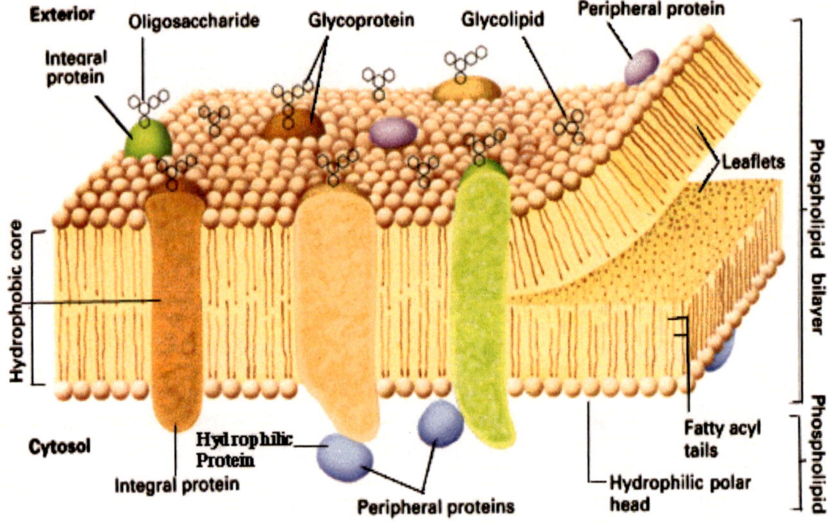

There are two types of proteins – Integral and Peripheral proteins;

Integralproteins are tenaciously held and can be removed only by destruction of the membrane by treatment with detergents. The lipid part of the membrane has fluidity. Hence the components can move around laterally. This fluidity appears to be essential for various membrane functions.

Cytoplasmic membrane acts as a hydrophobic barrier to most water soluble molecules. Specific proteins in the membrane facilitate the passage of small molecules across the membrane. Enzymes involved in respiratory metabolism and synthesis of capsular and cell wall components are present in the cytoplasmic membrane. It is impermeable to protons and act as the site of generation of proton motive force (pmf).

Peripheral proteins are loosely attached and can be removed by mild treatments like osmotic shock.

The cytoplasmic membrane of the mycoplasma contains sterols. It is a very important functional structure. Damage by physical and chemical agents can result in the death of the cell.

Properties

a) Semi-permeability b) Possess strong affinity for bacterial stains

D. Structures Internal to Cell Wall

1. Adhesion sites or Bayer junction:

In Gram-negative bacilli, the cell wall and cytoplasmic membrane are joined in number of areas and these areas are called adhesion sites or Bayer junctions. The polysaccharides are transported hrough these sites.

2. Mesosomes (Chondroids)

It is a specialized convoluted invagination of the cytoplasmic membrane. It is more prominent in Gram-negative bacteria and involved in DNA segregation during cell division.

There are two types - **Central mesosomes** and **Peripheral mesosomes**.

Central mesosomes penetrate deeply in to the cytoplasm. They are located near the middle of cell and appear to be attached to the nuclear material.

Peripheral mesosomes are shallow penetrations in to the cytoplasm. They are not restricted to central location and not associated with nuclear material.

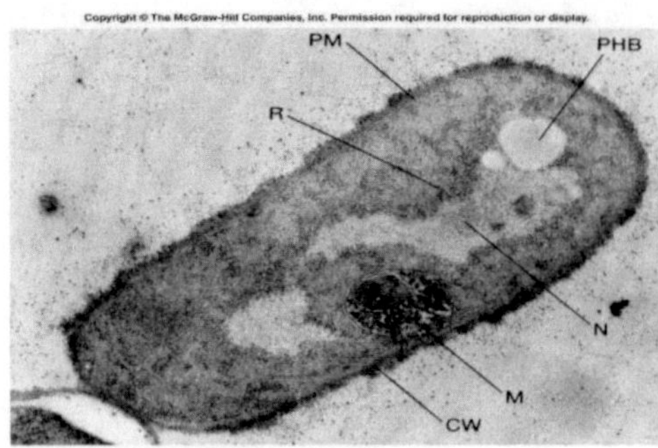

PM = plasma membrane, TR = ribosome
CW= cell wall, N = nucleoid
M = mesosome, PHB = inclusion body

3. Cytoplasm

Bacterial cytoplasm lacks endoplasmic reticulum. Cytoplasmic viscosity is very high in bacteria. It contains nucleus, ribosomes, plasmids, inclusions and vacuoles. It may be divided into:

1. Cytoplasmic area – granular in appearance.
2. Chromatic area – rich in DNA.
3. The fluid portion with dissolved substances.

a) Nuclear Region

The nuclear region is filled with DNA fibrils. The ds DNA is not bound by a nuclear membrane. The genome consists of a single molecule of dsDNA arranged in the form of a circle, about 1mm in length. The DNA is composed of deoxyribose, phosphoric acid and nitrogenous bases. The molecular weights of bacterial chromosome vary from $0.85x10^9$-$85x10^9$. The nucleic acids consist of long chains of nucleotides with alternating pentose and phosphate groups with a base attached to each sugar molecule. Such a nucleotide has an orientation, i.e. a polarity, with a phosphate group in the 5' position at one end, and a free hydroxyl group in the 3' position at the other end. The bacterial chromosome replicates and segregates between the daughter cells by a process called binary fission. The DNA encodes for about 4000 proteins.

b) Ribosomes

Ribosomes are complexes of ribosomal RNA (rRNA) and proteins. They are distributed throughout the cytoplasm. The bacterial ribosome has a sedimentation coefficient of 70 S and consists of two subunits, whose sedimentation coefficients are 50S and 30S.The 50S subunit contains one 23S rRNA molecule, one 5S rRNA molecule, and 32 different proteins. The 30S subunit contains one 16S rRNA molecule and 21 different proteins. Ribosomes are distributed throughout the cytoplasm and are the sites of protein synthesis. The amount of ribosomes will be high in fast growing cells. There will be about 5000 – 50,000 ribosomes. Chains of ribosome can be seen in electron micrographs like pearls of necklace called **polyribosomes or polysomes.**

c) Plasmids and Episomes

Plasmids are small extrachromosomal, self- replicating circular strands of DNA that replicates autonomously from the bacterial chromosome. There are usually multiple copies of each plasmid in the bacterial cell. The plasmids contain genes that code for 2-3 proteins. Episomes are plasmids that are capable of integrating into the bacterial chromosome.

Different plasmids provide a variety of functions;

- R (Resistance) factors: are plasmids that contain genes that code for antibiotic resistance. These factors are capable of transferring resistance between strains of the same bacteria and between closely related bacteria. They are responsible for 60-90% of the resistance of the gram negative bacteria and some gram positive bacteria. The R factor carries resistance to several antibiotics. (e.g. R factor is beta-lactamase, an extracellular enzyme that specifically hydrolyses the amide bond in the beta-lactam ring of penicillin, rendering the antibiotic inactive).
- F (Fertility) factors: are plasmids that promote the transfer of the donor bacterial chromosome at a high frequency of recombination (Hfr) into the recipient bacterial chromosome during conjugation.
- Colicinogenic (Col) factors are found in many species of enterobacteria and code for the enterotoxins, which are extracellular toxins that inhibit the growth of strains of the same or different species of bacteria. Colicines are the subset of bacteriocins produced by the coliform bacteria.
- Virulence factors may be encoded in the genetic material of plasmids and episomes. These factors include hemolysins, fimbrial adherence antigens and exotoxins.

d) Cytoplasmic Granules

The presence and amount of these storage particles vary with the type of the bacterium and its level of metabolic activity. Under appropriate conditions many microorganisms contain intracellular inclusions that can be regarded as reserve or storage materials. Among these are polysaccharides, lipids, polyphosphates and sulphur. These materials are accumulated when their precursors are present in the medium.

Volutin granules or metachromatic (Babes-Ernstgranules) granules - They are composed of polyphosphate. They stain intense reddish purple with methylene blue. In electron microscopy they appear as round dark areas. They serve as reserve source of phosphate. Eg: Corynebacterium

Poly - b –hydroxybutyrate (PHB) granules - They are found in aerobic bacteria. They are made up of chloroform soluble lipid like material which can serve as a reserve of carbon and energy source. PHB granules can be stained with lipid soluble dyes such as Nile blue. In electron microscopy they appear as clear round areas.

Polysaccharide granules – They are made up of glycogen. The granule can be stained with iodine and appear brown in colour. In electron microscopy they appear as dark granules.

Gas vacuoles are found in some bacteria living in aquatic habitats. They provide buoyancy. They appear as bright, retractile bodies in light microscope. In electron microscope they appear as hollow, rigid cylinders with more or less conical ends and have a striated protein boundary.

Intra cellular **globules of elemental sulphur** are found in some bacteria growing in hydrogen sulphide rich environments. Eg: Thiobacillus

Carboxysomes: These bodies are seen in bacteria that utilises carbon dioxide as their sole source of carbon. Eg. Cyanobacteria

L-Forms

Kleinberger-Nobel, studying cultures of *Streptobacillus moniliformus* observed swollen cells and other aberrant morphological forms and named them L-forms after the Lister Institute, London. These L-forms occur in the presence of inhibitors of cell-wall synthesis (eg) Penicillin, absence of nutrients necessary for cell wall synthesis.(eg) diaminopimelic acid and in the presence of destructors of cell wall (eg) antibody and complement.

These L-forms lack in cell wall either completely or the wall is weakened. They are soft protoplasmic bodies. They are unstable and revert to the normal form under favorable environment.

15

Staphylococci

First discovered by Scottish surgeon sir Alexander ogston (1880) in infected tissues. He named it as staphylococcus" (Greek staphyle, bunch of grapes; KOKKAS, berry).

Systematics

Domain : Bacteria

Phylum : Firmicutes

Class : Bacilli

Order : Bacillales

Family : Staphylococcaceae

Genus : *Staphylococcus*

Species : *Staph. aureus*

Staph. Intermedius

Staph. hyicus

Habitat and Ecology

Staphylococcus occurs worldwide in mammals although the spread of staphylococcal strains between different animal species is limited. They colonise the nasal cavity, skin and mucous membranes and can be transient in the intestinal tract. It is an opportunist type organism.

Morphology

Spherical cells, 0.8 to 1.0, micrometre in diameter on agar media the cocci are arranged in grape- like clusters. In broth they occur as small groups, pairs or short chains of not more than four numbers. They are gram positive, non-motile, non-acid fast and non-sporing and have no flagella.

Cultural Characteristics

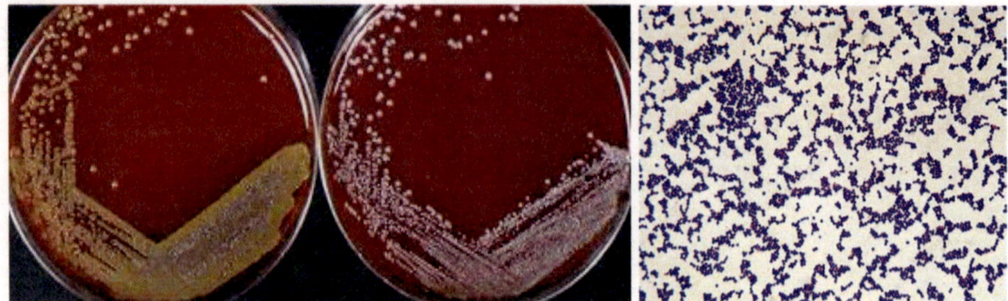

- They grow at an optimum temperature of 35-37°C.
- They are aerobic and facultatively anaerobic.
- Staphylococcus grows in the presence of 7-10% Sodium chloride. Hence a media-containing salt is selective for this organism (eg; Mannitol Salt agar) and it is highly inhibitory to many other bacteria particularly gram-negative bacteria.
- In the Nutrient agar plate the colonies are round, smooth, glistening, opaque, low, convex, entire edge and golden yellow or white colour. In nutrient broth a uniform turbidity is present with powdery sediment.
- Phenolphthalein diphosphate or tellurite agar selectively inhibits non-pathogenic strain.
- Haemolysis on blood agar. Capable of liberating 'V' factor into the medium, which favours the growth of Haemophilus organism.
- Purple agar, containing bromocresol purple as a pH indicator and 1% maltose, is used to differentiate Sta. aureus and Sta. intermedius.
- Most strains form pigments, hence previously classified as per the colour of the pigment produced.

  Eg: Staph.aureus : yellowish or golden orange pigment.

  Staph.albus : white colonies.

  Staph.citreus: lemon yellow colour pigment.
- Later on it was found that pigment formation is variable. Hence such classification was no longer followed.
- In sheep or rabbit blood agar plate "double hemolysis" around colonies are formed with incubation at 37°C it produces an incomplete haemolysis, which develops into a complete haemolysis when held at 4°C. This is called as hotcold lysis phenomenon.

Resistance

Temperature of 60°C for half an hour destroys all the cells. Death is accomplished by 1% phenol and 10% formaldehyde for 35 minutes and 10 minutes of exposure respectively.

Bio-Chemical Properties

Staphylococcus aureus produces acid from glucose, maltose, mannitol, lactose, and sucrose and not from salicin, raffinose & inulin. The organism is indole negative positive for NH_3, Methyl Red and Voges – Proskauer and catalase. Hydrolyses gelatin and coagulate serum. Negative for oxidase and H_2S production.

Antigenic Structure

It consists of;

1) Group antigen

a) Carbohydrates

The cell wall of *Staph. aureus* contains ribitol teichoic acid and *Staph. intermedius* contains glycerol teichoic acid.

b) Protein A. is present only in *Staph. aureus*

2) Type antigen

Proteins other than protein A

Virulence Factors

Three groups of virulence factors

1. Surface components

It consists of a capsule, Teichoic acid, which has antiphagocytic ability.

2. Exotoxins

Hemolysin

Four antigenically distinct hemolysin causes hemolysis of erythrocytes. They are Alpha, Beta, delta and Epsilon toxin, which produce partial hemolysis (hot

– cold hemolysis). The alpha toxin is the major toxin in gangrenous mastitis. It causes spasm of smooth muscle and is necrotizing and potentially lethal.

Leukotoxin

It has leucocidal activity and includes a and d toxins. By along so, the organisms may spread more easily to other parts where they develop secondary lesions.

Enterotoxin

It is seldom produced by animal strains. There are several antigenically distinct types of heat stable enterotoxins. They are not destroyed at 100°C for 80 minutes. They are responsible for food poisoning in man.

Exfoliative or Derma Necrotoxin

This toxin causes necrosis of skin by exfoliation and intraepidermal separation.

Toxic shock syndrome toxin (TSST)

Induce excessive lymphokine production results in tissue damage. Bovine and human strains of *Sta. aureus* produce TSST

3. Extracellular enzymes

Coagulase

On their ability to coagulate plasma they are classified as coagulase positive staph. (CPS) and coagulase negative staph. CPS is considered to be significant pathogens. It is resistant to heat. The coagulation of plasma produces a fibrin film on the surface of the organisms, which allows multiplying.

Hyaluronidase:

It hydrolyses hyaluronic acid, the mucoid ground substance of connective tissue.

Nucleases

Deoxyribonucleases (DNAse) hydrolyze DNA and Ribonucleases hydrolyze RNA.

Fibrinolysin

Fibrinolysin is commonly referred to as staphylokinase, is an activator of the plasma system leading to the breakdown of fibrin.

Lipases and esterases

They hydrolyze lipids.

Lysozyme

Hydrolyses the peptidoglycan in the cell wall of many bacteria.

PATHOGENESIS

Capsules
Normal inhabitant ⟶ Tissue damage ⟶ antiphagocytic
Protein A activity

Establishment
of staphylococci ⟵ Protect bacteria ⟵ Tissue damage
Coagulase Extracellular
Enzymes

Pathogenicity

Horse

Botryomycosis - Infrequent chronic granulomatous lesions involving the udder of the mare, cow and sow and the spermatic cord of horses.

Cattle

Mastitis - Staphylococcal bovine mastitis may be chronic, acute and peracute. Gangrenous mastitis due to a toxin is seen in postparturient cows.

Sheep

Tick pyemia in lambs occurs in 2-5 week old lambs, which is heavily infected Staphylococcal with *Ixodes ricinus*. Periorbital eczema is an infection due to abrasion. dermatitis is due to scratches from vegetation.

Poultry(Bumblefoot)

A pyogranulomatous process of subcutaneous tissue of foot that involves the joints. Staphylococcal arthritis and septicemia in turkeys, omphalitis – yolk sac infection, wing rot or gangrenous dermatitis infection in poultry

Pig

Exudative epidermitis (greasy pig disease) is an acute generalised infection of suckling and weaned pigs caused by *Sta. hyicus*. This disease is characterised by excessive sebaceous secretion, exfoliation and exudation.

Dogs and Cats

Pyoderma is one of the most common skin diseases of dogs. In addition to this, Otitis-externa and other suppurative conditions are caused by *Sta. intermedius*. Staphylococcal antigens produce intense inflammatory reactions and promote persistence of the bacteria.

Other Staphylococcal Organisms

S. aureus sub sp. *Anaerobius* causes caseous lymphadenitis. They are anaerobic and catalase negative.

S. caprae → Goat's milk

S. gallinarum and *S. arlettae* → Skin of chickens

S. lentus → Skin of sheep and goats

S. equorum → Skin of horses

S. simulans and → Cats

S. delphini → Skin of clophins

S. aureus → Staphylococcal scalded skin syndrome (SSSS) and Toxic shock syndrome (TSS) in humans

MRSA → Methicillin resistant Staphylococcus aureus

Diagnosis

- Direct microscopic examination
- Cultural characters in selective media.
- In case of food poisoning identification of the enterotoxin is diagnostic use
- CPS produces a typical reaction in the Hotis test of milk samples. These organisms produced green or greenish brown colonies with white centres after the milk samples were incubated for 16-20 hours.
- DNA tests and protein A tests are also employed.

Pathogenicity Test

Coagulase Test

When culture is added to Rabbit plasma, fibrinogen is converted to fibrin by coagulase enzymes. In this test, a suspension of staphylococci is mixed with rabbit plasma either on a slide or in a small tube.

The slide test detects the presence of a bound coagulase or clumping factor on the bacterial surface. A positive reaction is indicated by clumping of bacteria within 1 to2 min.

The tube test detects the free coagulase or staphylocoagulase, which is secreted by bacteria into the plasma. It is the definitive test for coagulase production and positive reaction is indicated by clot formation in the tube following incubation at 37^0C for 24 hrs.

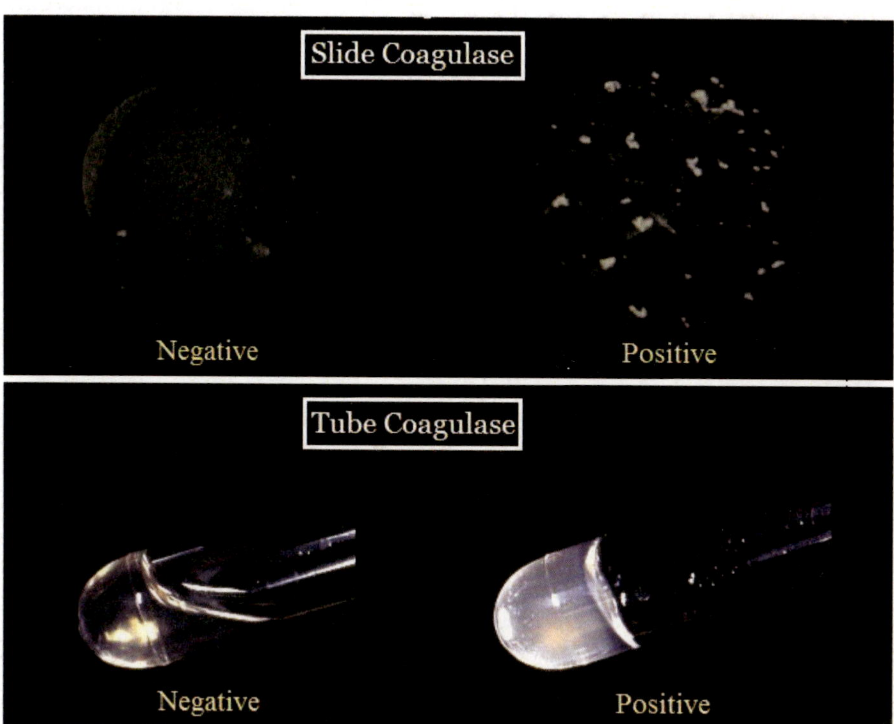

Phage Typing

Phage typing is carried out for epidemiological purposes, particularly for *Staph. aureus* strains of bovine mastitis.

Treatment

Penicillin is the drug of choice if strains are susceptible. New synthetic penicillins such as methicillin, oxacillin are effective. Tetracycline, bacitracin, nitrofuran, Trimethoprim, Cephalosporin, and enrofloxacin are effective.

Main differentiating characteristics of the gram-positive cocci

Organisms	Coagulase	Catalase	Oxidase
Pathogenic staph.	+	+	-
Non.Pathogenic Staph.	-	+	-
Enterococci	-	-	-
Streptococci	-	-	-
Micrococci	-	-	+

16

Streptococci

Systematics

Domain : Bacteria

Phylum : Firmicutes

Class : Bacilli

Order : Lactobacillales

Family : Streptococcaceae

Genus : *Streptococcus*

Species : *Str. agalactiae, Str.dysgalactiae, Str. equi subsp. zooepidemicus, Str. uberis, Str. equi subsp equi, Str.suis, Str. canis, Str. pyogenes* (human)

History

Rivolta (1873) described chain forming organisms in pus from a case of strangles in horses. In 1878-79, Pasteur recognized this organism as a pus-forming agent. In 1903, Hugo Schottmuller introduced blood to differentiate various types of hemolysis. In 1928, Rebecca Lancefield reported a serological method of grouping *Streptococci*.

Habitat

Streptococci are worldwide in distribution. Most of the *Streptococci* of Veterinary interest live as commensals in the mucosa of the upper respiratory and lower urogenital tracts. They do not survive for long away from the animal hosts.

Morphology

Streptococci are gram positive, spherical or ovoid cells, arranged in chains or pairs. Chain formation is due to the cocci dividing in one plane only and the daughter cells failing to separate completely. Each coccus is about 1 m in diameter. They are facultative anaerobes, catalase negative, oxidase-negative, and non-spore forming and non-motile with the exception of some of the *Enterococci*.

Capsulation is not a regular feature of *Streptococci* but some strains of *Str. pyogenes* and some group C strains have capsules composed of hyaluronic acid while polysaccharide capsules are encountered in members of group B and D. Protoplasts (L-forms) may be induced by penicillin or phage associated lysine and may be propagated on hypertonic media.

Classification of Streptococci

I. Several systems of classification have been employed. Based on growth characteristics, type of haemolysis and biochemical activities they can be divided into 6 principal categories.

1. Pyogenic *Streptococci*	:	*Str. pneumoniae, Str.pyogenes, Str.equi, Str. dysgalactiae*
2. Oral *Streptococci*	:	*Str.salivarius*
3. *Enterococci*	:	*Str.faecalis, Str.avium, Str. gallinarum*
4. Lactic acid *Streptococci*	:	*Str.lactis*
5. Anaerobic *Streptococci*	:	*Str. monbillorum*
6. Other *Streptococci*	:	*Str.bovis, Str.uberis, Str.equi subsp zooepidemicus*

II. The aerobic and facultative anaerobic *Streptococci* are classified, based on their haemolytic properties. Brown (1919) established this method by employing meat infusion peptone agar with 5% horse blood. He recognized three types of reactions.

beta-hemolysis alpha hemolsys gamma hemolysis (no hemolysis)

Alpha - haemolytic *Streptococci*

They produce a greenish discolouration with partial haemolysis around the colonies. The zone of lysis is small (1 or 2 mm wide), within which the partially lysed erythrocytes are seen. These ά - Streptococci are generally commensals in the throat. Eg. *Str.pneumoniae*

Beta haemolytic *Streptococci*

Beta haemolytic *Streptococci* produces a sharply defined, clear, colourless zone of haemolysis, 2 or 4mm wide, within which the red cells are completely lysed. Most of the pathogenic *Streptococci* fall into the beta group and are called the β haemolytic *Streptococci.*

Gamma or non-haemolytic *Streptococci*

Gamma or non-haemolytic *Streptococci* produces no change in the medium. The gamma *Streptococci* includes the faecal *Streptococci* (*Str. faecalis*) and related species. They are called the *Enterococcus* or indifferent *Streptococci.*

Haemolytic *Streptococci* of group A are known as *Str.pyogenes*. These may be further subdivided into types based on the protein (M, T and R) antigens present on the cell wall. The M protein is acid and heat labile and the T protein is acid labile and trypsin resistant. Some of the lancefield groups may be further subdivided by means of the agglutination test and designated by Arabic numbers-Griffth typing.

Another important way in which the beta haemolytic *Streptococci* were classified by Rebecca lancefield (1933) based on the nature of a carbohydrate (C) antigen on the cell wall. They are known as Lancefield groups, 19 of which have been identified so far and named by the capital letters A-U (without I and J)

Lancefield group	Species	Host	Disease
A	*Streptococcus pyogenes*	Humans	Scarlet fever, Septic sore throat, erysipelas, abscesses and rheumatic fever
B	*S. agalactiae*	Cattle, sheep & goats Human & dogs Cats	Chronic mastitis Neonatal septicaemia Kidney and uterine infections
C	*S. dysgalactiae*	Cattle Lambs	Acute mastitis Polyarthritis

Lancefield group	Species	Host	Disease
	S. dysgalactiae subsp. Equismilis	Horse	Abscesses, endometritis and mastitis
	S. equi subsp. Equi	Horse	Strangles, genital and suppurative conditions, Mastitis and purpura Haemorrhagica
	S. equi subsp. Zooepidemicus	Horse Cattle Pigs Poultry Lambs	Mastitis, abortion, secondary Pneumonia and navel infections Metritis & Mastitis Septicaemia & arthritis in 1-3 wk old piglets Septicaemia & Vegetative Endocarditis Pericarditis and Pneumonia
D.	*Enterococcus faecalis*	Many species	Opportunistic infections
	S. equines and Str. Bovis	Many species	Opportunistic infections
E. (P,U,V)	*Str. Porcinus*	Pigs	Jowl abscesses & lymphadenitis
G	*Str. Canis*	Carnivores Cattle	Neonatal septicaemia, genital, skin and wound infections Occasional mastitis
N	*Lactococcus lactis*	Cattle	unknown
Q	*Enterococcus avium*	Many species	unknown
R	*S.suis type 2*	Pigs (4 to 6 months)	Meningitis and arthritis
S	*S.suis type 1*	Pigs (2 to 4 wks old)	Meningitis and arthritis
Ungroupable	*S.uberis*	Cattle	Mastitis
	S.pneumoniae	Guinea pigs, rats and primates	Pneumonia

Haemolytic *Streptococci* of group A are known as *Str.pyogenes*. These may be further subdivided into types based on the protein (M, T and R) antigens present on the cell wall. The M protein is acid and heat labile and the T protein is acid labile and trypsin resistant. Some of the lancefield groups may be further subdivided by means of the agglutination test and designated by Arabic numbers-Griffth typing.

Cultural Characteristics

It is an facultative anaerobe, growing best at a temperature of 37°C. They grow best in enriched media. On blood agar after incubation for 24 hrs small, circular, semi transparent colonies with an area of clear haemolysis are produced. Virulent strains from fresh isolates produce 'matt' (finely granular) colonies and avirulent strains form 'glossy' colonies. Strains producing capsules form mucoid colonies.

In glucose or serum or Brain Heart Infusion broth, growth occurs as a granular turbidity with a powdery deposit. No pellicle is formed.

In Edward's medium (selective media) it produces dewdrops like black colonies. Edward's medium contains blood agar, crystal violet and aesculin (differentiate among different species of Streptococci which do or do not hydrolyse aesculin). *Str. pneumonia* is alpha haemolytic, produces mucoid or flat colonies with smooth borders and a central concavity after 48-72 hrs on blood agar (draughtsman colonies). Ability to grow in 0.1 % tellurite broth is characteristic of Str. faecalis.

Biochemical Properties

Streptococci are catalase and oxidase negative. Ferment several sugars producing acid but no gas. They ferment sorbitol, trehalose, lactose, maltose, dextrin, and mannitol. Gelatin not liquefied. Nitrates not reduced, Indole is negative. Solubility in 10% bile differentiates pneumococci from other *Streptococci* species.

Resistance

It is a delicate organism easily destroyed by heat (54°C for 30 minutes). It can survive in dust for several weeks, if protected from sunlight. It is rapidly inactivated by antiseptics. It is more resistant to crystal violet and susceptible to sulfonamides and other antibiotics. Sensitivity to bacitracin is employed as a convenient method for differentiating *Str. pyogenes* from other β haemolytic *Streptococci*.

Antigenicity

The hyaluronic acid capsule of *Str. pyogenes* inhibits phagocytosis. The cell wall is composed of an outer layer of fimbria containing proteins and lipoteichoic acid, a middle layer of group specific carbohydrate and an inner layer of peptidoglycan. The Cell wall polysaccharide has been shown to have a toxic effect on connective tissue in experimental animals.

The peptidoglycan is responsible for cell wall rigidity. Several protein antigens have been identified in the cell wall (M, T and R). They are responsible for type specificity in Str.pyogenes. Among these the M protein acts as a virulence factor by inhibiting phagocytosis.

Hair-like pili project through the capsule of group A streptococci. The pili consist partly of M protein and are covered with lipoteichoic acid, which is important in the attachment of Streptococci to epithelial cells.

Toxins and Other Virulence Factors

Streptococci form several exotoxins and enzymes, which contribute to its virulence

Extracellular Toxins

a) Hemolysins

Streptolysins O and S are produced by groups A, C and G.

1) **Streptolysin O:** It is so called because it is an oxygen labile. It is an antigenic protein and is active in the reduced form. On blood agar, its activity is seen only in pour plates and not in surface cultures. It is also a heat labile. It is lethal on I/V injection into animals and has a specific cardiotoxic activity. It is a general cytotoxin. Red cells of all animal species except mice are lysed. Streptolysin O is antigenic and antistreptolysin regularly appears in sera following Streptococcal infection. Estimation of this antibody (ASO) titre is a standard serological procedure for the retrospective diagnosis of infection with Str.pyogenes.

2) **Streptolysin S:** It is an oxygen stable haemolysin and so is responsible for the beta haemolysis seen around streptococcal colonies on the surface of blood agar plates. It is called Streptolysin S since it is soluble in serum. Addition of serum to broth increased the yield of haemolysin. It is protein but not antigenic. It has been shown experimentally to be nephrotoxic.

b) Erythrogenic toxin (Streptococcal pyrogenic exotoxins/ Dick toxin / erythrogenic toxin)

Four erythrogenic toxins are known and most strains of Str. pyogenes produce one or more. They are pyogenic and enhance susceptibility to lethal shock by endotoxin. The toxin is thermostable and antigenic. The intradermal injection in rats leads to development of erythema. This reaction is called the "Dick test" or Schultz-charlton reaction and it is useful for diagnosis of scarlet fever.

c) Enzymes

i) Streptokinase: (Fibrinolysin): Filtrate of Streptococci gp, A, E & G produces fibrinolysin. This toxin promotes the lysis of fibrin clots by activating a plasminogen. Fibrinolysin plays a biological role in streptococcal infections by breaking down the fibrin barrier around the lesions and facilitating the spread of infection.

ii) Deoxyribonucleases: (Streptodornase): This causes depolymerisation of DNA, pyogenic exudates containing large amounts of DNA, derived from the nuclei of necrotic cells. Streptodornase helps to liquefy the thick pus and may be responsible for the thin serous character of streptococcal exudates. Four antigenically distinct streptodornase (A, B, C & D) have been recognized.

iii) Hyaluronidase: This enzyme breaks down the hyaluronic acid of the tissues. This might favour the spread of infection along intercellular spaces. Streptococci possess a hyaluronic acid capsule and also elaborate a hyaluronidase- a seemingly self-destructive process. This is produced by group A, C, G and B Streptococci.

iv) Proteinase: This is another instance of an apparently self-destructive enzyme, since it is capable of breaking down the M protein, streptokinase and hyaluronidase. The enzyme is, however, produced only under special conditions such as an acid pH (5.5 –6.5). Such conditions may be produced by tissue destruction, as in abscesses. Most strains of *Str. pyogenes* forms proteinase.

v) Neuraminidase: This activity is detected in *Streptococci* groups A, B, C, G and L. This enzyme is a virulence factor for pathogens surviving on the mucosal surface. In addition to this M types of *Str.pyogenes* produce NADase and other many strains also produce esterase, amylase and N-acetyl glucosaminidase.

vi) Serum opacity factor: When gp. A. Streptococci grown in horse serum it produces an opalescence of the serum. This opacity factor is a protein. This is a lipoproteinase and opalescence is a result of an agglomeration of the bacterial antigen.

Pathogenesis

The natural habitat of the species of Streptococci are skin, nose, throat, digestive and urogenital tract of man and animals. Str.pyogenes is present in the human nose and throat without causing any disease, while Str.agalactiae and Str.uberis can exist in bovine udder causing mastitis.

Streptococcus → through teat end → Adherence (M protein) & multiplication of the organism on the epithelium of teat & duct sinuses
↓ Death of PMN
PMN attracted fibrin plug formation in the smaller milk ducts ← inflammation and fibrosis
Involution of Secretory Tissue ← PMN attracted fibrin plug formation in the smaller milk ducts
Chronic Mastitis & Loss of milk Producing capacity ← Involution of Secretory Tissue

Pathogenicity

Cattle

Bovine mastitis: It is caused by *Str. agalactiae* (group B), *Str. dysgalactiae* (group C), *Str. equi* subsp. *zooepidemicus* (group. C), *Str. uberis* (group C, D, E, P, V). Mastitis arises from the multiplication of Streptococci in the teat sinus and extends into the ducts.

It causes parenchymatous mastitis, which is characterised by progressively chronic conditions resulting in fibrosis. In acute stages milk is composed of purulent exudate, dead tissue cells, coagulated milk protein and bacteria.

Peptostreptococcus indolicus is an anaerobic streptococcus, which is responsible for summer mastitis in cattle in association with *Trueperella pyogenes.*

Horse

Str. equi and *str. equisimilis* is the main cause of strangles in young horses. It is characterised by a catarrhal discharge, with inflammation of the nasal mucous membranes, followed by swelling of pharyngeal LN's in which abscesses develop. The infection spreads through lymph channels. It also causes metritis and cervicitis in horses.

Purpura haemorrhagica, considered to be an immune mediated disease, occurs in horses 1 to 3 weeks after illness.

Bastard strangles – in which abscess developed in many organs. It is a very serious complication.

Chicken

Str. gallinarum causes typical acute septicemia with peritonitis in chicken.

Dogs

Str. canis is considered to be the cause of acid milk in puppies and canine tonsillitis. It is also associated with neonatal septicaemia and toxic shock syndrome.

Pigs

*Str. suis c*auses porcine cervical lymphadenitis and is also isolated from pneumonia, septicemia, arthritis, endocarditis, meningitis and reproductive tract infections. It also causes erosive arthritis in young pigs.

Diagnosis

It involves clinical, microscopical and bacteriological examination. Palpation of the udder and supramammary lymph nodes will be helpful in distinguishing the chronic and insidious form of mastitis produced by Str.agalactiae and Str. uberis.

In contrast Str.dysgalactiae and Str.zooepidemicus causes sudden onset of acute inflammation of one quarter only with an acute systemic disturbance followed by joint infections and lameness.

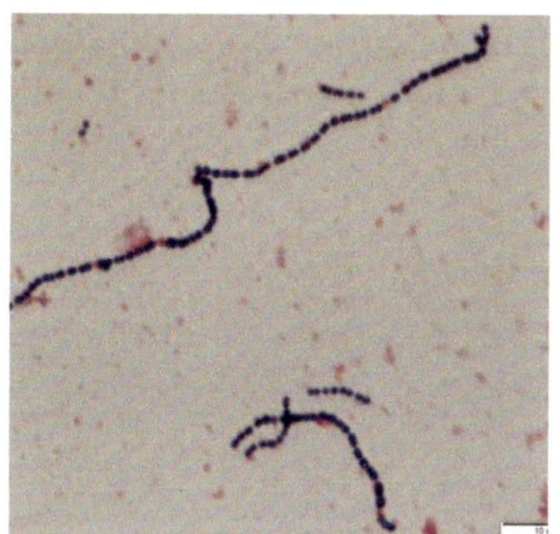

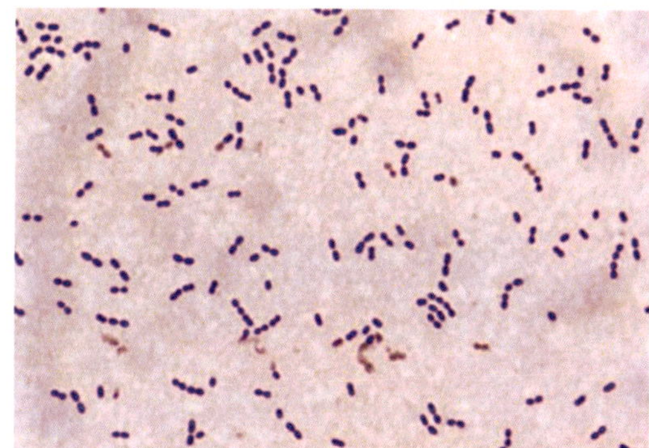

Microscopical examination

When long chains of organisms are detected in milk samples from chronic mastitis, it is caused by Str.agalactiae.

Bacteriological Examination

When 0.1 ml of secretions inoculated on Edward's medium (blood agar, crystal violet and aesculin) Str.agalactiae produces bluish-grey colonies and Str.uberis produces dark colour colonies.

CAMP test (Christie, Atkins, Munch and Peterson, 1944)

This test is based on the observation that ruminant red blood cells lysed by the beta toxin of staphylococci at 37°C are completely lysed in the presence of Str. agalactiae (group B).

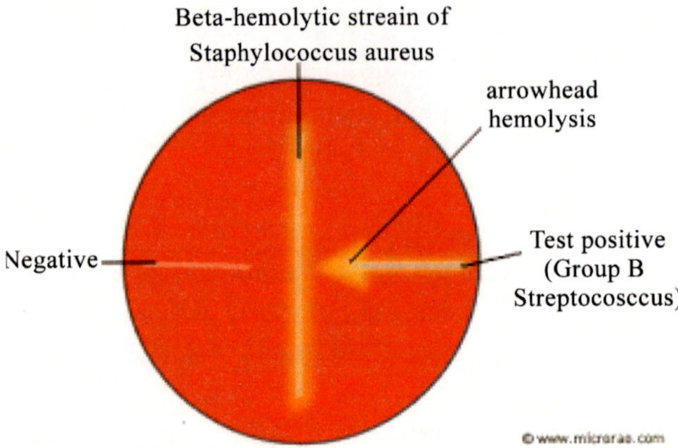

Bile solubility test

Differentiation between the pneumococcus and Str.viridans organisms can be achieved by bile solubility and the optochin test. Autolysis of pneumococcal cultures takes place within 15 minutes at 37°C in the presence of 10 per cent sodium deoxycholate. These substances have no effect on Str. viridans organisms.

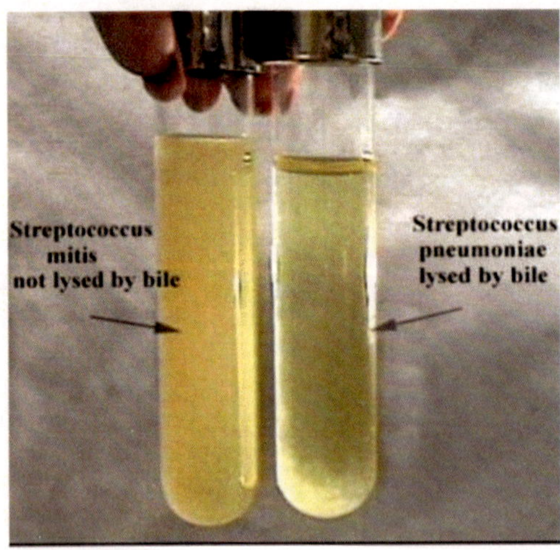

The optochin test

The majority of pneumococcal strains are sensitive to optochin (Ethylhydrocupreine hydrochloride), whereas Str.viridans organisms are not. This test consists of placing a small circular piece of filter paper, impregnated with 1:4000 aqueous solution of optochin, in the centre of a blood agar plate after inoculating the test cultures in streaks across the full width of the medium. The growth of pneumococcal strains will be inhibited to a distance of some 5mm from the circumference of the filter paper

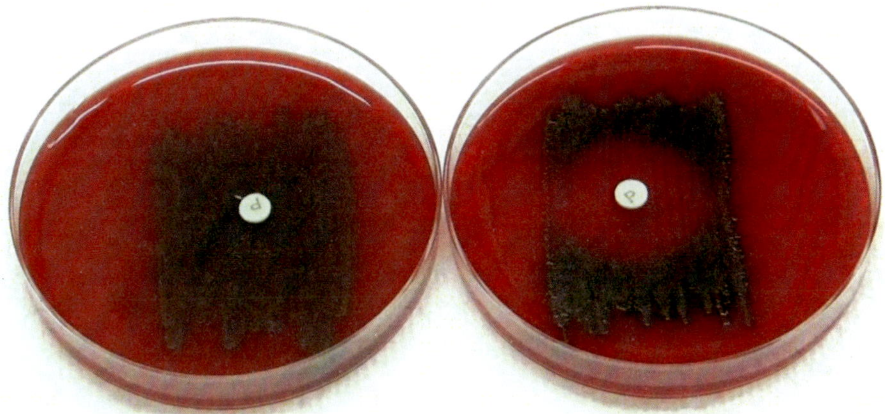

Control and Prevention

To treat the Streptococcal infections in animals, including mastitis, the most satisfactory method is antibiotic therapy using penicillin preparations to other substances because Streptococci develop resistance to penicillin comparatively infrequently.

Vaccines are of very limited value for the immunisation of animals against streptococcal infections, with the possible exception of equine strangles.

17

Bacillus

B.anthracis causes anthrax in animals and Wool sorter's disease, hide porter's disease, Malignant Pustule in humans. *B.cerus* causes food poisoning in humans. Other bacillus in this group is non-pathogenic and they are called as anthracoids. *B.licheniformis* is an emerging pathogen and it is implicated in sporadic abortions in cattle and sheep.

Systematics

Domain : Bacteria

Phylum : Firmicutes

Class : Bacilli

Order : Bacillales

Family : Bacillaceae

Genus : Bacillus

Species : *B. anthracis, B. cereus, B. subtilis, B. mycoides, B. megaterium, B.mesentricus*

Family Characters

They are gram +ve large rods, aerobic (facultative anaerobic), endospore forming, capsulated, mostly catalase positive and fermentative organisms. They are motile by peritrichous flagella.

Bacillus Anthracis

History

- Discovery of the anthrax bacillus is credited to Davaine and Rayer (1863 –1868).
- Considerable historic interest attached to anthrax bacilli;

- Pollender 1849 – Anthrax bacillus was the first pathogenic bacterium observed under the Microscope
- Davaine 1850 – first communicable disease shown to be transmitted by inoculation of infected blood
- Koch 1863 – first bacillus to be isolated in pure culture
- Pasteur 1881 – used for the preparation of attenuated vaccine.

Morphology

The anthrax bacillus is one of the largest pathogenic bacteria. They are gram+ve, straight, rod shaped, non-motile organisms measuring 4-8 um x 1-1.5um. In cultures bacilli are arranged end to end in long chains. The ends of the bacilli are truncated or often concave. So that chain of bacilli presents a bamboo stick appearance (also called Box - car bacillus - look like linked rail carriages).

In tissues or in blood smear, it is found singly, in pairs or in short chains, the entire bacilli being surrounded by capsule. The capsule is polypeptide in nature, being composed of a polymer of d-glutamic acid. Capsules are not formed under ordinary conditions of culture, but only if the media contains serum, albumen, charcoal, starch or bicarbonates with reduced partial pressure of carbon dioxide.

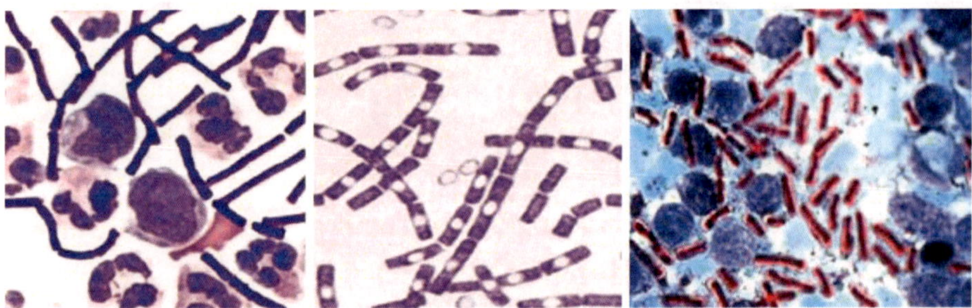

When blood films containing bacilli are stained with polychrome methylene blue for a few seconds and examined under the microscope, an amorphous purplish material is noticed around the bacilli. This represents the capsular material and is characteristic of the anthrax bacilli. This is called as McFadeyan'sreaction. This reaction depends on the degree of heat employed for fixation of a blood film.

Sporulation occurs readily outside the body in the presence of oxygen. Spores are formed in culture or in the soil, but never in the animal body during life. Sporulation occurs under unfavourable conditions for growth and is encouraged by distilled water, 2%NaCl or growth in oxalated agar. Sporulation takes place at an opt.temp.of 25-30^{0}C and in an atmosphere containing low partial pressure of oxygen.

Spores are central, elliptical or oval in shape and are of the same width as the bacillary body. So that they do not causes bulging of the vegetative cell. The spores do not stain by ordinary methods. But can be stained with Sudan black B.

Cultural Characters

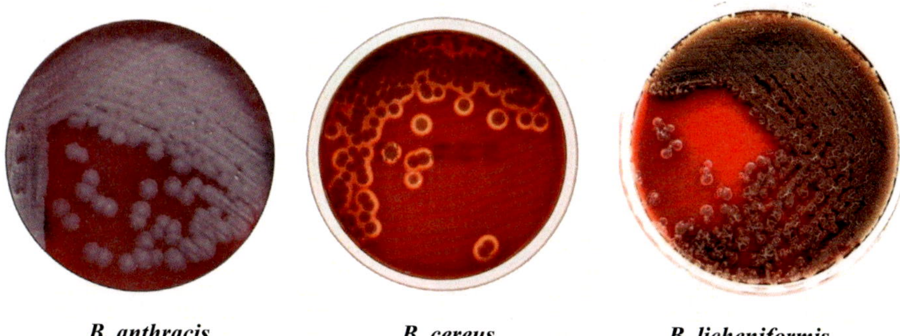

B. anthracis ***B. cereus*** ***B. licheniformis***

- Grows readily under aerobic and a facultative anaerobic at an opt.temp. of 35-37^{0}C
- On agar plates, irregular, round, raised, dull, opaque, greyish white, 2-3 mm in diameter frosted glass appearance colonies are produced. Under the low power microscope, the slightly serrated edge of the colony is composed of long, interlacing chains of bacilli, resembling locks of matted hair. This is referred to as medusa head or judges wig or women's curling hair type of growth.
- *B. licheniformis* produces characteristic hair-like outgrowths from streaks of the organisms on agar media. Colonies become brown with age. The name of this species derives from the similarity of its colonies to lichen.
- In medium containing iron salts, virulent *B.anthracis* produces pink or purple colured pigment colonies.
- Virulent capsulated strains form rough colonies, while avirulent attenuated strains form smooth colonies.
- In gelatin stab, fine filaments of growth develop laterally along the line of inoculum. The growth nearer to the surface of the medium is the longest and they progressively shorter. Where there is less oxygen resembling inverted fir tree appearance.

- On haemolysis, the *B.anthracis* produces slight haemolysis with capsule production compared with anthracoid organisms. B. cereus produces a wide zone of complete haemolysis around the colonies.
- A selective medium (PLET) consisting of polymyxin, lysozyme, EDTA and thallous acetate added to heart infusion agar are useful for isolating *B.anthracis* from mixtures containing other spore-bearing bacilli.

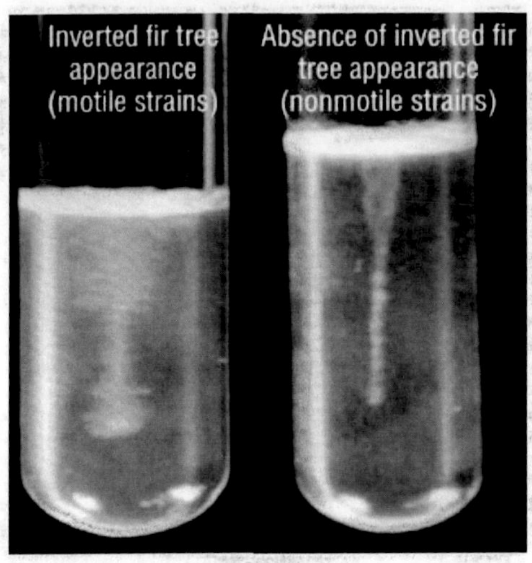

Biochemical Characters

They are Catalase +ve. It ferments Glucose, maltose, saccharose, trehalose and dextrin to produce acid but no gas. Nitrates reduced to nitrites. Indole is negative.

Resistance

The vegetative bacilli are destroyed at 60°C in 30 min. In the carcases of animals the bacilli remain viable in the bone marrow for a week and in the skin for two weeks. Normal heat fixation of smears may not kill the bacilli in blood film.

The spores are highly resistant to drying, heat, cold and disinfectants. Spores remain viable for many years in soil, water and animal hides and products. Spores have been isolated from naturally contaminated soil for as long as 60 years. They resist dry heat at 140°C for 2-3 hrs and boiling for 10 min. They survive on 5% phenol for weeks. Spores can be killed at 120°C for 10 min and 4% $KMNo_4$ treatment for 15 min. Destruction of the spores in animal products is achieved by HCHO. Treat 2% solution of HCHO at 39-40°C for 20 mins for disinfection of wool and as 0.25% at 60°C for 6 hrs for animal hair and bristles. This process is called duckering.

The anthrax bacillus is susceptible to sulfonamides, penicillin, erythromycin, streptomycin, tetracycline and chloramphenicol.

Antigens and Toxins

The complex antigenic structure includes a capsular polypeptide, a somatic protein and a somatic polysaccharide antigen. The capsule contains poly-d-glutamic acid, which is only detectable in virulent strains, having antiphagocytic properties. The somatic protein is a protective antigen present in the edema fluid.

B. anthracis produces an extracellular toxin which is composed of three components, Factor I, II & III namely oedema factor (EF), protective antigen (PA) and lethal factor (LF)respectively. Factor I consists of chelating compounds containing phosphorus with protein and carbohydrate moieties. Factor II & III consists of proteins. All three factors are essential to exhibit their toxic properties. They are not toxic individually but the whole complex produces local edema and generalised shock.

Other Toxins

- InH – immune inhibitor
- MprF- multiple peptide resistance factor
- Anthrolysins – similar to phospholipase C and cholesterol binding cytolysins – lysis of phagolysosome - liberate *B.anthracis*- lysis of cell membrane – liberate into extracellular environment
- DIp- iron binding protein
- AtxA – anthrax toxin activator
- AcpA- anthrax capsule activator

Pathogenesis

Spores are acquired from the environment. They are phagocytosed by macrophages. The spores germinate within the phagolysosome compartment. Vegetative bacteria produce the regulatory proteins AtxA and AcpA, which in turn regulate the production of LeTx, EdTx, capsule and other toxins. InH and MprF aid in the survival of *B.anthracis* while it is in the phagolysosome.

Anthrolysins permit escape from first the phagolysosome and then the macrophage itself. Production of lethal toxins triggers the hyper production of cytokines (IL-1, 6& 8, TNF) leading to multiple organ dysfunction, shock and depletion of clotting factors as well as apoptotic death of macrophages. Edema factor produces severe fluid and electrolyte imbalances, resulting in localised edema and systemic shock.

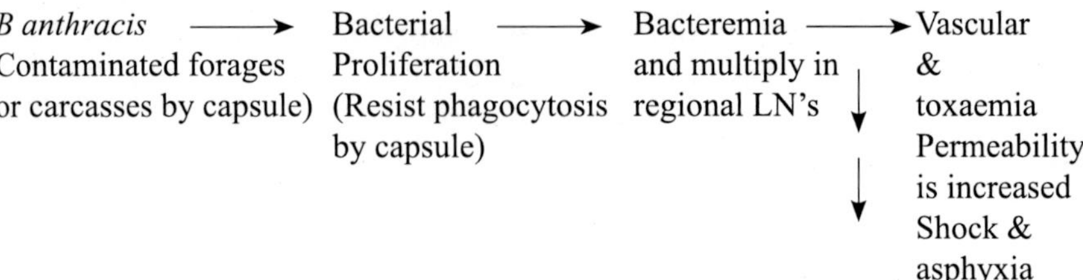

Symptoms

Horses

Acute form is very common and death may take place one day after edematous swelling of the throat and neck region. There may be symptoms of colic. In less acute cases, oedematous swelling becomes generalised and death occurs after 2-3 days.

Cattle

Bulls are more susceptible than cows. They have a mortality rate of 90%. There are three clinical causes of bovine anthrax. In peracute septicemia death occurs within 2 hours after an animal collapses with convulsions, sudden death in animals that appeared normal is common.

In acute septicemia death occurs within 48 to 96 hours clinical signs include fever, anorexia, ruminal stasis, hematuria and blood tinged diarrhoea. Pregnant animals may abort and milk production often abruptly decreases. Terminal signs include severe depression, respiratory distress and convulsions.

In chronic cases, clinical signs that manifest for more than 6 days are rare. *B.licheniformis* infection is associated with the feeding of contaminated silage and is responsible for abortion in cattle and sheep.

Sheep

Disease is acute and death occurs rapidly after convulsions.

Pigs

They have a greater natural resistance than herbivores. The usual signs are oedematous swellings in the region of the throat and neck interfering with normal respiration, enteritis and rise in temperature. The disease is not always fatal.

Dogs

The throat will be inflamed and swollen with gastroenteritis

Lesions

The carcass of animals will putrefy rapidly and develop incomplete rigour mortis. The blood is dark, (tarry coloured), clots poorly & exudes from the natural orifices. The spleen is greatly enlarged, dark and friable. The spleen reveals black cherry jam consistency. The LN (lymph node) at the region of initial infection site is hemorrhagic and edematous. Ecchymotic haemorrhages on the serosal surface of the abdomen, thorax, epicardium and endocardium are common. Subcutaneous edematous swellings are present on the ventral aspect of the neck

(Note: When suspected of anthrax, care should be taken not to open the carcass. Muzzle piece or earpiece is usually sent for examination)

Diagnosis

1. Clinical symptoms

2. Blood films from dead animals made by puncturing the superficial vein of the ear or in the region of the foot. Care should be taken to seal the injection site by placing cotton soaked in alcohol and ignited. The smears are heat fixed and stained by wright or giemsa's stain to reveal *B. anthracis* as large blue rods with characteristic dark pink or purple coloured capsules.In case of horse and pigs since peripheral blood contains fewer organisms, smears should be made from the edematous fluid or LN's.

3. Cultural examination:

- Swabs from blood are inoculated in blood agar plates and incubated at 37°C for 24 hrs and examined for their typical growth.
- Swabs are inoculated in agar enriched with blood or serum and incubated for 6 hrs at 37°C and examined by stained smears.

4. Bacteriological examination of hair, wool, hide, bone, bone meal & others

Samples are added to cold saline and shaken intermediately for 3 hours. The supernatant fluid is then heated at 70°C for 10 minutes. Then filtered through two layers of muslin cloth, added to melted agar, and poured into petri dishes, allowed to set and incubated at 37°C. After 12 hours incubation, plates are examined for characteristic colony morphology.

5. Ascoli's Precipitation test: (Ascoli 1911)

Grind up the organ or blood of the suspected animal and suspend it to 5-10 parts of saline and boil for 15 minutes. Filter through filter paper and allow cooling. Place 0.5 ml of anti-anthrax serum (1:50) in a small test tube and overlay with 0.5 ml of clear filtrate. Stand at room temperature for 15 minutes. A white ring of precipitation indicates a positive reaction.

6. String of Pearl's test: (Charlton, 1980)

B.anthracis produces swollen round cells in chains (string of pearls) when incubated for 3-6 hrs on tryptose agar containing 0.05 –0.5 IUof penicillin/ml

To 100ml of molten nutrient agar, add required sodium benzylpenicillin and mix carefully. Pour into petri dishes and allow to set. With a scalpel cut a block about 1.6cm^2 from the penicillin agar plate and place it on a microscopic slide in a petri dish containing a small piece of moistened absorbent cotton wool to prevent the agar drying out. Use a young colony to streak inoculates the centre of the agar block. Place a clean coverslip on the agar block and incubate the petri dish at 37^{0}C. After 2hrs, remove the slide and examine the inoculum microscopically by oil immersion for the string of pearls growth.

7. Fluorescent antibody technique.

8. Bacteriophage identification (Differentiation from non-pathogenic Bacillus sp.)

It is considered reliable but needs experience. Cherry gamma phage is used. It can be propagated on *B. anthracis* strain 14, which makes a suitable control. On one half of the blood agar plate, streak *B. anthracis* strains 14 cultures as control and on the other half streak the culture to be identified. Add drops of phage preparation (1:10 dilution) on both halves. Incubate the plates in the upright position for 8-12 hrs. Clear zones of clearing will be seen on control culture and on the suspected half if it is *B. anthracis*

9. Animal Inoculation:Guinea pigs & mice are highly susceptible. Materials are inoculated by dermal scarification, subcutaneous or intramuscular route. Death occurs in 2-3 days and the organisms will be readily identified in the blood & tissues.

Apart from this laser pyrolysis, gas-liquid chromatography, mass spectrometry is used for detection of anthrax toxin productions.

Treatment

The organism is susceptible to penicillin-G, tetracyclines, erythromycin and chloramphenicol.

Immunity

Prevention of anthrax in animals is aided by active immunisation. Initially oedema fluid and tissues obtained from anthrax lesions, which had been freed from viable organisms, had protective properties. He termed the active substances as "aggressins". Pasteur's vaccine was anthrax bacillus attenuated by growth at 42 –43°C.

As the spore is the common infective form in nature, vaccines consisting of spores of attenuated strains were developed. The Sterne vaccine contained spores of a non capsulated avirulent mutant strain. It should be given 1 month before anticipated outbreaks. The Mazucchi vaccine contained spores of stable attenuated carbazole strain in 2% saponin. It gives good protection when given subcutaneously. Protective humoral antibodies develop in 7-10 days against Factor II of the exotoxin complex and last about one year.

Public Health Aspects

There is a need for great care in performing necropsy on animals. Infections most often result from spores entering through injuries to the skin causing cutaneous anthrax. Spores are present in soil, hair, hides, wool (wool sorter's disease-Pulmonary anthrax), faeces, milk, meat and blood products. The skin lesion (cutaneous anthrax) is usually solitary, painless, seropurulent, necrotizing, hemorrhagic and ulcerous. It leaves a black scar (anthrax=coal), which accounts for the name malignant pustule.

Difference between *B. anthracis* & Anthracoid organisms in culture

Sl.no.	*B. anthracis*	Anthracoid organisms
1.	Non motile	generally motile
2.	Capsulated	non capsulated
3.	Grows in long chains	Grows in short chains
4.	No turbidity in broth	Turbidity in broth
5.	Inverted fir tree in gelatin	Atypical or absent
6.	Methylene blue reduced weekly	Reduced strongly

7.	Haemolysis weak	Strong
8.	Liquefaction of gelatin is slow	Rapid
9.	Lecithinase reaction is weak	Rapid
10.	Ferments salicin slowly	Rapid
11.	Produces toxin neutralised by	not neutralised
	B. anthracis antitoxin	
12.	Pathogenic to g.pigs & mice	non pathogenic
13.	Susceptible to gamma phage	not susceptible.

Difference between *B. anthracis* & Anthracoid organisms in blood smear stained with Polychrome methylene blue stain

	B. anthracis	Anthracoid organisms
1.	Dark pink or purple coloured capsules	No capsule
2.	Organism rod stained blue	Organism rod stained blue
3.	Ends truncated	Ends bulged and rounded
4.	Single, pair or short chain	Usually long chain
5.	Absence of spores	Spores may be present

18

Clostridium

The genus Clostridium consists of gram positive, spore forming anaerobic bacilli. The spores are wider than the bacillary bodies, giving the bacillus a swollen appearance, resembling a spindle. Hence, the name clostridium (Kloster- meaning spindle) was given.

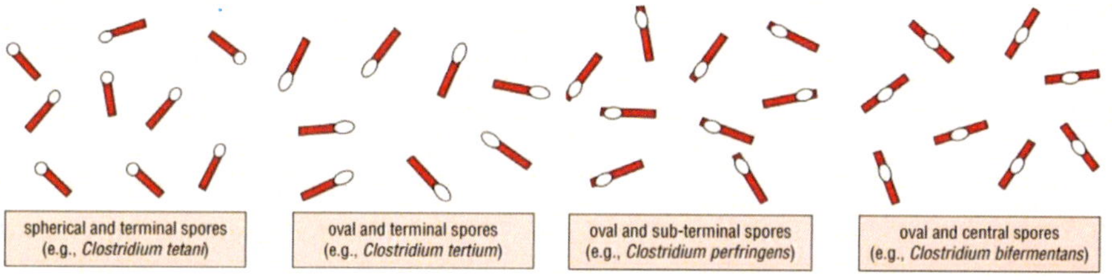

Systematics

Domain : Bacteria

Phylum : Firmicutes

Class : Clostridia

Order : Clostridiales

Family : Clostridiaceae

Genus : Clostridium

The clostridia can be divided into four major groups according to the kind of disease they produce. They are as follows.

1. The Histotoxic clostridium causes a variety of tissue (often muscle) infections frequently following wounds or other trauma (eg).

Cl. chauvoei	- Cattle, sheep (pigs)	- Black quarter (Black leg).
Cl. septicum	- Cattle,	- Malignant edema
	- Sheep	- Braxy
	- Chicken	- Necrotic dermatitis
Cl. novyi type A	- Sheep	- Big head of rams
	- Cattle and Sheep	-Gas gangrene
Type B	- Sheep (Cattle)	- Black disease (necrotic hepatitis)
Type C	- Water buffaloe	- Osteomyelitis

2. Hepatotoxic clostridia produce their toxins in the liver, thus resulting in the disease Bacillary haemoglobinuria and Black disease

Eg;

Cl. haemolyticum (*Cl. novyi* type D)	- Cattle, (sheep)	- Bacillary haemoglobinuria
Cl. sordellii	- Cattle, Sheep, Horses	- Gas gangrene.
Cl.colinum	- Birds	- Quail disease,
		Ulcerative enteritis
Cl.piliforme	- foals, laboratory animals Calves, dogs & cats	- Tyzzer's disease (Hepatic necrosis)

3. The Enterotoxigenic clostridium produces mainly enterotoxaemia and food poisoning although they are occasionally histotoxic (Eg).

Cl. perfringens (Cl.welchi)		
Type A	- Humans	- Food poisoning, gas gangrene
	- Lambs	- Enterotoxaemic Jaundice (Yellow lambs disease)
	- Broiler chickens	- Necroticenteritis
Type B	- Lambs	- Lamb dysentery (Under 3 weeks old)
Type C	- Piglets, lamps, calves and foals.	- Hemorrhagic enterotoxemia (Clostridial enteritis)
	- Broiler chickens	- Necroticenteritis
	- Adult sheep and goat	- Struck
Type D	- Sheep	- Pulpy kidney disease (except neonates)
Type E	- Calves and lambs	- Enterotoxaemia

4. The Neurotoxic clostridia cause the disease by the production of the potent exotoxins (Neurotoxins)

Eg.	
Cl. tetani	- Tetanus
Cl. botulinum	- Botulism

Family Characters

They are pleomorphic, rod shaped; long filaments and involution forms are common. Spore formation occurs with varying frequency in different species. The shape and position of the spores vary in different species. The clostridia are motile with peritrichous flagella except *Cl.welchi* and *Cl.tetani* type VI. *Cl.welchi* is capsulated, while others are not.

Clostridia are anaerobic. *Cl.odematiens* is strict anaerobes and is dying on exposure to oxygen. *Cl.histolyticum* and *Cl.welchi* are aerotolerant and may even grow aerobically. The clostridia are fermentative, oxidase negative and catalase negative organisms.

A very useful media for isolation of clostridia is Robertson's cooked meat broth. Clostridia grow in the medium, rendering the broth turbid most species produces gas.

- Saccharolytic species turn the meat pink – *Cl.odematiens, Cl.septicum, Cl.chauvoei and Cl.welchi*
- Proteolytic species turn the meat black - *Cl.tetani, Cl. Botulimum,* and produce foul and pervasive odour *Cl.hamolyticum*
- In litmus milk medium, the production of acid, clot and gas can be detected.

Uninoculated/Negative growth

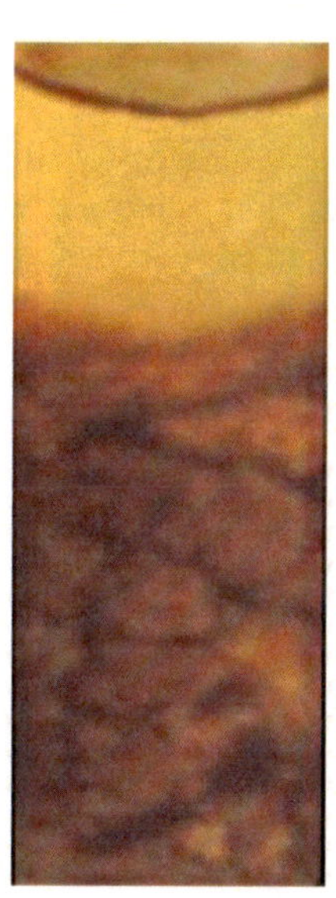

Saccharolytic Growth

Proteolytic Growth

Neurotoxic Clostridia

Clostridium Tetani

It is the causative organism of tetanus in Man and Animals.

History

- Tetanus has been known from very early times, having been described by Hippocrates. But the knowledge of the disease was achieved only in 1884.
- Rosenbach –1886 - demonstrated a slender bacillus with round terminal spores in a case of tetanus.
- Kitasato –1889 – isolated *Cl.tetani* in pure culture and reproduced the disease in animals by inoculation of pure culture.
- The Greek term "tetanus" which means 'contracture' has been taken from the Latin medicine "rigor".

Natural Habitat

Soil, especially that contaminated by animal faeces, is the natural habitat as *Cl.tetani* is often transient in the intestines of horses and other animals. It is ubiquitous and has been recovered from a wide variety of other sources, including street and hospital dust, cotton wool, bandages, catgut, plaster of paris, clothing etc. It may occur as an apparently harmless contaminant in wounds.

Morphology

Cl.tetani is a straight, slender, gram positive rod that characteristically produces terminal, spherical endospores that bulges the cell giving the characteristic drumstick appearance. (The young spore may be oval rather than spherical). It occurs singly and occasionally in chains. It is non-capsulated and motile by peritrichous flagella.

Cultural Characters

Very strict anaerobe grows at an opt.temperature of 37°C and pH7.4. It grows on ordinary media and the growth is improved by blood and serum and not by glucose.

Surface colonies are very difficult to obtain as the growth has a marked tendency to swarm over the surface of the agar especially if the medium is moist. The swarming nature /spreading can be inhibited by increasing the concentration of agar upto 3% (stiff agar). In this stiff agar individual rhizoid colonies are formed.

On blood agar, it develops partially translucent, greyish colonies with filamentous edges giving a fuzzy appearance. On horse blood agar, alpha haemolysis is produced, which later develops into beta haemolysis due to the production of haemolysin (tetanolysin).

Cl.tetani grows well in Robertson's cooked meat broth, with turbidity. The meat is not digested, but is turned black on prolonged incubation. In gelatin stab cultures a fir tree type of growth occurs, with slow liquefaction. A greenish fluorescence is produced on media containing neutral red.

Biochemical Reactions

Cl.tetani has a feeble proteolytic, so it does not ferment any sugars. It forms indole. It is MR and VP negative, nitrates not reduced.

Resistance

The endospores are highly resistant and while boiling kills the spores of most strains in 15 mins. Autoclaving at 121^0C for 15 mins and dry heat temp of 150^0C for more than one hour is completely sporicidal. Spores are able to survive in soil for years and they are resistant to most antiseptics. They are not destroyed by 5% phenol or 0.1% mercuric chloride solution in two weeks or more. Iodine (1% aqueous solution) and H_2O_2 kill the spores within a few hours.

Antigenicity

Ten serological types have been recognized based on the flagellar antigen (types I to X). Type VI contains non flagellated strains. All types produce the same neurotoxin- tetanospasmin. This can be neutralised by one common antitoxin.

They have a common heat stable somatic antigen shared by all types. A second somatic antigen is shared by type II, IV, V and X.

Toxins

Cl.tetani produces at least two distinct toxins.

- Tetanolysin
- Tetanospasmin

They are antigenically and pharmacologically distinct and their production is mutually independent.

Tetanolysin

Tetanolysin is a cholesterol binding cytolysin. It binds to cholesterol-containing rafts in the eukaryotic cell membrane. Once bound it forms a pore resulting in the death of the cell. Tetanolysin has no known pathogenic significance in the production of tetanus.

It causes lysis of rabbit and horse RBC'S. It is a heat and oxygen labile similar to those of streptolysin O, ä toxin (*Cl. Oedematiens*) &ϕ toxin (*Cl. Welchii)*

Tetanospasmin

Tetanospasmin is a 'di-chain' molecule consisting of a light chain with (zinc endopeptidase activity), a heavy chain composed of a translocation domain (responsible for forming a pore through which the light chain passes), and a binding domain (responsible for binding to nerve cells).

Tetanospasmin is a zinc endopeptidase, which hydrolyzes the docking proteins required by neurotransmitter-containing vesicles to fuse with the presynaptic membrane. Tetanospasmin hydrolyzes the docking proteins VAMP- vesicle associated membrane protein, also known as synaptobrevin.

Pathogenesis

Cl.tetani has little invasive power. The endospores enter traumatised tissue or surgical wounds, especially after castration or docking, via the umbilicus or into the uterus following dystocia in cattle and sheep. The spore implanted in a wound can germinate and multiply only if the conditions are favourable.

Destruction and necrosis of tissue, lack of drainage in the area, presence of extraneous matter especially of soil, all create anaerobic conditions and favour germination of *Cl.tetani* spores. The resultant vegetative cells multiply at the site and produce the potent tetanospasmin. This travels via peripheral nerves or blood stream and attaches to ganglioside receptors on the nearest cholinergic nerves and is internalised within a vesicle (by receptor mediated endocytosis). Then travels retrograde inside the axons to the cell bodies in the ventral horns of the spinal cord thus affecting many groups of muscles at various levels.

Symptoms

It is influenced by several factors, such as the site and nature of the wound, the dose and toxigenicity of the contaminating organism. The incubation period is variable from 2 days to several weeks but is commonly 6-12 days.

Initial symptoms include mild stiffness and unwillingness to move. This may proceed with the head, neck and tail becoming rigid. Mild twitching of muscles develops into obvious spasms of muscles, which can occur in response to sudden noises, animals fall over to one side and unable to rise.

In the terminal stages the rigidity of muscles extend from the limbs to the trunk, nostrils get dilated, ears erect, nictitating membrane protrusion and mastication becomes impossible because the mouth cannot be opened – hence called "LockJaw".Respiration becomes shallow and rapid before final respiration failure.

Lesions

No characteristic lesion for this disease but there may be a superficial wound which has developed from accidental injury or from surgery.

Diagnosis

In tetanus, the diagnosis is often based on the history and on the characteristic clinical signs without reference to laboratory tests.

Direct Microscopy

Demonstration of characteristic drumstick spores of *Cl.tetani* by gram stained smears of material from a wound (but it is not confirmative, because *Cl. tetanomorphum* and *Cl.tetanoides* also produce drumstick spores).

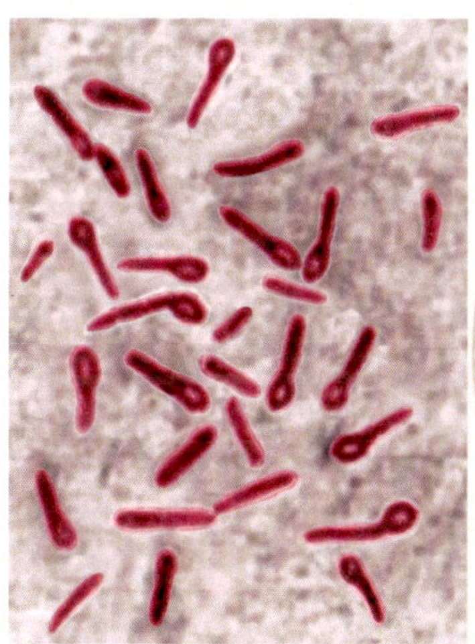

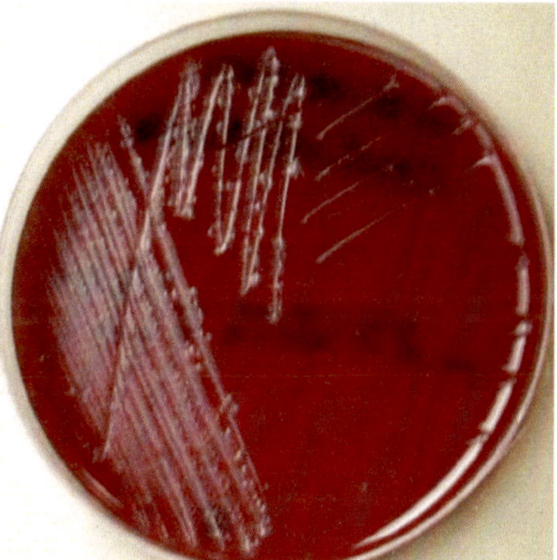

Isolation

Necrotic tissue from a wound or wound exudates can be heated to 80^{0}C for 20 min and used to inoculate a blood agar plate and another blood agar plate containing stiff agar. A tube of thioglycollate medium or cooked meat broth could also be inoculated and subcultured into blood agar. The plates are to be incubated anaerobically for 2-3 days. Growth is noticed using a hand lens as a filamentous growth spreading throughout the medium. The edges of this growth give pure culture on subcultivation.

Confirmation is done by identification of toxins. The toxin present in animal's serum or in filtrate from cooked meat broth or thioglycollate medium can be inoculated into mice S/C or I/M and identified by neutralisation or protection tests using specific antitoxin.

Control and Prevention

The disease is due to the action of the toxin, and hence, the obvious and most dependable method of prevention is to build up antitoxic immunity by active immunisation.

The available methods of prophylaxis are;

1. Surgical attention
2. Antibiotics
3. Immunisation – passive, active or combined.

Surgical attention aims at removal of foreign bodies, necrotic tissue and blood clots, in order that an anaerobic environment favourable for the tetanus bacillus is not provided. Flushing with hydrogen peroxide in the wound area produces aerobic conditions.

Tetanus can be prevented by antibiotics (Large doses of penicillin) when administered 4 hrs after infection but not after 8 hrs. Hence, prompt administration is essential. Bacitracin or neomycin may be applied locally. Penicillin can be given as both injections and orally till healing is established. Antibiotics have no action on the toxin.

Antitoxin should be administered promptly, either I/V or into the subarachnoid space, for three consecutive days to neutralise unbound toxins. Toxoid may be given subcutaneously to promote an active immune response even on those animals, which have received antitoxin. For prevention, the farm animals should be vaccinated routinely with tetanus toxoid.

Clostridium Botulinum

Cl. botulinum denotes a group of bacteria that produce extremely potent neurotoxins. These toxins cause botulism, a disease characterised by flaccid paralysis in many animals and humans. Botulism is most common in water birds, ruminants, horses, mink and poultry.

Botulism in animals has been called a variety of names;

Horses : Spinal typhus / Shaker foal syndrome

Cattle : Lamsiekte, loin disease and contagious bulbar paralysis

Water fowl : Limberneck, alkali poisoning and western duck sickness

(Note: Botulism is rare in domestic cats. Pigs and dogs are relatively resistant)

History

The name botulism is derived from sausage (botulus, latin for sausage), an article of food that used to be associated with the type of food poisoning. *Cl.botulinum* was first isolated by VanErmengam (1896) from a piece of ham that caused an outbreak of botulism.

Natural Habitat

The endospores are widely distributed in soils and aquatic environments throughout the world. The diseasebotulism is mainly due to the ingestion of preformed toxin. Germination of the endospores, with growth of vegetative cells and production of toxin, occurs in anaerobic situations such as contaminated cans of meat, fish or vegetables, carcases of invertebrate and vertebrate animals, rotting vegetation and baled silage.

Morphology

Gram +ve, straight rods, occurring singly, pairs or occasionally in chains, Spores are oval, wider than the bacilli, situated centrally, terminal or subterminal, Non-capsulated, motile by peritrichous flagella.

Classification

Type	Toxin produced	Most susceptible animals	Sources of toxin	Disease
A	A	Humans, chickens, pigs	Vegetables, fruits, meat and fish	Food borne botulism
Cα	C_1	Waterfowl	Invertebrate carcases, rotting vegetation and material on refuse dumps	Limberneck in long necked birds
Cβ	C_2	Cattle, horses, mink, dogs	Carcases, baled silage, chicken manure as feed supplement	Forage poisoning
D	D	Cattle, sheep	Eating contaminated bones and carcass of small mammals (Phosphorus deficiency-Pica)	Lamsiekte

Cultural Characters

It is a strict anaerobe, opt.temp is 30-37°C and pH is 7-7.6. Good growth occurs on ordinary media. Surface colonies are large, irregular, and semitransparent with fimbriate borders. Spores are produced consistently when grown in alkaline glucose gelatin media at 20 to 25°C.

In horse blood agar, large transparent colonies with irregular edges are developed with a narrow zone of haemolysis. On sheep blood agar *Cl.botulinum* produces beta-haemolysis, the colonies are slightly domed with a ragged edge.

In cooked meat medium, the proteolytic strains type A, B and F produce blackening of meat and non-proteolytic strains C, D and E do not blacken meat. Gelatin liquefied rapidly by type A&B and slowly or not at all by type C, D and E.

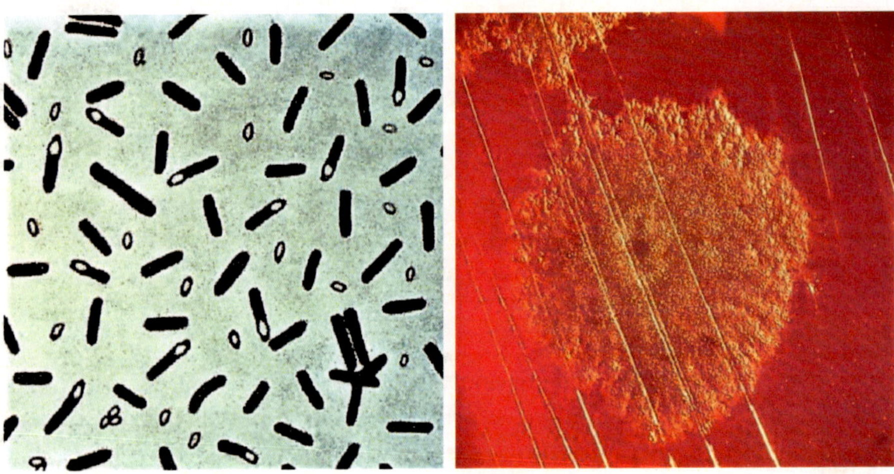

Resistance

Spores are highly resistant, surviving several hours at 100°C and for up to 10 mins at 120°C. But spores can be killed at 121°C for 15 mins, while the toxins are destroyed at 100°C for 20 mins.

Antigens and Toxins

Cl.botulinum possesses a number of H, O and spore antigens. There are seven types of botulinum toxin. Based on the toxin production *Cl.botulinum* is classified into 8 types (Type A to G). The toxins differ from other exotoxins in that they are not released during the life of the organism. It appears in the medium only on the death and autolysis of the cell. It is believed to be synthesised initially as a nontoxic protoxin or progenitor toxin. Trypsin and other enzymes activate progenitor toxin to active toxin.

These neurotoxins are identical in pharmacological action but differ in potency, distribution and antigenicity. They are neutralised only by the homologous antiserum. All seven types of toxins are zinc endopeptidases. They hydrolyze the docking proteins required by neurotransmitter-containing vesicles to fuse with the presynaptic membrane. Though the end result is the same (blockage of the release of neurotransmitter), the various types of toxins hydrolyze different docking proteins.

Comparison of the toxins of *Cl. tetani* and *Cl. botulinum*

Characteristics	*Cl.tetani*	*Cl.botulinum*
Site of toxin production	Wounds	Carcases, decaying vegetation and occasionally wounds and intestine
Mode of action	Centrally by blocking synaptic inhibition	Peripherally by blocking neuromuscular transmission
Type of paralysis	Spastic paralysis	Flaccid paralysis
Antigenic types of toxin	Tetanospasmin (one antigenic type)	Eight different toxins produced by types A-G.

Pathogenicity

Ingested botulinum toxin absorbed from the glandular stomach and anterior small intestine and distributed via the bloodstream. It binds to receptors and enters the nerve cell after receptor mediated endocytosis. Vesicles containing toxin remain at the myoneural junction. A light chain of the toxin translocates across the vesicle membrane into the cytosol of the nerve cell and subsequently hydrolyzes docking proteins. The synapse degenerates and flaccid paralysis results due to the lack of neurotransmitter- acetylcholine. When this affects muscles of respiration, death due to respiratory failure occurs. *Cl.botulinum* is non-invasive and virtually non infectious. Botulism is of three types

1. **Food borne botulism:** It is due to the ingestion of preformed toxin in foodstuffs. The toxin is adsorbed from the intestinal tract and is transported via the bloodstream to peripheral nerve cells. Where it binds to susceptible cells and suppresses the release of acetylcholine at the myoneural junctions. This results in flaccid paralysis, death being caused by circulatory failure and respiratory paralysis.
2. **Wound botulism:** The spores are introduced into wounds where they germinate. Toxin is formed at this localised site and spreads through the body. The shaker foal syndrome in horses is thought to be caused in this way.

3. **Infant botulism:** It occurs when spores are ingested in food and germinate in the intestines when the normal flora has not been fully established. This form is seen in human infants (Floppy baby syndrome) and as rare epidemics of type C in broiler chickens and turkey poults.

Symptoms

In cattle, the incubation period varies between 2-10 days depending upon the dose of toxin ingested. Initially there will be excitation, followed by incoordination and paralysis of the hind limbs. There will be paralysis of muscles in the mouth, pharynx and neck resulting in the animal being unable to swallow and the tongue protruding from the mouth. This is followed by death. In South Africa this condition is termed as lamsiekte in cattle caused by type D especially in the phosphorus deficient animals.

In poultry it results with the ingestion of type C toxin and the disease is known as duck sickness or western duck disease and Limberneck(drooping head posture)in chicken. The symptoms include paralysis of the wings, legs and neck, protrusion of nictitating membrane, diarrhoea and comatose before recovering in 5-6 days time.

In horses the incubation period varies between 12 hours to ten days. The symptoms include generalised muscle weakness, slow eating and a shuffling gait with toe dragging. Sluggish pupillary light response may be detected within 6 to 18 hours after toxin ingestions. The tongue tone, eyelid and tail tone are weak.Prehension of food and ability to swallow are usually affected in horses with botulism. Affected one walk with a stiff, stilted gait and short “choppy” strides and oftentimes have muscle fasciculations progress to recumbency within 12-24 hours

Lesions

Pathological changes in the CNS especially the brain stem and 3rd ventricle, catarrhal gastroenteritis, hepatitis and nephrosis

Diagnosis

The diagnosis of botulism is based on history, clinical signs and demonstration and identification of toxin in serum of moribund or recently dead animals as well as the detection of toxin and /or *Cl.botulinum* in the suspected foodstuff.

Toxin Demonstration

Serum or centrifuged serum exudates from animals can be directly inoculated I/V (0.3 ml) or I/P (0.5 ml) into mice. If toxin is present the characteristic wasp waist appearance in the mice will be seen in a few hrs or up to 5 days. The appearance is due to abdominal breathing because of paralysis of respiratory muscles.

Extraction of toxin in foodstuffs is accomplished by grinding the material in saline. The suspension is centrifuged and the supernatant is filtered through a 0.45μm filter. As the toxin can be in a protoxin form 9 parts of filtrate are treated with one part of 1%trypsin solution and incubated at 37^{0}C for 45 mins. Mice or guinea pigs are inoculated intraperitoneally.

Toxin Identification

Mouse (or guinea pig) neutralisation test using a polyvalent antitoxin initially followed by monovalent antitoxin is used to identify the toxin.

Isolation of *Cl.botulinum* from Foodstuffs

Several samples of the foodstuffs are macerated in a small amount of physiological saline. The suspension is heated at 65-80^{0}C for 30 mins to kill most of the contaminating organisms and to induce the *Cl.botulinum* spores to germinate. Blood agar plates are inoculated with the suspension and incubated under CO_2 at 35^{0}C for upto 5 days. To determine whether the isolate is a toxin producing strain, a cooked meat broth is inoculated and incubated at 30^{0}C for 5-10 days. Filtrates are prepared and lab. Animals can be used for demonstrating and identification of the toxin.

Treatment

Neutralisation by polyvalent antiserum is effective. Therapeutic agents such as triethylamine and guanidine hydrochloride, which enhance transmitter release at neuromuscular junctions, may be of value when given intravenously.

Control and Prevention

Immunisation is not followed. In South Africa attempts were made by giving two injections of types C&D toxoid at an interval of several weeks. Bivalent & Trivalent antitoxins are available commercially.

Clostridium Perfringens (Clotridium Welchi)

History

The bacillus was originally cultivated by Achalme (1891), but it was first described in detail by Welch and Nuttal (1892)-who isolated it from the blood and organs of cadaver.

Natural Habitat

Type A occurs in the intestinal tract of humans and animals and in most soils. Type B to E is more adapted to survival in the intestines but in outbreaks of disease they survive long enough in soil to infect other animals.

Morphology

Gram +ve, bacillus, straight, parallel sides, rounded or truncated ends, occur either in singly or in chains or small bundles. It is highly pleomorphic, filamentous and involution forms are common. It is capsulated and non-motile. The spores are oval, subterminal and bulge the mother cell. They are rarely produced and their absence is one of the characteristic morphological features of *Cl.welchii.*

Classification

Based on the type of toxins, they are classified into 5 types

Cl.perfringens types	Major toxins	Host	Disease
A	Enterotoxins	Human	Food poisoning
	Alpha	Lambs Broiler chickens	Enterotoxaemic jaundice Necrotic enteritis
B	Beta and alpha	Lambs under 3 weeks old	Lamb dysentery
C		Piglets 1-3 days old	Haemorrhagic enteritis (Clostridial enteritis)
	Beta and alpha	Broiler chickens (2 weeks old)	Necrotic enteritis
		Adult sheep and goats	Struck
D	Epsilon	Sheep all ages (except neonates)	Pulpy kidney disease (overeating disease)
E	Iota and alpha	Calves and lambs	Enterotoxaemia (Haemorrhagic enteritis)

Cultural Characters

It is an anaerobe, but can also grow under micro aerophilic conditions. Oxygen is not actively toxic to the bacillus and cultures do not die on exposure to air. Though this bacillus is grown at 37^{0}C, pH 5.5-8.0, the temp of 45^{0}C is optimal for many strains. The generation time at this temperature is 10 minutes only.This property can be utilised for obtaining pure cultures of *Cl.welchii.* Robertsons cooked meat broth inoculated with mixtures of *Cl.welchii* and other bacteria and incubated at 45^{0}C for 4-6 hrs serves as enrichment. Subcultures from this onto blood agar plates yield pure or predominant growth of *Cl.welchii*

Good growth occurs in Robertson's cooked meat medium. The meat turns pink but it is not digested. In litmus milk, fermentation of lactose leads to formation of acid, which is indicated by the change in the colour of litmus from blue to red. The acid coagulates the casein (acid clot) and the clotted milk is disturbed due to the vigorous gas production. The paraffin plug is pushed up and shreds of clot are seen sticking to the sides of the tube. This is known as stormy fermentation.

After overnight incubation on rabbit or sheep blood agar, colonies of most strains show a "target haemolysis" resulting from a narrow zone of complete haemolysis due to theta toxin and a much wider zone of incomplete haemolysis due to alpha toxin. This double zone haemolysis pattern is characteristic for *Clostridium welchii.* All the five types of *Cl.welchii* (A –E) produce alpha toxin. This is a lecithinase C, which, in the presence of calcium and magnesium ions splits lecithin into phosphotidyl choline and diglyceride. This specific lecithinase effect can be demonstrated by Nagler's reaction.

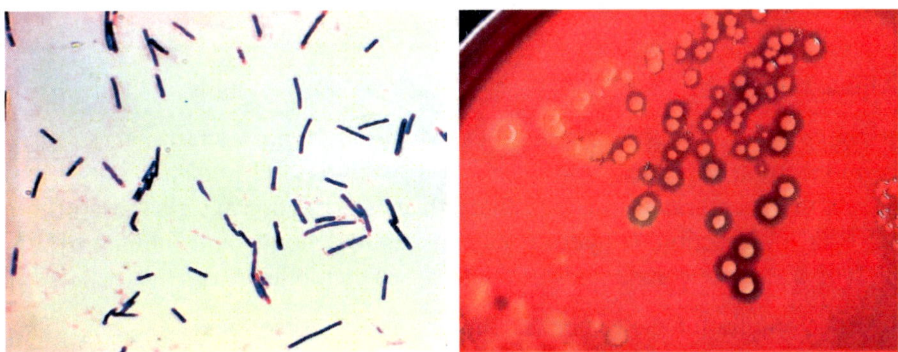

Type A antitoxin is spread over half of an egg yolk agar plate and allowed to dry. The suspect *Cl.perfringens* is streaked across both sides of the plate. All the types of *Cl.perfringens* produce the alpha toxin that is a lecithinase. On the half of the plate without the antitoxin, the lecithin in the medium is attacked causing opalescence around the streak. The lecithinase reaction is neutralised on the other half of the plate with the antitoxin but the growth of *Cl.perfringens* is unaffected.

Biochemical Reactions

Cl.welchi ferments several sugars (glucose, maltose, lactose and sucrose) and produces acid and gas. Indole –ve, MR +ve, VP-ve, H_2S +ve.

Resistance

Spores are usually destroyed within 5 minutes by boiling, but those of the food poisoning strains of type A and type C strains resist boiling for 1-3 hrs. Autoclaving temp is lethal.

Toxins

Cl.welchii is one of the most prolific of toxin producing bacteria forming at least 12 different toxins, besides many other enzymes and biological active substances.

Cl. perfringens	Major toxins			
Type	Alpha	Beta	Epsilon	Iota
A	+	-	-	-
B	+	+	+	-
C	+	+	-	-
D	+	-	+	-
E	+	-	-	+

Alpha Toxin

Alpha toxin is produced by all types. Mostly by type A strains. It is lethal, dermonecrotic and haemolytic. This is a lecithinase C (phospholipase) that attacks cell membranes causing cell death and destruction and also responsible for Nagler's reaction. It hydrolyzes phosphatidylcholine and sphingomyelin, both of which are constituents of the host cell membrane.

It is haemolytic for the red cells of most species except horse and goat. This toxin gives a zone of partial haemolysis on blood agar. The haemolysis is of hot-cold variety being best seen after incubation at 37^{0}C followed by chilling at 4^{0}C.

Beta Toxin

Beta toxin is lethal and necrotising. It is a pore forming toxin, which damages host target cells such as intestinal epithelial cells and endothelial cells. It is sensitive to trypsin and this explains the prediction of types B and C for neonates as colostrum has anti trypsin activity. It is a labile toxin and may be destroyed if there is a delay in small intestinal contents, containing the toxin, reaching the laboratory.

Epsilon Toxin

Epsilon toxin is secreted as a protoxin (proto toxin) and is activated in the intestines by proteases such as trypsin. Pulpy kidney disease is not usually seen in neonatal lambs as colostrum contains an antitrypsin factor that can prevent the epsilon toxin being activated. The toxin itself increases gut permeability, assuring absorption of the toxin into the blood stream. It damages vascular endothelium (including blood vessels in the brain) leading to fluid loss and edema. This epsilon toxin can be regarded as an enterotoxin and neurotoxin.

Iota Toxin

Iota toxin is also produced as a protoxin and is not unique to *Cl.perfringens* type E as it is also formed by *Cl.spiroforme* and *Cl.difficile.*It is a binary toxin composed of a binding portion that binds the toxin to target epithelial cells, and an enzymatically active portion. It is also a pore forming toxin.

Perfringolysin O (also known as theta toxin)

It binds to cholesterol containing rafts in the eukaryotic cell membrane, and forms a pore, which results in the death of the cell. It is also responsible for lysis of phagolysosomal membranes.

Enterotoxin

It is produced during sporulation of Cl.perfringens. When the endospore is released, enterotoxin is also released in the surrounding media. It is a bifunctional toxin, first forming a pore in the apical portion of small intestinal epithelial cells resulting in fluid and electrolyte abnormalities, as well as providing access to tight junction proteins results in further losses in control of fluid and electrolytes.

Besides, several minor toxins are produced such as the haemolysin, Kappa (collagenase), lambda (Proteinase), Mu(hyaluronidase) and Nu(DNase) – all these may contribute to tissue damage.

Pathogenicity

The enterotoxemia are often precipitated by certain husbandry and environmental factors such as abrupt changes in feeding usually to a richer diet and overeating and voracity on high protein and energy rich feeds. This leads to slowing of peristalsis with retention of bacteria in the intestines, absorption of toxins, inadequately digested carbohydrate and the provision of a rich medium for the proliferation of *Cl.perfringens*

The bacterium inhabits the large intestine in normal animals, but if overgrowth occurs *Cl.perfringens* can spill over into the small intestine with the production of a large amount of toxin and enterotoxaemia

Cl.perfringens Type A

Tissue destruction is probably due to the membrane active toxins (alpha and perfringolysin O), and the toxins that affect connective tissue (collagenase, hyaluronidase, and sialidase)

Cl.perfringens Type B

Beta toxin is considered the principal factor producing hemorrhagic enteritis affecting the small intestine. In lambs the signs are depression, anorexia, abdominal pain and diarrhoea. The course is rapid with mortality rate approaching 100%. A chronic form occurs in older animals. The characteristic intestinal lesion is haemorrhagic enteritis.

Epsilon toxin being a permease increases intestinal permeability, ensuring its absorption into the circulation where it affects vascular endothelium, leading to fluid loss and edema, as well as damage to kidney function. Congestion, edema, serosal effusions and haemorrhages in various organs are seen.

Beta and epsilon also affect the nervous system which leads to severe depression and high mortality.

Cl.perfringens Type C

Beta toxin is considered the principal factor producing hemorrhagic enteritis affecting the small intestine. In neonates the signs are depression, anorexia, abdominal pain and diarrhoea. The course is rapid with mortality rate approaching 100%. The characteristic intestinal lesion is haemorrhagic enteritis. The signs and lesions are mainly due to alpha, beta and perfringolysin O.

Cl.perfringens Type D

The key epsilon toxin is secreted as protoxin activated by intestinal proteases. Epsilon toxin increases intestinal permeability, ensuring its absorption into the circulation where it damages vascular endothelium, leading to fluid loss and edema. When toxin levels are high, affected capillary endothelial cells in the brain are damaged, and the resultant edema greatly increases the intracranial pressure. When the amount of toxin is lower, it damages the capillary endothelial cells in the brain and leads to focal symmetrical encephalomalacia. Epsilon toxin also causes cAMP related hyperglycemia and glycosuria.

It is common in suckling lambs 3 to 10 weeks old and in feedlot animals. Lambs may die without premonitory signs. Gross lesions may be absent. Convulsions and diarrhoea may present. Cattle and older sheep show neural manifestations.

Affected sheep display blindness, head pressing and anorexia. In goats, diarrhoea is common. Catarrhal, fibrinous or hemorrhagic enterocolitis is a consistent lesion in goats. Death rates are high in lambs.

Disease	Clinical and postmortem signs
Food poisoning	Sudden onset, diarrhoea, abdominal pain and nausea. But vomiting is uncommon. Short course and rarely fatal
Enterotoxaemic jaundice	Depression, anaemia, icterus, haemoglobinuria and lambs die within 6-12 hrs of first signs known as the yellows or yellow lamb disease
Lamb dysentery	A haemorrhagic and rapidly fatal enterotoxaemia. Lambs are often found dead
Hemorrhagic enterotoxemia (Clostridial enteritis)	Dysentery, collapse and death. Small intestine is dark red and has gas bubbles in mucosa
Necrotic enteritis in broilers	Depression, diarrhoea, death in a few hrs. Mortality 2-50%. Mucosa of the small intestine has a brown pseudomembrane. Most common in deep litter units.
Struck	Sudden death due to an enterotoxemia
Pulpy kidney disease (Overeating disease)	Edema of brain, glycosuria, sudden death. Excess fluid in body cavity, focal symmetrical encephalomalacia occurs in well-grown lambs.

Diagnosis

Gram stained smears can be made from the mucosa of the small intestine of a recently dead animal. Large numbers of gram-positive rods are suggestive of *Cl.welchii.*

Histopathology on brain sections helps to demonstrate focal symmetrical encephalomalacia in pulpy kidney disease. Rapid kidney autolysis, pulpy cortical softening and Glucosuria are suggestive of pulpy kidney disease.

Demonstration of Toxin in the Small Intestine

Collect 20-30ml of ileal contents from a recently dead animal and send it to the laboratory as soon as possible. The ileal contents are centrifuged and the clear supernatant is tested for toxin. In ileal contents, the epsilon and iota toxins are usually in the active form.

To demonstrate the toxin 0.4ml of the clarified ileal contents can be inoculated intravenously into mice. If a mouse dies within 5 mins this is probably due to shock, death from toxin usually occurs within 10hrs. Identification of toxin in the clarified ileal content is carried out by a neutralisation test by using suitable antitoxin.

Control and Prevention

Before the lambing season the ewes are vaccinated with formalised whole culture or alum precipitated vaccine. The resulting passive immunity, which unweaned lambs, derived from colostrum, protects lambs for the first 3 weeks of life.

Similarly alum precipitated trypsin –treated toxoid is also satisfactory. Lambs can also be vaccinated by giving the first dose within 72 hrs of birth and repeated at 4 weeks of age. Immunity may not last more than 6-12 months unless a booster dose is given.

Histotoxic Clostridia

Histotoxic clostridial organisms typically are involved in infections of muscle, liver or heart.

Cl. chauvoei	- Cattle, sheep (pigs)	- Black quarter (Black leg)
Cl. septicum	- Cattle,	- Malignant edema
	- Sheep	- Braxy
	- Chicken	- Necrotic dermatitis
Cl. novyi type A	- Sheep	- Big head of rams
	- Cattle and Sheep	- Gas gangrene
Type B	- Sheep (Cattle)	- Black disease
		(Infectious necrotic hepatitis)
Type C	- Water buffaloe	- Osteomyelitis
Type D	- Cattle, (sheep)	- Bacilliary haemoglobinuria
(Cl. haemolyticum)		
Cl. sordellii	- Cattle, Sheep, Horses	- Gas gangrene
Cl.colinum	- Birds	- Quail disease
		Ulcerative enteritis
Cl. piliforme	- foals, laboratory animals	- Tyzzer's disease
	Calves, dogs& cats	(Hepatic necrosis)

Clostridium Novyi (Clostridium Odematiens)

Natural Habitat

Clostridium novyi is world wide in distribution. The principal habitat of *Cl.novyi* is soil and the intestine of animals. Type A commonly found in soil. Type B is rarely found in soil, and it is common in the normal intestinal tract of herbivores. Strains of type A and B are recovered from the livers of normal animals. Type D *(Cl. haemolyticum)* - is found in the ruminant digestive tract, liver and in the soil. Based on toxin production *Clostridium novyi* classified into four types (A-D).

Morphology

Large, pleomorphic, gram+ve rods with oval to cylindrical subterminal spores. There is little or no swelling of the mother cell, non-capsulated, motile by peritrichous flagella.

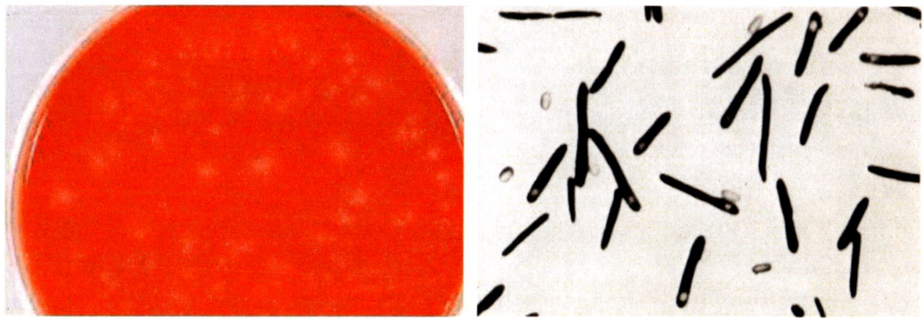

Cultural Characters

Clostridium novyi type B and *Cl.haemolyticum* are very demanding in both their anaerobic and nutritional requirements. Very strict anaerobic procedures are necessary and media containing cysteine should be used. These clostridia can die within 15 mins of being exposed to atmospheric O_2

These organisms are difficult to grow in primary culture and the growth is enhanced by agar enriched with glucose or freshly prepared blood or fresh brain infusions. On blood agar *Cl.novyi* produces characteristic large, irregular colonies with a rhizoid edge and a large zone of clear haemolysis.

On the moist surface of solid media after 3-4 days incubation colonial motility develops which is characterised by the movement of daughter colonies moving away from parent colonies in spirals or arcs and few return and fuse to the parent colony.

In horse blood agar, the colonies are haemolytic, small and usually rhizoidal in nature. Areas of hemolytics develop beneath the colonies and develop into wider zones after 48-72 hrs incubation.

In Sheep blood agar very slight haemolysis develops. In Robertson's cooked meat medium *Cl.novyi* type D is very strong proteolytic. Type A, B and C are saccharolytic. The lecithinase activity of beta toxin of type B and D, and Gamma toxin of type A are producing quite distinct opacity changes on egg-yolk agar.

Cl.novyi type A exhibits lipase activity on egg-yolk agar. It will produce a characteristic iridescent pearly layer on the surface of the colonies extending on to the surface of the medium immediately surrounding them. *Cl.novyi* type A is the only species among clostridia that produces both a lecithinase and a lipase.

Biochemical Characters

Saccharolytic types ferment glucose and maltose but not lactose.

Resistance

Spores of most strains survive heating to 95°C for 15 minutes, and are killed at the autoclaving temperature. Spores are resistant to 5% phenol, 10% formalin or 0.1% merthiolate. They are killed rapidly by exposure to hypochlorite. Spores remain viable for years in soil.

Antigens and Toxins

They possess several somatic and flagellar antigens, which are not of much importance. Based on toxin production *Cl.novyi is classified* into 4 types. *Cl.novyi* synthesises fivemajor toxins, α, β, γ, δ and ε.

Type of organisms	Alpha	Beta
Cl. novyi type A	+	-
Type B	+	+
Type C	-	-
Type D *Cl. haemolyticum*	-	+++

Type A produces all toxins except beta. Type C isolates are non toxigenic. In addition to this type B also produce zeta, eta and theta toxin.

- Alpha toxin produced by type A and B is lethal, necrotizing, causes increased capillary permeability, and is toxic to several tissues including muscle, heart and liver.
- Beta toxin is a lecithinase that hydrolyzes phosphotidyl choline and sphingomyelin and it is produced by type B and by *Cl.haemolyticum* (type D) in greater amounts. This may account for the haemolytic crisis and death in bacillary haemoglobinuria.
- Gamma toxin is a necrotising phospholipase D.

- Delta toxin is an oxygen labile haemolysin - Novyilysin
- Epsilon toxin is a lipolytic enzyme.
- Zeta toxin is partly haemolytic.
- Theta toxin is a lipase.
- Eta toxin is a tropomyosins, which degrades tropomyosin and myosin and may play a role in destruction of infected muscles.

Pathogenesis

In black disease (infectious necrotic hepatitis) and bacillary haemoglobinuria, the spores, normally present in the intestine, may reach the liver and remain dormant in the kupffer cells. Any destruction of liver tissue could be the inciting factor. The tissue damage is usually due to migration of immature liver fluke (Fasciola hepatica), and anaerobic conditions permit germination of spores, growth of vegetative cells and subsequent production of toxins (alpha and beta).

Alpha toxin produced, in the local area of necrosis, in the liver is absorbed into the circulation and results in systemic effects. Death may be sudden or within 2 days. Signs include depression, anorexia and hypothermia. Necropsy reveals edema, serosal effusion and liver necrosis. Subcutaneous venous congestion secondary to pericardial edema darkens the underside of the skin, suggesting the name black disease In case of bacillary haemoglobinuria the dominant toxin is beta toxin (phospholipase C). It causes haemolytic crisis and death within hours. Big heads in rams develop when s/c tissues are traumatized during fights and are subsequently invaded by *Cl.novyi* type A. Toxic endothelial damage (by alpha toxin and novyi lysine) produces edema involving head, neck and cranial thorax. Death occurs in 2 days. The yellow tinge of the edema fluid which is clear and gelatinous with little haemorrhage is a postmortem change.

Symptoms

Big head

Edematous swelling occurs in the head, face and neck. It will be followed by collapse and death of animals. The mortality rate may be more than 90%.

Black disease

Acute toxaemia leads to sudden death. The signs include rapidly /decreasing ability to move, unsteady gait and collapse.

Bacillary haemoglobinuria

Common in summer months, affected animals suffer from fever, pale, icteric mucous membranes, anorexia, agalactia, abdominal pain, port-wine coloured urine, diarrhoea, haemoglobinuria and hyperpnea. The mortality rate is 90%.

Lesions

Black disease

Number of clearly defined grey-yellow foci (necrotic areas) in the liver. The lesion consists of a central core of necrosis surrounded by a zone of leukocytes in which there will be masses of *Cl.novyi.* Excess fluid in body cavities. Straw-coloured exudates will be present in pericardial and peritoneal cavities.

Extensive subcutaneous and bloodstained edema can be noticed in the carcass. Venous congestion occurs that darkens the skin. (Black disease)

Bacillary haemoglobinuria

There are a number of typical anaemic infarcts in the liver. Pale and raised surrounded by a blue-red zone. There will be blood stained intestinal contents, dark coloured urine in the bladder, marked icterus of the carcase, widespread edema and haemorrhages in the myocardium.

Diagnosis

1. Based on history
2. Direct gram stained smears

   Presence of characteristic liver lesions together with large number of gram +ve rods in liver impression smears from a recently dead animal is suggestive of the disease.

3. FAT is useful for the identification of *Cl.novyi* type B and *Cl.haemolyticum* in acetone fixed liver impression smears.
4. Isolation of organism from affected tissue (as like other clostridial infections) and by characteristic cultural characters.
5. Animal inoculation

Toxin in the liver can be demonstrated by i/m injection of homogenates into guinea pigs. The pathogenicity is enhanced if the homogenate is added to an equal amount of 5% $Cacl_2$ solution before inoculation. The guinea pigs die in 1-2 days with very extensive subcutaneous edema. Specific antitoxin is not readily available for neutralization tests.

Control and Prevention

1. Elimination of liver flukes through destruction of snail
2. Aluminum hydroxide adsorbed formalized whole culture vaccines are available.
3. Outbreaks of the disease may be controlled by the prompt injection of hyper immune sera.

Clostridium Septicum

Clostridium septicum is very closely related with *Clostridium chauvoei*, hence it is called Clostridium *chauvoei* type A.

Clostridium septicum causes

- Malignant edema in Cattle, sheep and pigs
- Braxy (Bradsot) in Sheep
- Necrotic dermatitis in Chickens

History

The bacillus was first described by Pasteur and Jourbert (1887) and named it as "Vibrion Septique".

Normal Habitat

It is found in soil and the intestine of animals

Morphology

Gram +ve, highly pleomorphic, characteristic long filamentous forms are seen in stained smears of affected muscle. Cigar shaped rods and citron forms are more common. Spores are oval, central or subterminal. Non-capsulated, motile by peritrichous flagella.

Cultural Characters

Strict anaerobe, growth at an opt.temp of 37^{0}C, growth is promoted by glucose. On ordinary media, the colonies are irregular and transparent initially, turning opaque (large, greyish white on continued incubation). The colonies are swarming and spreading over the entire surface.

On stiff agar, the colonies are irregular with a rhizoid edge. Some strains produce smooth, round colonies.

In cooked meat medium meat turns pink with rancid odour, and produces abundant gas (because it is saccharolytic). Like *Cl.perfrigens,* the *Cl.septicum* inoculated into litmus milk produces the classical stormy clot or stormy fermentation reaction.

Biochemical Reactions

Ferment glucose, lactose, maltose and salicin but not sucrose. Acid and gas are not produced.

Resistance

Spores are killed by steam in 40-50 mins and by autoclaving. Spores are also susceptible to 3% formalin for 15 mins.

Antigens and Toxins

- Six groups have been recognized, based on somatic and flagellar antigens. *Clostridium septicum* produces at least fourdistinct toxins and fibrinolysin.
- The α toxin is oxygen stable haemolysin, dermonecrotic and lethal. It is a pore forming toxin. The cell bound proteolytic enzyme, furin, cleaves it, resulting in fragments that insert into the membrane forming pores leading to death of the cell.
- The β toxin is leucotoxin and DNAse
- The γ toxin is a hyaluronidase
- The δ toxin is an oxygen labile haemolysin- streptolysin O
- α toxin has a direct effect on cardiac muscle and is capable of causing capillary damage. Iron is required both for growth of bacteria and for production of α toxin

Pathogenesis

Braxy or Bradsot in sheep

The disease is more common in winter months. When ingestion of large volumes of frozen grass, the spores that are present in the soil are ingested with feed. The mucosa of the abomasum is damaged due to cold conditions from an adjacent rumen. Any *Cl. septicum* spores present can germinate and replication of the bacterium leads to toxin production, toxaemia and rapid death.

Malignant Edema (Anaerobic cellulitis)

In this exogenous gas gangrene infection, spores are introduced into wounds where they may germinate in the anaerobic necrotic material and toxin is produced by the vegetative cells.

Symptoms

Braxy usually occurs in well-nourished one-year-old sheep, ailing animals showing signs of abdominal pain and diarrhoea. Death occurs within a few hours.

In malignant edema, an infected wound, which becomes gas gangrenous. Fever, soft swelling around the wound and spreading to muscles. Swelling edematous and wet with much exudates and gas. Muscles appear dark red to black color.

Lesions

In braxy, the lesion may be confined to the abomasums; there will be the characteristic area of hemorrhagic inflammation in the wall of the abomasum. There will be an extensive quantity of blood stained fluid in the peritoneal cavity.

Diagnosis

Based on history, symptoms and characteristic lesions are strongly suggestive of this disease.

FAT is to be employed for differentiate it from *Cl.chauvoei*

Identification of specific Clostridial toxin by mixing 1.2ml of culture fluid with 0.3ml of specific antitoxin, allowing the mixture to stand for 30 minutes at 37^0C and inoculate two guinea pigs i/peritoneally. If a specific toxin is present, it will be neutralised.

Control and Prevention

Penicillin alone or with hyper immune serum can be used to treat infections. Sheep can be effectively immunised against the disease and a multi component vaccine is used.

Clostridium Chauvoei

Synonyms

- *Cl. chouvoei type B, Cl. feseri*
- It causes **black quarter or black leg** in Cattle & Sheep

Natural Habitat: Worldwide distribution in soil and pastures

Morphology

Gram positive, rod shaped with rounded ends 3-8m in length & 5m in width. Sometimes pleomorphic, large cigar shaped rods or citron forms occur. Non-capsulated and motile by peritrichous flagella. Non-motile variants do occur. Spores are oval and located centrally or subterminal. Old cultures stain gram negative.

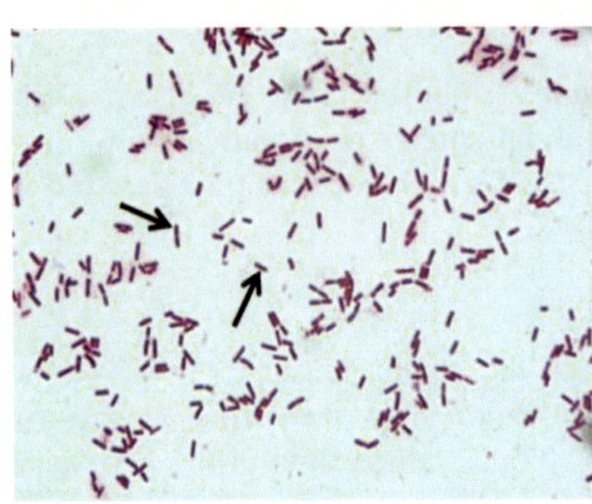

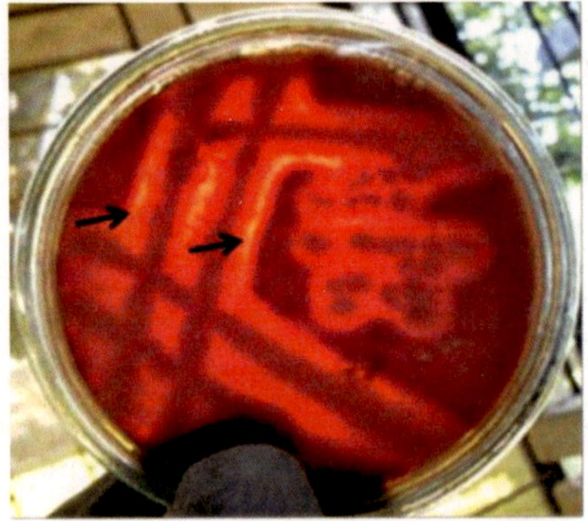

Cultural Characters

It is a strict anaerobe. Growth occurs at an optimum temperature of 37^0C. Growth enhanced by the addition of liver extract or glucose. In blood agar whitish grey colonies with irregular edges develop surrounded by a zone of haemolysis. In cooked meat medium growth is slow and meat is turned pink with sour odour

Biochemical Characters

Cl. chauvoei ferments glucose, lactose, sucrose, maltose with acid & gas, but not salicin.

Resistance: Similar to *Clostridium septicum*

Antigens and Toxins: *Cl. chauvoei* has somatic and flagellar antigens and produces 4 toxins

Alpha toxin - Oxygen stable haemolysin, pore forming necrotoxin that causes dermonecrosis and fibrinolysis

Beta toxin – DNA ase

Gamma toxin – hyaluronidase

Delta toxin – Oxygen – labile haemolysin. This toxin is lethal for mice & Guinea pigs when given I/v.

Chauveolysin- it binds to cholesterol containing rafts in th cell membrane. Once bound, it forms a pore resulting in the death of the cell.

Neuraminidase – Sialidase- removes sialic acid residues from glycoconjugates on cell walls resulting in disruptions of the intercellular matrix.

Pathogenesis

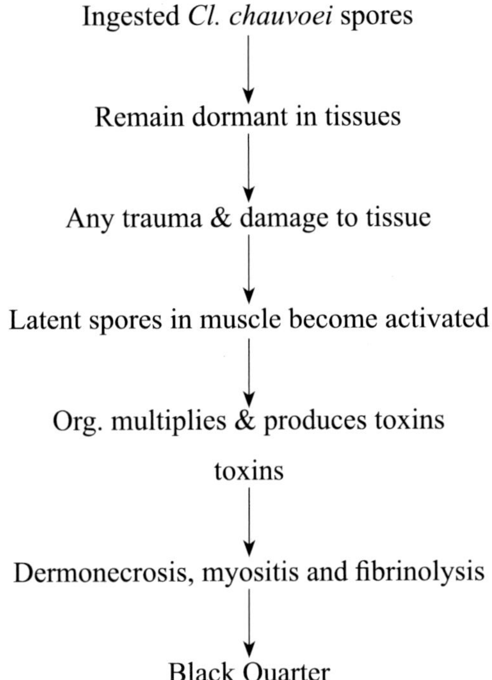

Symptoms

The disease usually occurs in young cattle of 6 months to about 2-3 years of age. High fever, anorexia and lameness are common. The most obvious sign is crepitus swelling particularly in the hind or fore quarter which crackles when rubbed with the fingers as a result of gas production.

The affected animal will become lame and the affected muscles show trembling with violent twitching. Death usually occurs within 24 hours. In sheep an acute febrile condition develops within 1-2 days following an injury and a typical black quarter lesion can be observed at the site. Death occurs suddenly

Lesions

In the central part of the lesion there is usually a well-defined area of muscle, which is dark red in colour, dry, necrotic and filled with small gas bubbles (emphysematous), which give a swollen appearance to the muscles. The lesion has a characteristic rancid butter odour. Surrounding this area of muscle there will be yellowish or blood stained oedematous fluid. In ewes there will be necrosis of the vaginal mucosa and skin with extensive oedema involving the hind limbs and thigh muscles.

Diagnosis

1) History
2) Symptoms
3) Smears prepared from the lesions and oedematous fluid reveal gram positive rod.
4) Isolation can be done from the centre of the lesion, oedematous fluid and from heart blood & spleen.
5) FAT used to differentiate from *Cl. septicum*
6) Broth cultures or oedematous fluid from the lesions can be tested for toxicity and specific neutralisation by antitoxin in mice or Guinea pigs.

Control and Prevention

The most reliable results are obtained from using a formalised alum precipitated whole culture that confers immunity against the bacteria as well as the toxin. For economic reasons a multi component vaccine containing Cl. *chauvoei, Cl. septicum, Cl.welchii type D and Cl. tetani* is used.

A stronger immunity is stimulated by two doses of vaccine at a time interval of at least 2-3 weeks. Hyper immune serum (HIS) is used to control explosive outbreaks. Penicillin along with HIS is used to treat the disease. Oxytetracycline & Chlortetracycline can also be employed effectively in early stages.

19

Listeria

Listeria has been divided into seven species with two distinct groups. Among which the *Listeria monocyotgenes* and *Listeria ivanovii* are haemolytic and pathogenic for animals. The *Listeria murrayi* and *Listeria grayi* are nonhaemolytic, rarely isolated and considered to be non pathogenic. Among which the genus *L. monocytogens*, being the cause of septicaemia, abortion and CNS infections in a wide range of animal species including humans.

Host and disease syndrome of pathogenic listerias

Species	Host(s)	Disease
L.monocyotgens	Young animals of many species, including lambs and calves. Birds can be affected	Visceral listeriosis (septicaemic) Myocardial degeneration in birds
	Sheep, goats, cattle	Neural listeriosis- circling disease
	Sheep, goats, cattle	Abortion
	Cattle	Iritis (Endopthalmitis)
L.ivanovii	Sheep and cattle	Abortion

History

L.monocytogenes first described by Murray (1926) who named it as bacterium monocytogenes because of characteristic monocytosis infection in laboratory animals. It was renamed *Listerella hepatolytica* by Pirie (1927) and the present name given by him in 1940. The *Listeria momnocytogenes* was first isolated by Gill (1929) from sheep.

Natural Habitat

Listeria species are widely distributed in the environment and can be isolated from soil, faeces, plants, decaying vegetation and silage (pH 5.5) in which the bacteria can multiply. Silage is commonly implicated in outbreaks of listeriosis in cattle and sheep. In poor quality silage the listerial numbers may reach 10^7 cfu/kg of silage. Asymptomatic faecal carriers occur in man and many animal species. *L.monocytogens* can be excreted in bovine milk. Human foods associated with listeriosis in man include soft cheeses, milk and poultry meat.

Morphology

L.monocytogenes are medium sized, G+ve rods, non-spore forming and non-acid fast. Old cultures stain gram –ve. From rapidly growing cultures or animal tissues the cells can appear coccal. They are multiple by few (1-5) peritrichous flagella. They are motile at room temperature, but not at 37^{0}C.

Cultural Characters

L.monocytogenes is able to grow at temperature ranges from 4 to 45^{0}C and grow at pH range of 5.5 to 9.6. It is relatively resistant to high salt (10%) concentrations. They are facultative anaerobes. The growth is enhaced by agar enriched with glucose, blood, liver extract and by 10% CO_2. They grow on nutrient agar, blood agar but not on MacConkey agar. Small transparent colonies with smooth borders appear on blood agar in 24hrs, becoming grayish white in 48hrs.

L.monocytogenes and other non-pathogenic listeria produce narrow zones of beta haemolysis, often only under the colony itself. *L.ivanovii* produces a comparatively wide zone of haemolysis and is very similar in appearance to beta haemolytic streptococci. *L.monocytogenes* produces a CAMP reaction with the haemolysis of *Staphy.aureus*. In contrast; *Listeria ivanovii* is negative in the CAMP reaction with *Staphy.aureus*.

Oxford medium or PALCAM mediumcan be used as selective media. Oxford medium contains lithium chloride, acriflavin, colistin sulphate, cefotetan, cycloheximide, phosphomycin, ferrous ions and aesculin. Listeria hydrolyses aesculin to aesculetin and dextrose. Aesculetin reacts with ferric ions and produces black zones under and around the colonies. Small black, grey or brown colonies that are surrounded by a black zone is a typical colony morphology of Listeria in this agar plate.

PALCAM medium contains Polymyxin-Acriflavin-Lithium chloride-Ceftazimide- Aesculin – Mannitol. Listeria hydrolyses aesculin to aesculetin and dextrose. Aesculetin reacts with ammonium ferric citrate and forms a brown-black complex seen as a black halo around grey-green colonies. On TSA or BHI agar, these colonies have a characteristic blue-green sheen when light is reflected obliquely at a 45^0 angle off their surface. In fluid medium, slimy tenacious precipitate forms after incubation for several days.

L.monocytogenes, particularly shows the characteristic tumbling motility when a 2-4hr broth culture, incubated at 25^{0}C, is examined by the hanging drop method. This motility is an end-over-end tumbling of individual cells with periods of quiescence. When grown in semisolid motility media the Listeria spp. give an unusual umbrella shaped growth in the subsurface.

Biochemical Characters

All the listeria spp hydrolyse aesculin, CAMP +ve, Catalase +ve, Oxidase –ve, Indole –ve. Acid production from glucose and rhamnose, but not from xylose and mannitol. Nitrates not reduced.

Resistance

It is killed by moist heat at 55^{0}C for 40minutes and is readily susceptible to the lethal effects of disinfectants. Under natural conditions, in summer they survive for 1month and in winter for 3-4 month.

Antigens and Toxins

Based on somatic and flagellar antigens, sofar 16 serovars have been identified. Of these 16 serovars, all cases of animal and human infections are caused by 3 serotypes. ½ a, ½ b and 4 b. Numbers indicate O antigen and alphabet indicate H antigens.

Toxins

- **Act A**: Act A protein is important in intracellular movement by actin polymerization and is also thought to play a role in tropism and invasion.
- **Cell wall:** The lipoteichoic acid and peptidoglycan of the cell wall intract with macrophage cells resulting in the release of proinflammatory cytokines.
- **Internalins:** Internalins are surface proteins responsible for adhesion and entry into target cells.
- **Listeriolysion O: I**t is a pore forming cholesterol dependant cytolysin. It releases *L.monocytogenes* from the phagosome into the cytosol following phagosome acidification. It also induces apoptopsis in hepatocytes.
- **Phospholipase C** – important in membrane lysis.

Pathogenesis

In both cattle and sheep, listeriosis can manifest itself in four ways;

- As a CNS infection (meningo encephalitis in adults and meningitis in the young)
- As abortion
- As a generalized septicaemia with involvement of the liver and other organs
- As mastitis in dairy cattle.

Silage is commonly implicated in outbreaks of Listeriosis in cattle and sheep. Most pathogenic bacteria require the availability of iron in the host for metabolic activities. High iron levels in silage that lead to elevated tissue concentrations of iron may predispose cattle and sheep fed on silage to Listeriosis. Poor quality silage with a pH greater than 5.5 is favourable for multiplication of *Listeria monocytogenes*.

Exposure to listeria occurs via oral route. Most listeria are destroyed by gastric acids. Use of antacids and H2 blockers are considered as risk factors for infection. Intestinal translocation appears to be a passive process that involves both intestinal epithelial cells and M cells.

After penetrate the epithelial barrier in the intestine and multiply in hepatic and spleenic macrophages aided by the haemolysin named listeriolysin O. Then the listeria escapes from the phagosome (with Listeriolysin O) become associated with actin filaments in the cytoplasm, and propels itself to the cell's plasma membrane via polar assembly for actin filaments (with Act A). In this way, it is able to pass to neighbouring cells in plasma membrane protrusions and thus avoid host defense mechanisms.

If the pathogen penetrates through damaged mucousal surfaces (oral, nasal or ocular) to the CNS, via the neural sheath of trigemianl nerve or an alternate route it may penetrate through the dental pulp (when sheep are cutting or losing teeth) to the CNS results in neural form.

Symptoms

Four syndromes;

a) Subclinical:infections are the most common form of infection. Usually outbreak occurs when fed with poor quality, high pH silage, particularly during cold weather.

b) Neonatal infections:Characterised as visceral infections with a septicemia. Often gastroenteritis and bilateral meningitis. Deaths are frequent in neonatal animals.

c) Listerial abortion is a sporadic condition in cattle & sheep. Abotion is usually late term – after 7 months in cattle and 12 weeks in sheep. The fetus may be macerated or delivered weak and moribund. Retained placenta and metritis are common.

d) Neural Listeriosis: Encephalitis (circling disease):

The incubation period is ranges from 14 to 40 days.The disease is more common in winter or early spring. The clinical presentation of meningoencephalitis in adult ruminants may begin with signs of depression and confusion. The ears droop; animal holds it head to one side.

Protrusion of the tongue and salivation are common and twitching or paralysis of the facial and throat muscles may occur. When the animal moves, it tends to be in a single direction, giving rise to the common name of circling disease. In the terminal stages, the animal may fall and be unable to rise.

In poultry, there are signs of torticolis, weakness, incordination of legs and sudden death in young birds. The disease is usually fatal in sheep, pigs and horses.

Lesions

Microabscess in the brain stem, usually unilateral, together with perivascular cuffing is very characteristic of listeriosis. The lesions are most common in the midbrain, pons and medulla oblongata. In addition to this there will be generalized septicaemia, focal necrosis of the liver and spleen will be seen.

Diagnosis

Specimens to be Collected

Visceral form : Material from lesions in liver, kidneys or spleen

Neural form: Spinal fluid, brain stem, and tissue from several sites in the medulla oblongata

Abortion: Placenta (cotyledon), foetal abomasal contents and /or uterine discharges.

1. Stained smears from lesions may reveal gram +ve rods (often coccobacillary)
2. Histopathological examination of brain tissue can often give a presumptive diagnosis of neural listeriosis

3. Isolation and Identification: Inoculation of specimens on selective media includes blood agar with an antibiotic supplement or blood agar containing 0.05% potassium tellurite (inhibitory to gram –ve). Specimens from the visceral form of the disease or from abortion cases are inoculated directly onto the laboratory media. A cold-enrichment procedure is necessary for brain tissue from neural listeriosis. Small pieces of spinal cord and medulla are homogenized and a 10% suspension is made in a nutrient broth. The broth suspension is placed in the refrigerator at 4^0C and sub cultured on to blood agar once weekly for upto 12 weeks.
4. Inoculation in developing chicken embryos causes development of focal necrotic lesions on the chorio allantoic membrane (CAM).
5. Anton's test: Inoculation of live bacterial suspension into the conjunctiva of a rabbit or guineapig only *L.monocytogenes* causes a purulent keratoconjuctivitis within 24-36hrs of inoculation.

Treatment

Ruminants in early stages of septicaemic listeriosis respond to systemic therapy with ampicillin or amoxicillin. Response to antibiotic therapy may be poor in neural listeriosis although prolonged higher doses of ampicilins or amoxicilin combined with an aminoglycoside may be effective. Ocular listeriosis requires treatment with antibiotics and corticosteroids injected subconjuctivily.

Control and Prevention

Poor quality silage should be avoided. Vaccination with killed vaccines, which do not induce effective cell-mediated immune response, is not protective because *L.monocytogenes* is an intracellular pathogen. Live, attenuated vaccines, which contain serovars 1/2a, 1/2b, and 4b are reported to reduce the prevalence of listeriosis in sheep.

20

Erysipelothrix

Systematics

Main host and diseases of *Erysipelothrix rhusiopathiae*

Main host (s)	Disease syndrome
Pigs	Swine erysipelas * Acute septicaemic form (Pregneant sows may abort) * Urticarial form (Diamond skin disease) * Vegetative endocarditis and Polyarthritis (Chronic form)
Sheep	* Poly arthritis in lambs * Post-dipping lameness * Valvular endocarditis and pneumonia
Turkeys, Geese and other birds	* Acute septicaemia (Turkey erysipelas) * Vegetative endocarditis and arthritis (Chronic form)
Human	Erysipeloid (localized cellulitis)

Natural Habitat

The bactrerium is widespread in nature and has been recovered from a wide variety of wild and domestic animals including mammals, fish (both fresh and salt water), birds, reptiles and amphibians. It is present in the soil and can survive for 20 days or longer in alkaline soil. The major source of infection for swine and turkeys is carrier animals of the same species. It is reported that 30-50% of pigs carry the bacterium in their tonsils, other lymphoid tissues. It is present in slurry of piglets and can be recovered from the faeces of carrier pigs.

Morpholgy

Erysipelothrix rhusiopathiae (previously named *Erysipelothrix insidiosa)* from S (Smooth) -form colonies and usually from acute syndromes is a gram-positive rod, the R (rough) form colonies usually from chronic disease is a gram-positive filament. The organism is non-motile, non-spore forming, non-acid fast, capsulated occur either in singly, in groups or in chains.

Cultural Characters

It is a facultative anarobe, but growth is enhanced by 10% CO_2. It is able to grow in a temperature range of 5⁰C to 42⁰C, within a pH range 6.7 to 9.2 and 8% $NaCl_2$. Growth occurs on nutrient agar but is improved by the addition of serum or blood. It will not grow on Mac Conkey agar. Media contain either sodium azide (0.1%) or crystal violet (0.001%) may be used as selective media.

On blood agar, non-haemolytic pinpoint colonies (0.5 mm) appear at 24hrs incubation. Colonial variation becomes obvious at 48hrs incubation when a zone of greenish haemolysis often develops under and just around the colonies. The smooth form colonies are convex, circular with an entire edge. The large rough form colonies are flatter, more opaque and have an irregular edge.

A characteristic reaction is produced when triple sugar iron agar is stab inoculated. When incubated at 37⁰C for 24hrs. H_2S is produced as a thin, black line just along the inoculation stab. The R forms gives a bottlebrush or pipe cleaner type of growth in stab cultures of gelatin incubated at 21⁰C for 5 days.

Biochemical Characters

The bacterium is coagulase positive, catalase negative and oxidase negative. It does not hydrolyse aesculin or produce urease. *Erysipelothrix rhusiopathiae* usually ferments lactose, glucose, levulose and dextrin. But the acid production is poor. To obtain the good result, carbohydrate tests can be carried out in peptone water with added sterile horse serum (5-10%) with phenol red as the indicator. Indole, Methyl red and Voges proskauer tests are negative.

Resistance

Erysipelothrix rhusiopathiae is resistant to several chemicals including sodium azide, and to drying, pickling, salting and smoking. It is capable of surviving for nearly a year in putrefying meat. But they are susceptible to caustic soda and hypochlorites. They are readily killed in moist heat at a temperature of 55⁰C for 10mts.

Antigens and Toxins

Based on heat labile and heat stable antigens sofar 23 serotypes have been identified. Strains of serotype 1 are subdivided into 1a and 1b. Serotypes 1a, 1b and 2 are most frequently involved in disease in swine.

Cell wall

The lipoteichoic acid and peptidoglycan of the cell wall interact with macrophage cells resulting in the release of proinflammatory cytokines.

Neuraminidase

It is responsible for adherence to cell surfaces. It causes cleavage of sialic acid residues on endothelial cells leads to thrombus formation.

Hyaluronidase and Coagulase

These enzymes are also produced by some strains.

Pathogenicity

The carrier animals are an important source of the organism. Entry of the organism may be by the oral, cutaneous or respiratory route. Ingestion of contaminated feed or water or contaminations of abraded skin are the most common means of infection in swine.

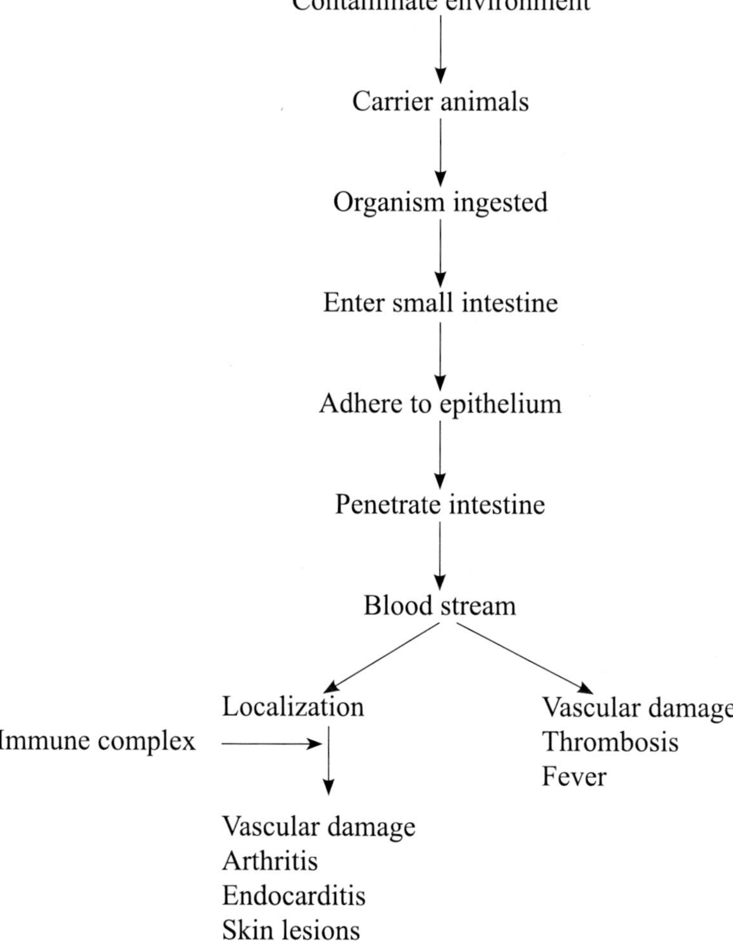

Erysipelothrix rhusiopathiae is able to adhere to epithelial cells, and that they invade the blood stream and cause localization. The more virulent strains produce high levels of neuraminidase. That cleaves sialic acid present on cell surfaces leads to vascular damage and thrombus formation. Congestion of dermal capillaries results in diamond skin disease.

Localization of E. rhusiopathiae in joints of swine leads to fibrinous exudation and pannus formation. Subsequent damage to the articular cartilage and prolonged retention of bacterial antigen in the synovial tissue and chondrocytes are responsible for chronic infection.

Symptoms

Erysipelas occurs in pigs of all ages, but pigs from 2 months to one one year age are highly susceptible.

Four forms of clinical disease in swine have been described. These may occur alone or in combination. They are;

- Acute septicaemia
- Urticarial or diamond skin lesions
- Vegetative endocarditis
- Arthritis

Swine erysipelas manifest in three forms - Acute, subacute or chronic course. The acute disease is characterized by high fever, inappetance, depression, a rapid course of illness, and death within 2-3 days in untreated animals. Some animals may show a stiff gait and reluctance to stand or move, and urticrial cutaneous lesions may develop. The urticaria may be pink or purplish especially in the abdomen, thighs, ears and tail. In severe cases the skin become necrotic and is sloughed. The diamond shaped raised skin lesions is pathognomonic. Pregnant sows may abort. A vegetative endocarditis is manifested by signs of cardiac insuffiency or sudden death. Subacute disease is similar to the acute except that it is less severe and animals are likely to recover within 5-7 day.

In the chronic form arthritis is more common. The hock, stifle, elbow and carpal joints are most likely to be affected resulting in severe lamness. The mitral valves are involved in valvular endocarditis. Valvular vegetations are due to fibrin deposition and connective tissue proliferation.

Turkeys develop a cyanotic skin, become droopy, and may subsequently die. A swollen cyanotic snood and diffuse reddening of the skin are pathognomonic.

Diagnosis

Diamond shaped skin lesions are pathognomonic.

Specimens to be Collected

It includes liver, spleen, heart valves or synovial tissues. Organisms are rarely recovered from skin lesions or chronically affected joints.

Smear examination:

1. Gram-positive rods in acute cases and gram-positive filaments in chronic cases.
2. Based cultural characters and biochemical tests.
3. Serological tests are not applicable for diagnosis.

Treatment

In addition to hyper immune serum, treatment with antibiotics such as penicillin and tetracyclines are effective.

21

Mycobacteria

Mycobacteria are slender rods of varying lengths that sometimes show branching filamentous form resembling 'fungal mycelium'. Hence, the name mycobacteria, meaning fungus like bacteria.

Although cytochemically gram positiive, the Mycobacteria do not take up the dyes of the gram stain because the cell walls are rich in lipids – Mycolic acid. Once a dye has been taken up by the cells they are not easily decolourised, even by acid-alcohol. Mycobacteria are therefore called as acid-fast bacilli.

The genus includes animal and human pathogens as well as saprophytic members often referred to as atypical, anonymous, opportunistic, tuberculoid and MOTT (Mycobacteria other than typical tubercle) bacilli.

Clasification of *Mycobacteria* (Tubercle Bacilli)

I. Slowly growing *Mycobacteria*

- *Mycobacterium tuberculosis* causes human tuberculosis in human and dogs.
- *Mycobacterium bovis* causes bovine tuberculosis in many animal species and also cause tuberculosis in human
- *Mycobacterium africanum* causes human tuberculosis.
- The human type (*Mycobacterium tuberculosis*) is primarily a pathogen for man. But can cause disease in cattle, pigs, dogs, monkeys, parrots and other species. The bovine type (*Mycobacterium bovis*) is a common cause of disease in domestic animal particularly cattle, pigs, cat, dogs and horse. The avian type (*Mycobacterium avium*) is primarily a pathogen for birds. But can cause disease in cattle, sheep, goat and pigs.

II. Atypical Mycobacteria

Runyon (1959) grouped the atypical mycobacteria on the basis of pigmentation, colonial morphology and growth rate. The photochromogens will produce pigment only if exposed to light. The scotochromogens are those that produce yellowish orange pigments in the dark. The slow growing mycobacteria are those that require over 7 days incubation. And rapid growers are those requiring less than 7 days.

1. Slowly growing photochromogens

   Mycobacterium kansasi, Mycobacterium marinum, Mycobacterium simiae

2. Slowly growing scotochromogens

   Mycobacterium gordonae (tap water scotochromogens)

3. Slowly growing non-chromogens

   Mycobacterium avium (Avian tuberculosis)

   Mycobacterium intracellulare

   Mycobacterium paratuberculosis

   (Johne's disease – chronic hypertrophic enteritis in cattle)

   Mycobacterium lepraemurium (Feline leprosy)

4. Rapid growing mycobacteria

   Mycobacterium phlei (timothy grass bacillus)

   Mycobacterium smegmatis

III. Non-cultivable Mycobacteria: *Mycobacterium leprae*

Addition to this the unspecified acid-fast bacilli such as *Mycobacterium senegalense* and *Mycobacterium farcinogens* were isolated from Bovine farcy.

History

- The generic name mycobacterium (fungus bacterium) was proposed by Lehmann and Neumann (1896).
- The first member of this genus to be identified was the lepra bacillus discovered by Hansen (1868) – Hansen bacillus.
- Koch (1882) isolated the mammalian tubercle bacillus and proved its causative role in tuberculosis by satisfying Koch's postulates.
- The acid-fast property of Mycobacterium was discovered by Ehrlich (1882).
- Johne (1895) described Johne's bacillus - *Mycobacterium paratuberculosis*.

Natural Habitat

It has a worldwide distribution. The usual habitats of the great majority of the cultivable mycobacteria are water and watery habitats, marshes, wet soil, streams, lakes, rivers. The source of the pathogenic mycobacteria is usually infected animals. *Mycobacterium bovis* is excreted in respiratory discharges, faeces, milk, urine and semen. *Mycobacterium avium* and *Mycobacterium paratuberculosis* are shed in faeces and *Mycobacterium tuberculosis* mainly in respiratory discharges.

The atypical mycobacteria are widespread in soil, pastures, grass and water. A few are commensals in animals and may infect them.

Morphology

They are acidfast organisms, which are usually straight or slightly curved rod occurring singly, pairs or in small groups. The morphology varies from cells of species to species. *Mycobacterium tuberculosis* is often arranged in serpentine cords (Smear prepared from liquid medium). *Mycobacterium kansasi* is distinct banded or beaded appearance, while *Mycobacterium avium* is often almost coccoid. In clinical materials M. avium *subsp.paratuberculosis* may appear as bundle of faggots. They are non-motile, non-sporing and non-capsulated.

Cultural Characters

A comparatively slow growth rate is characteristic of the mycobacteria, with generation time ranges from 14-20hours. Colonies appear only in about two weeks and sometimes may be delayed upto 6-8 weeks. Optimum temperature is 37^0C and pH is 6.4 –7.0.

Mycobacterium tuberculosis is an obligate aerobe while *Mycobacterium bovis* is microaerophilic. Growth is stimulated by 5-10% CO_2. Tubercle bacilli do not have exact growth requirements. But they are highly susceptible to even traces of toxic substances like fatty acids in culture media. The toxicity is neutralized by addition of serum, albumin or charcoal.

Several media both solid and liquid are available. The egg based Lowenstein Jensen medium and Stone Brinks medium are most commonly used. Malchite greendye (0.025g/100ml) is commonly used as the selective agent. *Mycobacterium tuberculosis, Mycobacterium avium* and many of the atypical mycobacteria require glycerol for growth. However glycerol is inhibitory to *Mycobacterium bovis*, while sodium pyruvate enhances its growth. On Lowenstein Jensen medium (i.e. glycerol containing media), Mycobacterium tuberculosis giving the characteristic rough, tough and buff colonies – is known as eugonic. On Lowenstein-Jenson (LJ) slants, M. tuberculosis typically has a crumbly, “bread-crumb” appearance, whereas M. avium complex has a bright yellow, smooth, characteristic colony (so-called “condom colony”).

The growth of *Mycobacterium avium* in this media also described as eugonic. *Mycobacterium bovis* has sparse, thin growth on glycerol containing media that is called dysgonic. *Mycobacterium bovis* however grows well on pyruvate containing media without glycerol (i.e. stone brink medium)

Pigment formation is tested with young, well-developed colonies on Lowenstein Jensen medium. The cultures are exposed to a 100-watt, clear electric bulb, at a distance of 50cm, for atleast an hour and then incubated again in darkness for a further 1-3 days. After this treatment the photochromogens will develop pigment. Many of the mycobacteria produce yellow/orange pigments while *Mycobacterium tuberculosis, Mycobacterium bovis* and *Mycobacterium avium* are non-chromogenic.

In liquid media, the growth begins at the bottom, creeps up the sides and forms a prominent surface pellicle (mould like pellicle) that may extends along the sides above the medium. Smears made from liquid medium (Virulent stains) gives characteristic longserpentine cords, while avirulent strains grow in a more dispersed fashion.

Supplementation of media with mycobactin (extracted from non-mycobactin dependant isolates of *M.avium subsp. paratuberculosis)* is required for *M.avium subsp. paratuberculosis.*

Biochemical Characters

They are oxidative. Atypical mycobacteria are catalse positive, while tubercle bacilli are peroxidase positive. Niacin production and nitrate reduction is only by *Mycobacterium tuberculosis*. Urease is reduced by *Mycobacterium tuberculosis* and *Mycobacterium bovis* but not by avian strain.

Resistance

The Mycobacteria are resistance to physical influences and will retain their viability in soil and particles of dried faeces for many months. They are not specifically heat resistant; being killed at 60^{0}C in 15-20mts. Cultures may be killed by exposure to direct sunlight for two hours. Bacilli in sputummay remain alive for 20-30hrs and in droplet nuclei for 8-10 days.

They are relatively resistant to disinfectants i.e. exposure to 5% phenol, 15% H_2SO_4, 3% nitric acid, 5% oxalic acid and 4% NaOH. It is destroyed by tincture of iodine in 5mts and by 80% ethanol in 2-10mts.

Antigens and Toxins

Many antigens have been identified in mycobacteria. Group specificity is due to polysaccharide and type specificity is due to protein antigen.

They do not produce any exotoxins. The cellwall of the mycobacterium is composed of peptidoglycan, arabinogalactan and mycolic acid. In addition to this it contain wide range of lipids. The outerlayer layer of the cellwall is composed of mycosides (Peptidoglycolipids or Phenolic glycolipids).

Mycosides are responsible for the control of cellular permeability, resistance to action of water-soluble enzymes, antibiotics and disinfectants.

Cord factor (Trehalose–6, 6' dimycolate) and Wax D- inhibits chemotaxis, leukotoxic, responsible for delayed hypersensitivity.

Sulfatides- sulfur containing glycolipids – promote the survival of virulent tubercle bacilli within macrophages by inhibiting phagolysosome formation and avoiding exposure to hydrolytic enzymes present in the lysozomes.

Virulence appears to reside in the lipids of the cell wall. Mycosides, phospholipids and sulpholipids are protecting the tubercle bacilli against Phagocytosis.

Pathogeneis

Infection is usually by inhalation and ingestion. The mucociliary clearance by mucus and epithelial cilia in the upper respiratory passages provides defense against infection. However, microorganisms on small particles (1-4 μm in size), such as, dust and water droplets reach alveolar spaces.

The infectious process begins with deposition of tubercle bacilli in the lung or on pharyngeal or intestinal mucous membranes. In previously unexposed animals, local multiplication of the mycobacteria occurs and the resistance to phagocytic killing allows continued intra cellular and extra cellular replication. Infected host cells with mycobacteria can reach local lymphnodes and from there may pass to the thoracic duct with general dissemination. Haematogenous spread may produce milliary tuberculosis (in deer). This involves multifocal tubercle formation in an organ.

Lesions

Cattle

Catlle are usually infected with *M. bovis*. Infection enters usually by the respiratory tract. Tuberculosis consists of a characteristic lesion – the tubercle. This is an avascular granuloma composed of a caseous necrosis in a central area encircled by a zone of epitheloid cells, and a peripheral zone of lymphocytes, granulocytes and fibroblasts. Calcification may be present in the necrotic centers. An outer boundary of fibrous tissue is usually present between the lesions and normal tissue.

Tubercle lesions are more commonly present in the lymphnode, lungs and pleura. Haematogenous dissemination leads to similar lesions in liver, kidney etc. Similar lesions are seen in sheep and goats.

Horses

Horses are infected more often with *M. avium* than with *M. bovis*. Infetion enters usually by the alimentary tract. Primary complexes related to pharynx and intestine. Secondary lesions may be in lung, liver, spleen and serous membranes.

Pigs

Swine can be infected mostly by *M. bovis* via alimentary tract. *M. avium* infections are predominat in many countries. Tubercles are most commonly seen in the skeletonespecially vertebrate and long bones are common sites. Meninges and viscera are also commonly affected.

Birds

They are primarily susceptible to M.avium via alimentary tract. The grayish white granulamatous lesions are found in the liver and they are also present in the intestines, spleen and bone marrow. *Mycobacterium genavense* affects canaries and parrots.

Diagnosis

Specimens

Specimens from live animals include aspirates from cavities, lymphnodes, biopsies, tracheobronchial lavages and centrifuged deposit from about 50 ml of milk in the case of suspected tuberculous mastitis. With dead animals, collect fresh and fixed (10% formalin) samples of lesions.

Diagnostic Techniques

1. Based on history, signs and post mortem lesions

2. Direct microscopy

The Ziehl-Neelsen (acid-fast) stain is used to stain smears from lesions and other specimens. Organisms are appearing as slender, often beaded, red staining rods against a blue background. Smears stained by fluorescent dyes (auramine, acridine orange or fluorochrome) allow the mycobacteria to be seen more easily if relatively small numbers are present.

3. Isolation

Several preliminary procedures are necessary in order to recover the comparatively slow growing mycobacteria

a) Selective decontamination to reduce significantly the number of fast growing contaminating bacteria.

b) Digestion or liquefaction of mucus is necessary. Mucin-trapped mycobacteria in specimens, such as broncho-tracheal exudates, may not be available for growth in cultures.

c) If mycobacteria are present, they must be concentrated by centrifugation.

For deconatmination, the ground-up specimens must be treated with decontaminating agents, such as 5% oxalic acid, 4% sodium hydroxidefollowed by neutralization of the acid or alkalior20% antiformin can be used and subsequently cultured in stone brink or Lowenstein Jensen media.

4. By animal inoculation

Inoculation of suspected material	*Mycobacterium tuberculosis*	*Mycobacterium bovis*	*Mycobacterium avium*
Rabbits (I/V)	+	++	++
Guinea pig (S/C)	++	++	-
Chicken (I/V)	-	-	++

5. Bacteriophage typing: A, B, C, I

6. Tuberculin test (Intra dermal, double intra dermal, ophthalmic tests in cattle and wattle test in case of poultry)

7. Gamma interferon assay, Gas liquid chromatography, PCR, ELISA for detecting circulating atnibodies, FAT, LTT assays can also be used in the diagnosis

Control and Prevention

Treatment and vaccination are inappropriate in control programmes for cattle. In many countries, tuberculin testing followed by isolation and slaughter of reactors has been implemented as the basis of national eradication schemes.

Paratuberculosis (Johne's Disease)

Johne's disease or paratuberculosis is caused by *Mycobacterium avium subsp paratuberculosis* (also referred as *Mycobacterium johnei*). This disease is caused by chronic, contagious fatal enteritis, which can affect cattle, sheep, goats, camels and wild ruminants.

Note:*Mycobacterium avium subsp paratuberculosis* infection in human called as Crohn's disease (Chronic enteritis in human).

Morphology

Mycobacterium avium subsp paratuberculosis are acid-fast organism. They are short rods measuring 1-2µm in width with rounded edges. It is motile and does not form spores. On artificial media the organism tends to be shorter club form.

Cultural Characters

It requires mycobactin- Killed extracts of *M.phelei* or other killed acid-fast organism- enriched media for growth. Slants of Herrold's egg yolk mediumwith mycobactin are highly sutibale for isolation of organism from specimens.

The slants are incubated at 37^{0}C for upto 16 weeks and examined weekly for evidence of growth. They produce minute grayish white, friable irregular colonies, less than 1mm in d.m., in 5-16 weeks. Isolates from sheep may be pigmented.

Pathogenesis

The organism shed in the faeces, milk and semen of infected animals. They remain viable in the environment for upto one year under suitable conditions. Calves under one month of age are highly susceptible and they are developed clinical disease than animals infected later in life.

Infection is acquired mainly through ingestion. The organism is an intra cellular pathogen and cell mediated reactions are mainly responsible for the enteric lesion. Ingested mycobacteria, engulfed by macrophages in which they survive and replicate are found initially in Peyer's patches. As the disease progress, an immune mediated granulomatous reaction develops, with marked lymphocyte and macrophage accumulation in the lamina propria and submucousa. The resulting enteropathy leads to loss of plasma proteins and malabsorption of nutrients and water.

Symptoms

Clinical signs developed after prolong subclinical phase of infection. Affected cattle are usually more than 2 years of age when signs are first observed. In cattle, the disease is characterized by diarrhoea, initially intermittent, dark and semisolid, but becoming persistent and profuse. Progressive weight loss results without loss of appetite, emaciation and eventually death. The mortality rate may approach 100%. Asymptomatic carrier cattle have an increase incidence of mastitis and infertility. In sheep and goats, the disease is clinically evident only in mature animals. The diarrhoea is less marked and may be absent.

Lesions

Chronic catarrhal inflammation of the intestine is characteristic. In cattle, the mucousa of affected areas of the terminal small intestine and the large intestine is usually thickened and folded into transverse corrugation. The mesentric and ileocaecal lymphnodes are enlarged and odematous. Thickening of the intestinal mucousa is less marked in sheep, and necrosis and caseation may be present in the regional lymphmnodes.

Diagnosis

1. Microscopical examination by staining the faecal smears with acid-fast stain.
2. Bacteriological examination: Materials decontaminated with 0.3% bezalkonium chloride and concentrated by centrifugation and subsequentlu cultured in Herrold's egg yolk medium and incubated at 37^0C for upto 16 weeks.
3. Basen on PM lesions
4. Serological tests: Complement fixation test can be used, but are laborious and relatively insensitive. Agar gel precipitation test has been used for confirming clinical infection. ELISA, using serum absorbed with a M.Phlei, may detect subclinically infected animals.
5. Johnin test
    a. Intra dermal Johnin test: Inoculate Johnin PPD into the skin of the neck region. The delayed hypersensitivity reaction is measured after 48 hours. Peak response usually develops a month or so after infection.
    b. Intra venous Johnin test: The intravenous Johnin test reaction is measured by increase in body temperature following intra venous Johnin PPD injection.

6. DNA probes, which are highly sensitive, are being used to detect organisms in faeces.

Control and Prevention

Inactivated adjuvanated vaccines are available. A live vaccine consists of nonpathogenic strain of *Mycobacterium avium subsp paratuberculosis* is inoculated subcutaneously into calves soon after birth and before 4 weeks of age. It reduces the incidence of Johne's disease in the herd.

22

Paratuberculosis (Johne's Disease)

Corynebacteria are gram positive, non-acidfast, non-motile, non-capsulated, small pleomorphic rods. They frequently occur in rods, coccoid, club and filamentous shape. The term coryneform which refers to the pleomorphic club shape of these gram-positive bacteria. (From Coryne – meaning club). The major pathogen is *Corynebacterium diptheria*. Which causes diptheria in children. Corynebacteria associated with animals are called diphtheroids.

Systematics

The main diseases, hosts and natural habitats of the Corynebacteria are;

Spcies	Main host (s)	Diseases	Natural habitat
C. pseudotuberculosis (Corynebacterium ovis or Preisz Nocard Bacillus)	Sheep and Goats (Non-nitrate reducing biotype)	Caseous lymphadenitis	Skin, mucous membrane and G.I. tract
	Horses & Cattle (Nitrate reducing biotype)	Ulcerative lymphangitis Pigeon fever or Breast bone fever	
Corynebacterium renale	Cattle	Pyelonephritis and cystitis	Prepuce and semen of asymptomatic bulls, vaginal mucous membrane of heathy cows
	Pigs	Kidney abscess	
	Male sheep	Balanoposthitis (Pizzle rot)	
Corynebacterium cystitidis	Cattle	Severe cystitis, rarely Pyelonephritis	Male genital tract
Corynebacterium pilosum	Cattle	Pyelonephritis	Male genital tract and urine
Corynebacterium bovis	Cattle	Subclinical mastitis	Udder and teat canal of cows
Rhodococcus equi (Corynebacterium equi)	Foals (2-4 months)	Suppurative bronchonephritis	Soil and Faeces of foals and other herbivores
	Pigs	Cervical lymphadenitis	Soil

Morphology

They are gram-positive slender rod with a tendancy to clubbing at one or both ends; they are non-sporing, non-motile, non-capsulated and non-acidfast.

They have granules composed of (high energy phosphate stores) – polymetaphosphate. The granules are more strongly gram positive than the rest of the bacterial cell. Stained with Loeffler's methylene blue, the granules take up a reddish purple color and hence they are called metachromatic granules. They are called as volutinor Babes Ernst Granules. They are often situated at the poles of the bacilli and are called polar bodies. Special stains, such as Albert's, Neisser's and Ponder's have been devised for demonstrating the granules clearly.

Stained smears from animal tissues often reveal groups of cells in parallel (Palisades) or cells at sharp angles to each other (Chinese letter or Cuneiform arrangement). This is due to the incomplete separation of the daughter cells after binary fission.

Rhodococcus equi can appear as a gram-positive coccus or a rod or club shaped form arranged in clusters. It is capsulated and sometimes weakly acid fast.

Cultural Characters

Growth is scanty on ordinary media. Enrichment with blood, serum or egg is necessary for good growth. The optimum temperature for growth is 37^0C and optimum pH is 7.2. It is an aerobe and facultative anaerobe. Sheep or ox blood agar is used routinely along with Macconkey agar to detect any gram-negative contaminants that may be present.

On blood agar, *Corynebacterium ovis* colonies are small, white, dry and non-hamolytic at 24hr incubation. A narrow zone of haemolysis occurs at 72 hrs incubation. After several days incubation the colonies can reach 3mm in d.m. and appear dry, crumbly and cream in color.

Corynebacterium bovis colonies are small, white, dry and non-haemolytic. As it is a lipophilic corynebacterium, they grow very well on media enriched with 0.5 –1% tween 80.

Rhodococcus equi colonies are small, smooth, shiny and non-haemolytic after 24hrs incubation. But on 4-day culture, the colonies become larger, mucoid and salmon-pink in color. The salmon-pink pigmentation is not easily seen against a red background. So, the mucoid colonies and salmon-pink pigmentation can easily be demonstrated on nutrient agar (4-day culture). Nutrient agar enriched with yeast extract and glucose is useful for enhancing the salmon pink pigmentation.

The renale groups are non-haemolytic. On nutrient agar, after 48 hrs incubation, *Corynebacterium renale* produces dull yellow colonies. The *Corynebacterium pilosum* produces distinct yellow and *Corynebacterium cystitidis* exhibit white colonies. On milk agar *Corynebacterium renale* show casein digestion, while *Corynebacterium pilosum* and *Corynebacterium cystitidis* do not give this reaction.

On CAMP tests, the *Corynebacterium ovis, Rhodococcus equi* and *Corynebacterium renale* interacting with beta haemolytic of *Staphylococcus aureus* and gives the following results.

Organisms	Staphylococcal β haemolysis
Corynebacterium ovis	Inhibition
Rhodococus equi	Enhancement
Corynebacterium reanale	Enhancement

Biochemical Characters

They are catalase positive, oxidase negative. Except *Corynebacterium bovis* others are urease positive. The renale group is very strong urease positive (less than one hour). All diptheroids ferments sugar except *Rhodococcus equi. Corynebacterium bovis* and *Corynebacterium renale* ferments both glucose and maltose.

Two biotypes of *Corynebacterium ovis* are recognised. The ovine/caprine strains lack nitrate-reducing capacity, while the equine/bovine strains usually reduce nitrate.

Resistance

Diptheroids are readily destroyed by heat, 60^{0}C for one hour. They are highly susceptible to disinfectants. It is more resistant to the action of light, dessication and freezing. *Rhodococcus equi* is resistant to 2.5% oxalic acid for one hour.

Antigens and Toxins

The diphtheroids antigen and toxins are not well documented. *Corynebacterium ovis* produces a filterable toxin similar to that produced by *Corynebacterium diptheria*. It is a haemolytic toxin, which has phospholipase D activity. The lipoteichoic acids and peptidoglycan of the gram positive cell wall interact with macrophage cells resulting in the release of proinflammatory cytokines. The high concentration of lipids contributes to intraphagocytic survival and leukotoxicity.

In *Corynebacterium renale*, the pili is antigenic, it express fibrillar protein adhesin on their surfaces. Renalin- a *Corynebacterium renale* extracellular protein may play a role in lysis of cell membranes. The lipoteichoic acids and peptidoglycan of the gram positive cell wall interact with macrophage cells resulting in the release of proinflammatory cytokines. Urease – plays major role in pathogenesis.

Rhodococcus equi produces diffusible *Rhodococcus equi* factors (Phospholipase C and Cholestrol oxidase) and these as well as the capsule and cell wall constituents probably play a major role.

Pathogenesis

The diphtheroids infections are characterized by the development of suppurative lesions and clinical manifestations do not develop in the absence of predisposing factors.

Corynebacterium ovis

The prevalence of caseous lymphadenitis may be as high as 50% in adult sheep. Some clinically normal sheep may carry the organism in the digestive tract, excrete in the faeces and contaminate the environment. When bacteria enter the host via skin wounds (or tick bite), multiply and are phagocytosed. Phagosome-lysosome fusion takes place. But *Corynebacterium ovis* multiplies in the phagolysosome and phagocytic cells die. Permeability of local blood vessels increases, encouraging the spread of infection from the initial site to other locations, often-regional lymphnodes and produces toxin – Phopholipase-D. Abscesses may develop at either primary or secondary sites, eventually rupturing and discharging thick, caseous pus containing large numbers of viable bacteria. In some instances, lesions become metastatic and, as they increase in number, the thin ewe syndrome develops, resulting in progreesive debilitation and death.

Corynebacterium Renale

Corynebacterium renale is a normal flora in the lower urogenital tract. This group possess fimbriae which allow attachment to the urogenital mucousa. The major predisposing factors that put a cattle at risk are the shortness of the female urethera and the effects of preganancy and parturition, thus, disease occurs most frequently in mature cows. The vulva may be an important portal entry for *Corynebacterium renale* into the urinary tract. Bacteria grow readily in urine and ascend (through vesiculourethral reflex) to the kidney. *Corynebacterium renale* has high urease activity. The urease is nephrotoxic and produces pyelonephritis.

Rhodococcus Equi

Rhodococcus equi may be a commensal in the intestine of horses and it is largely a soil organism. The soil enriched with equine faeces and summer temperatures are favours the rapid multiplication of this bacterium. The disease is usually seen in 2-4 month old foals, possibly due to the decline in maternal antibody at about 6 weeks of age. The main route of infection is by inhalation. *Rhodococcus equi* is a facultative intracellular pathogen. Its ability to survive, persist in and eventually to destroy alveolar macrophages is the basis of its pathogencity. It causes granulomatous inflammation and abscesses in the lung tissue. Heavily infected sputum may be swallowed by the affected foal leading to ulcerative colitis and mesentric lymphadenitis.

Symptoms

Corynebacterium Ovis

Caseous lymphadenitis in sheep is characterized by skin wounds, enlarged lymph glands and abscess distributed throughout the body. But the disease can be a mild infection and it is unnoticed until postmortem is done.

Ulcerative lymphangitis in horses are similar to glanders. Initially there is a pain and swelling of the hind limbs and enlargement of lymphatic vessels. Ulcers develop, starting at the fetlock, discharging purulent material. Its progress towards the inguinal region is marked by swelling and abscesses. In severe cases it spreads to abdomen, forelegs and neck leads to death.

Pectoral abscess is also called pigeon fever and breast bone fever. C. ovis causes abscesses, usually in the muscles of the chest and caudal abdominal region of horses. Signs such as swelling, pain and lameness depend on the location and size of the abscess may be observed. Septicaemia may result in renal abscess, debilitation and death.

Corynebacterium renale

In pyelonephritis, characteristically frequent passage of turbid or blood stained urine by animals, which are pregnant or recently calved. Restlessness and kicking at the abdomen may indicate renal pain. The urine contains red blood cells, pus cells and albumin.

Ovine posthitisorUlcerative balanoposthitis (Pizzle rot), particularly in common in Merino sheep and Angora goats, is characterized by necrotizing inflammation and ulceration around the preputial orifice, with a brownish crust developing over the lesion. Disease develops in the presence of the urealytic agent in an area constantly irrigated with urine. Pizzle rot typically occurs in animals on rich legume pasture that is high in proteins, which increase urea excretion, and estrogens, which cause preputial swelling and urine retention in the sheath.

Rhodococcus Equi

In suppurative bronchopneumonia, the foals develop cough, fever and increased respiratory rate. In severe cases purulent discharge from nose just before death.

Lesions

Corynebacterium ovis

In caseous lymphadenitis the superficial lymphnode contains a mass of greenish yellow caseation in concentric layers, which have an onion ring appearnce in, cross section (hence, named it as Corynebacterium pseudotuberculosis). In advanced cases, similar lesions are seen in lungs, kidney, liver and spleen.

Corynebacterium renale

The kidneys are enlarged, necrosis and suppuration in the medulla. Wedge shaped suppurative foci in the cortex. The kidney pelvis contains blood and pus. The bladder contains blood and pus with petechial haemorrhages and ulceration of the tract.

Rhodococcus equi

Characterized by pyogranulamatous lesions with abscesses in the lung, associated lymphnodes and pus in the bronchi are very characteristic.

Diagnosis

Specimens

Pus or exudates are collected from suppurative conditions and mid stream urine for isolation of the *Corynebacterium renale*. A tracheal wash technique with infusion of saline can be used for the recovery of *Rhodococcus equi* from affected foals. Diagnosis is mainly based on history, symptoms and lesions.

Isolation and identification of the organism

Based on microscopical appearance, colonial morphology on blood agar, CAMP tests and Biochemical tets are used for identification of the organism.

Control and Prevention

Diphtheroids are susceptible to penicillin, tetracycline, erythromycin, lincomycin, neomycin and gentamicin.

23

Actinomycetes

The actinomycetes comprise a heterologous group of prokaryotes that have the ability to form gram positive, branching filaments of less than 1 μm in d.m. The main animal pathogens in the actinomycetes are in the genera Actinomyces, Arcanobacterium, Actinobaculum, Nocardia and Dermatophilus. Non-pathogenic, prolific producers of antimicrobial substances – streptomyces are also included in Actinomycetes.

Actinomyces

Natural Habitat

The actinomyces species are present on mucous membrane of the host animal, often in the oral cavity, tonsils, and nasopharynx. The soil is the natural habitat of many actinomyces species.

History

The generic name Actinomyces was first used by Harz (1879). Boestrom (1891) isolated Actinomyces bovis. Cummins (1962) clearly demonstrated Actinomyces were bacteria and they are distinct from other branching genera.

Diseases caused by the pathogenic actinomycetes;

Actinomycete	Host (s)	Disease
Actinomyces bovis (Syn: Ray fungus)	Cattle	Bovine actinomycosis (Lumpy jaw)
	Horses	Poll evil and Fistulous withers (occur as a mixed infection with Brucella species)
Arcanobacterium pyogenes (Actinomyces pyogenes)	Cattle, Sheep and Pigs mainly	Chronic or acute suppurative mastitis, suppurative pneumonia, septic arthritis, vegetative endocarditis (Cattle), endometritis, umbilical infections, wound infections and Seminal vesiculitis (Bulls and Boars). Summer mastitis – a mixed infection with *Peptostreptococcus indolicus*

Actinomycete	Host (s)	Disease
Actinomyces viscous	Dogs	Canine actinomycosis 1. Localised cutaneous granulamatous abscess and/or 2. Pyothorax and granulomas in the thoracic cavity
Actinomyces isralii	Human	Human actinomycosis
Actinobaculum suis (Actinomyces suis)	Pigs	Pyogranulamatous mastitis, ascending pyelonephritis, cystitis.

Morphology

The organisms show considerable pleomorphism. Actinomyces species are usually long and fialmentousalthough short V, Y, and T configuration also occur. In lesions of actinomycosis, the pus contains small pale yellow granules referred as sulfur granules. The sulphur granule is composed of bacterial filaments and mineralized calcium phosphate of host origin. When the granules are crushed and gram stained, a mass of gram-positive branching filaments about 1μm in width, short rods, and cocci are evident. Around this mass, a circle of club shaped bodies with their narrow ends pointing towards the centre-staining gram negative. Hence, called ray fungus. They are non-acid fast, non-spore forming, nonmotile, non-capsulated and do not form endospores or conidia.

In case of *Arcanobacterium pyogenes* infections the pus or mastitic milk does not contain any granules. Gram stained smears reveal large numbers of small, highly pleomorphic, gram-positive rods, cocci and pear shaped cells. Occasionally short branching typical Chinese letter appearances are also seen.

Cultural Characters

They cannot grow on Sabouraud dextrose agar. Actinomyces require enriched media for growth. They grow well on sheep or ox blood agar. *Actinomyces bovis* is a capnophilic (i.e. required 5-10% CO_2 for its growth). *Arcanobacterium pyogenes* and *Actinomyces viscous* will grow aerobically but 5-10% CO_2 will enhance their growth. *Actinomyces bovis* and *Actinomyces viscous* usually require 2-4 days but the growth of *Arcanobacterium pyogenes* can usually be seen in 24 hrs.

Actinomyces bovis colonies are non-haemolytic, very small (< 1nm), white, rough or smooth and adhere tenaciously to solid medium. Gram stained smears show gram positive, slightly branched filaments or short forms. On subculture, the bacterium may become diphtheroidal or coccobacillary. *Actinomyces bovis* grows well in thioglycollate medium giving a characteristic diffuse growth in about 7-10 days. In broth cultures, it grows in coarse aggregates, which in some cases may result in a granular deposit with a completely clear supernate.

Arcanobacterium pyogenes produce a hazy- haemolysis after 24hrs incubation along the streak lines. At 48 hrs incubation, the colonies are surrounded by a narrow zone of complete haemolysis. *Arcanobacterium pyogenes* has the ability to pit a loeffler serum slope in 24-48 hrs. (i.e. A loopful of pureculture of the medium is taken and a heavy inoculum is made in a small area in the center of the slope, taking care not to break the surface of the medium. The medium is incubated at 37^{0}C for 24 –48 hrs).

Arcanobacterium pyogenes will give positive CAMP test with Staphylococcus aureus (i.e. enhancement of staphylococcal haemolysis). In litmus milk, the organism produces acid and clot after 3 days growth. *Actinomyces viscous* commonly produces two colonial forms, one being smooth, entire, convex and glistening and the other is smaller, rough dry and irregular. Neither is haemolytic. The larger colonial type yields gram-positive diphtheroided forms and the smaller colony has short branching filaments.

Biochemical Tests

Both *Arcanobacterium pyogenes* and *Actinomyces bovis* are catalase negative, ferments several sugars and produce acid. Reduction of nitrate is negative. *Actinomyces viscous* is catalse positive.

Resistance

Actinomyces are killed at a moist heat temperature of 60^{0}C for 20 mts and they are susceptible to various disinfectants.

Antigens and Toxins

With the exception of *Arcanobacterium pyogenes,* Actinomyces species have not been shown to produce any toxin. *Arcanobacterium pyogenes* produces a haemolytic exotoxin (Pyolysin O), which is dermonecrotic and lethal and it also produces a protease and an extracellular neuraminidase.

Pathogenesis

Actinomyces bovis, present as part of the normal flora of the mouth. Trauma to the tissues is the initiating event in disease and may occur as a result of shedding of teeth or as a result of coarse feed. Whenever there is a trauma, the organism invades a variety of tissues and often produces lesions on bone. Growth of the organism may involve maxillary bone, tongue, pharynx, lungs, lymphnodes and S/c tissues of the head and neck.It initiates rarefying osteomyelitis and soft tissue reaction, the condition being referred to as lumpy jaw.Granulation, mononuclear infiltration and fibrosis occur in the lesions with sinus tracts leading to the outside. Exudate from the tracts contains pus with sulphur granules. *Arcanobacterium pyogenes* is a commensal on the exposed mucosal surfaces of cattle, sheep and swine.

Arcanobacterium pyogenes infection is often a sequel to earlier tissue injury, or to infection with other bacteria. (i.e. Fusobacterium necrophorus, Peptostreptococcus indolicus). It produces toxins and established mastitis with abscess formation. The acute bovine mastitis is refered as summer mastitis. *Actinomyces viscous* serotype 1 appears too responsible for disease in dogs. Two syndromes can occur, either separetly or together. One is a localized granulomatous lesion involving skin and subcutis; the other is a pyothorax, with granulomas in the thoracic cavity and often a large accumulation of sanguinopurulent pleural fluid containg soft white granules.

Symptoms

In case of lumpy jaw in cattle there is marked swelling associated with suppurative (sinus tract containing pus) and proliferative chronic rarefying osteomyelitis in the region. There may be dislodgement of teeth, inability to chew and mandibular fractures. *Arcanobacterium pyogenes* infection occurs most frequently in heifer and dry cows during summer months. Hence, it is named as summer mastitis. The affected quarter become enlarged and firm. It often takes a destructive course, causing abscess formation and sloughing. Animals have fever with general toxaemia.

Lesions

Area of suppuration accompanied by granulation tissues is seen in lumpy jaw. Rarefying osteomyelitiss leads to replacement of normal bone by porous bone (irregularity and honey combed) with sinus tract containing pus. The pus characteristically contains small sulphur granules. In summer mastitis, abscess develops at any site containing greenish yellow foul smelling pus.

Diagnosis

1. Direct microscopy: The pus or exudate is placed in a Petridish and washed carefully with a little saline to expose the yellowish sulphur granules of *Actinomyces bovis* or the softer greyish white granules of

Actinomyces viscous. If it is stained with Gram's, the ray fungus can be demonstrated.

2. Isolation and Identification of organism
3. FAT
4. Pitting of loeffler serum slope and CAMP test in case of *Arcanobacterium pyogenes*

Control and Prevention

Actinomycetes are highly sensitive to tetracycline, chloramphenicol and pencillin including benzyl pencillin and ampicillin.

24

Nocardia

Nocardia has the ability to form gram-positive, branching filaments of less than 1um in diameter. It is closely related to Corynebacterium, Mycobacterium and Rhodococcus species

Diseases caused by the pathogenic Nocardia

Species	Host (s)	Disease
Nocardia asteroides	Dogs / Cats Cattle	Canine nocardiosis 1. Localised cutaneous granulamatous abscesses 2. Pyothorax and granulomas in the thoracic cavity Chronic granulamatous mastitis
Nocardia farcinica	Cattle	Bovine farcy in tropical regions
Nocardia nova	Dogs and cats	Mycetoma

History

Nocard described this organism in 1888, following its isolation from a case of bovine 'farcy', hence the name of the type species: Nocardia farcinica.

Natural Habitat

Nocardia species are soil borne saprophytes

Morphology

They are gram-positive, obligate aerobes ability to form branching filaments. Some produce true mycelia and some strains are acid-fast. All species are non-motile. Gram stained smears from lesions revealed gram-positive branching filaments that often show fragmentation into coccobacillary elements. The modified ZN stained smears exhibit a similar morphology but most of the filaments retain the carbol-fuchsin dye and stain red.

Cultural Characters

The Nocardia species grow very well in blood agar incubated aerobically at 37^0C for upto 7 days. Inoculate the suspected colonies from blood agar into Sabouraud dextrose agar (SDA) and incubated at 37^0C for upto 10 days.

Colonial Morphology

The colonies on blood agar often a vivid white and powdery, if aerial filaments and spores are formed. Occasionally the colonies are smooth, heaped and variably pigmented. Both types of colonies are firmly adherent to the agar surface. The colonies on SDA are dry, wrinkled and yellow, becoming deep orange color with age.

Microscopic Appearance

Gram-stained smears from colonies show gram-positive branching filaments that characteristically break up into rods or coccobacillary elements with age. An MZN - stained smear from young culture reveals red staining, branching filaments.

There are three morphological forms:

a. Group I strains have limited mycelia development due to early fragmentation of hyphae into coccoid forms within 2 to 14 hours of incubation;

b. Group II strains produce mycelia, which fragment in about 18 to 20 hours after incubation, though these mycelia break up into mycelial fragments within two days of growth

c. The pathogenic Nocardiaspecies belong to Group III. The colonies are usually leathery in appearance and pigmented. Extensive mycelium produced because fragmentation does not begin until after 5 days incubation.

Biochemical Characters

To differentiate Nocardia species tests such as decomposition of casein, hypoxanthine, tyrosine, urea and xanthine are useful. They are oxidative and Catalase positive. Reduced nitrates to nitrites. Gelatin not hydrolyzed.

Toxins

The cell wall of nocardia is typical of Gram-positive bacteria, with the addition of arabinogalactan, mycolic acids and high concentration of lipids. No exotoxins are known. But a superoxide dismutase acts as a virulence factor.

Pathogenesis

Nocardia are aerobic and essentially saprophytic. They cause suppurative and pyogranulamatous reactions in immunocompromised hosts or animals that have been exposed to large doses of the bacterium. The pathogenic nocardia survive within phagocytic vacuoles by preventing phagolysosome formation.

This is probably due to the surface lipids as Nocardia species have a cell wall similar to the mycobacteria. Other cell wall lipids may provoke the characteristic granulamatous reaction. Exudates are sanguineopuruelnt and can sometimes contain soft granules consisting of bacteia, neutrophils and debris. They lackthe microstructure of the sulphur granules produced by some of the Actinomyces species.

In bovine, the nocardial mastitis onset is sudden with fever, anorexia and abnormal milk secretion. Typically infection is introduced during the dry period with intramammary mastitis therapy. Bovine farcy is a chronic suppurative infection usually starting at the lower limbs. It involves lymphatics of the extremities or head region and the associated lymphnodes. Ulcers and discharging sinuses form along the path of infection.

Laboratory Diagnosis

Based on Direct Microscopy

Soft granules are not common in exudates from N. asteroids infections. Smears made from exudates, aspirates, granulomatous tissue and from centrifuged deposits of bovine mastitic milk are stained by Gram and MZN stain.

Gram-stained smears revealed gram-positive branching filaments that often show some fragmentation into coccobacillary elements. The MZN stained smears exhibit a similar morphology but most of the filaments retain the carbol fuchsin dye and stain red.

Based on Isolation and Identification / Biochemical reaction

- Characteristic colonial morphology on blood and SDA agar
- Microscopic appearance and
- Biochemical reaction.

Differentiation with Actinomyces

Actinomyces infections respond well to penicillin and other commonly used antibiotics, nocardial infections are often refractory to treatment and N. *asteroides* is susceptible only to limited range of antimicrobial agentssuch as Trimethoprim-sulphamethoxazole or erythromycin.

Characters	Nocardiosis	Actinomycosis
Granules in exudates	Not common	Usually present
Filaments MZN positive	+	-
Fragmentation of filaments	+	-
Growth on SDA	+	-
Powdery, white colonies (aerial hyphae)	+	-
Susceptibility to penicillin	-	+

Note

- Infection with Nocardiais by inhalation from the environment, while Actinomycesinfection begins in the host as a normal flora invading damaged tissues.
- Disseminated disease caused by Nocardiais more common in the dogs, while granuloma formation is the rule with Actinomycesinfection and spreads by local extension.
- When there is a doubt as to whether an animal has actinomycosis or nocardiosis, precautionary measures to preserve the anaerobic Actinomycesshould be institute, including prompt delivery to the laboratory under anaerobic condition, culturing on brain heart infusion agar blood plates and incubating under anaerobic, microaerophilic, and aerobic conditions. While Actinomycesfails to grow on Sabouraud agar,Nocardiagrows uninhibited.
- Acid-fast staining procedure of the sample exudate before culturing will also be helpful in the presumptive identification of the infecting agent.

25

Dermatophilus

Dermatophylaceae is a group of bacteria with mycelial filaments which divide transversely and in at least two longitudinal planes to form masses of coccoid or cuboidal cells, which characteristically become motile. They are gram positive, non acid fast, aerobic and produce aerial mycelium when their growths are stimulated by 10% CO_2. *Dermatophilus congolensis*, *Dermatophilus dermatonomus* and *Dermatophilus pedis* are the pathogens causing variety of skin lesions in mammals, including man. *Dermatophilus congolensis* causes very severe clinical disease and its infection is most common in tropical and subtropical regions.

Diseases caused by Dermatophilus species

Dermatophilus congolensis mainly affects Cattle, horses, sheep and goats, but many animal species and man can be infected.The disease has many names;

Cattle : Streptothricosis or Dermatophilosis

Horse : Skin Funk (Rain Rot, Rain Scald, Dew Poisoning), Grease heal

Sheep: Mycotic dermatitis (general infection), lumpy wool (wool- covered skin) and strawberry foot rot (skin of lower leg and coronet

Natural Habitat

Dermatophilus congolensis is the only species in the genus is thought to maintain itself in small foci of infection on a carrier animal or within scab particles in dust. It can survive in scab material for period's upto 3 years.

History

Bovine disease was first described by Van Saceghem in 1915 in the Congo now known as Zaire in Africa.

Morphology

*D. congolensis*is gram positive, non acid fast and aerobic to facultatively anaerobic.It is filamentous and branching. The reproductive unit of D. congolensis is the motile coccoid zoospores, about 2 um in diameter. Upon germinating, zoospores sprout a germ tube, about 1 um thick, which elongates and thickens, dividing both transversely and longitudinally (resulting in a 'tram-track'-like appearance) to form masses of coccoid or cuboidal cells, which characteristically become motile. If the flakes of scab (collected from infections) are treated too roughly, when the smears are made, the filaments will disintegrate and only Gram-positive cocci (zoospores) will be seen.

Developmental cycle of *D. congolensis*:

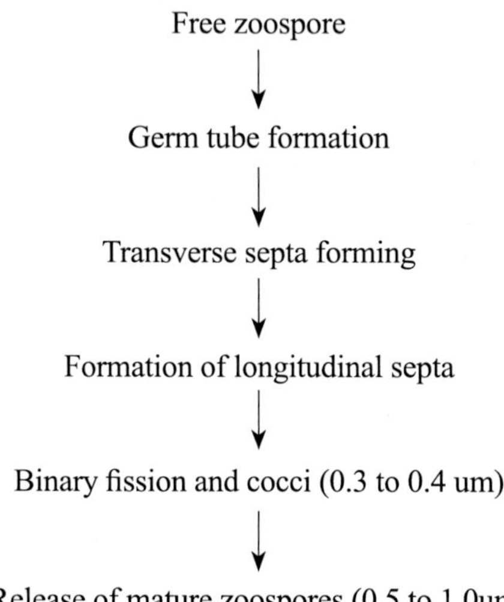

Transverse, horizontal and vertical septa form in the immature filaments dividing it into coccal zoospores. When mature, these zoospores are motile by polar flagella and are infective.

Cultural Characters

The bacterium is comparatively easy to culture and grows well on sheep or ox blood agar. An atmosphere of 5-10 per cent CO_2 enhances the growth of the organism especially on primary isolation. The inoculated plates are incubated at 37°C (Optimum temp. 37°C.) for up to 5 days, although colonies may be seen after 24-48 hours incubation.

Colonial Morphology

Small (about 1 mm) greyish-yellow,distinctly haemolytic colonies can be seen after 24-48 hours incubation. They are firmly adherent to the medium and are embedded in the agar. After 3-4 days, isolated colonies can be 3 mm in diameter and are rough, wrinkled and a golden-yellow colour. Older colonies can become mucoid. No growth occurs on Sabouraud dextrose agar.

Microscopic Appearance

Gram-stained smears from colonies do not show the characteristic 'tram-track' appearance seen on direct microscopy. Usually the smears reveal uniformly staining, Gram-positive, branching filaments but sometimes coccal forms predominate.

Biochemical Characters

D. congolensis is catalase-positive, urease-positive, gelatine - positive and produces acid from glucose, fructose and maltose. It is indole-negative, does not reduce nitrate and non-fermentative although acid is produced from certain carbohydrates.

Pathogenesis

The bacterium produces disease in many animal species. It is also a zoonosis. The common name for the disease is dermatophilosis or streptothricosis. As far as is known the bacterium is considered an obligate parasite, living only on animals.

Infection is spread by contact, biting insects, fomites and by other unknown means. Moist conditions and high relative humidity are known to promote the prevalence of the disease

D. congolensis causes skin infections most commonly seen in cattle, sheep, goats, horses and polar bears in zoological collections. The infection is characterised by the formation of thick crusts which come away easily with a tuft of hair, leaving a moist, depressed area with bleeding points from capillaries.

Infections can be localised but have a tendency to spread over large areas of the body and the morbidity and mortality can be high, especially in tropical regions.

The position of the lesions varies with the predisposing conditions. In periods of high rainfall the lesions tend to occur along the backs of animals. Where there is a heavy infestation with *Amblyomma* ticks the lesions are present in the predilection sites of the ticks: dewlap, axillae, udder and scrotum. In the dry season, in tropical regions, when feed is scarce the lesions are on the muzzle, head and lower limbs due to the animals foraging in thorn-covered scrub.

Though the disease does not lead to death in the adult cattle, deaths in young goats and cattle have been reported.

In sheep, the disease is called mycotic dermatitis and is seen in three forms:

1) Dermatitis of the wool-covered areas of the body or lumpy wool;
2) Dermatitis of the face and scrotum; and
3) Dermatitis of the lower leg and foot, which may result in severe ulcerative dermatitis referred to as "strawberry foot rot".

Laboratory Diagnosis

Based on Direct microscopy

Small pieces of material are shaved from the scab with a scalpel and the flakes of scab are softened in a few drops of distilled water on a microscope slide. A smear is made, taking care to leave a few flakes of scab material intact. The smear can be stained by either Giemsa or Gram stains. Giemsa is the better stain to show the characteristic morphology of the bacterium. Segmenting filaments and coccoid spores stain deep purple. The spores are seen in packets.

Based on Isolation and Identification

The organism grows well on blood agar plates within 24 to 48 hours as small grayish white colonies turning yellow to orange upon further incubation. Although the isolation of D. *congolensis* may not be necessary for a diagnosis of streptothricosis, Scab material contains many contaminants and Haalstra's method was developed to overcome this problem.

Haalstra's Method for the Primary Isolation of Dermatophilus congolensis;

1. Grind up a small amount of scab material and place a little in 2 ml distilled water in a container and incubated for 3.5 hours at room temperature.
2. Place the container, with lid removed, in a candle jar at room temperature for 15 minutes.
3. The motile zoospores are chemotactically attracted to the carbon dioxide-enhanced atmosphere in the candle jar and move to the surface of the distilled water. Remove a loopful of fluid from the surface and inoculate a blood agar plate. Incubate the inoculated plate at 37°C for 72 hours under 5-10 per cent CO_2.

Identification of this bacterium does not seem to pose any problems. Lesion and typical morphology are quite diagnostic.

Immunity

There is no known vaccine for immunization against the disease produced by Dermatophilus. Recovery from the disease seems to confer permanent immunity to reinfection.

26

Escherichia Coli

Introduction

Escherichia coli (commonly abbreviated as *E. coli*) is a Gram-negative, facultative anaerobic, rod-shaped bacterium that is commonly found in the lower intestine of warm-blooded organisms (endotherms).

Escherichia coli are common inhabitants of the terminal small intestine and large intestine of mammals. They are often the most abundant facultative anaerobes in this environment. They can occasionally be isolated in association with the intestinal tract of non-mammalian animals and insects. The presence of E.coli in the environment is usually considered to reflect faecal contamination and not the ability to replicate freely outside the intestine.

Morphology

E.coli is Gram-negative, facultative anaerobic and non-sporulating. Cells are typically rod-shaped, and are about 2.0 micrometers (μm) long and 0.25-1.0 μm in diameter, with a cell volume of 0.6–0.7 μm. It can live on a wide variety of substrates. Strains that possess flagella are motile. The flagella have a peritrichous arrangement. It is motile by peritrichous flagellae, though some strains are non-motile.Spores are not formed. Capsules and fimbriae are found in some strains.

Cultural Characteristics

Escherichia coli or *E.coli* cells may grow on a solid or in a liquid growth medium under a laboratory condition. Solid and liquid media may have exactly the same composition except that the solid medium contains an extra 1.5% agar. Different *E.coli* clones may have different properties. Colonies growing on solid media represent different clones. It is an aerobe and a facultative anaerobe. Optimal growth of *E. coli* occurs at 37°C (98.6°F) but some laboratory strains can multiply at temperatures of up to 49°C (120°F).

On Nutrient agar, colonies are large, thick, greyish white, moist, smooth, opaque or translucent discs. The smooth (s) form seen in fresh isolation is easily emulsified in saline, whereas the rough (R) form often auto agglutinates in saline.Some strains may form “mucoid”colonies. On MacConkey agar medium, colonies are bright pink due to lactose fermentation.

On selective media (Desoxycholate citrate agar-DCA; salmonella shigella-SS medium) used for the isolation of salmonella, their growth is inhibited; however their colonies are pink on DCA as it contains lactose and neutral red. In broth, there is generalized turbidity and deposit which disperses on shaking.

Temperature 37°C for 24 hrs	MacConkey Agar	Eosin-methylene blue Agar
Size in mm	1 mm	1 mm
Shape	Circular	Circular
Colour	Pink	Metallic sheen
Margin	Complete	Complete
Elevation	Slightly Raised	Convex
Opacity	Opaque	Translucent
Consistency	Soft	Soft

Biochemical Reactions

E. coli uses mixed-acid fermentation in anaerobic conditions, producing lactate, succinate, ethanol, acetate and carbon dioxide. Since many pathways in mixedacid fermentation produce hydrogen gas, these pathways require the levels of hydrogen to be low, as is the case when *E. coli* lives together with hydrogen consuming organisms, such as methanogens or sulphate-reducing bacteria.

Glucose, lactose, mannitol, maltose are fermented with acid and gas production, but sucrose is not fermented by typical strain of E. coli. In Triple sugar iron (TSI), acid and gas are produced. The four biochemical tests widely used for enterobacteriaceae classification are Indole (I), Methyl Red (MR), Voges Proskauer (VP) and Citrate (C) utilisation which are referred to by the mnemonic IMViC. Escherichia coli is Indole and MR positive VP and citrate negative (IMViC ++--), H_2S is not formed and urea is not hydrolysed.

Test	Reactions
Catalase	+
Oxidase	–
Urease	–
TSI	Acid butt, with gas, acid slant
MR	+
VP	–
Nitrate	+
Citrate	–
Indole (TW)	+
Gelatin	–

Key: + = reaction positive
– = reaction negative

Antigens and Toxins

Surface Antigens of *E.coli*

The capsular (K) antigens are polysaccharides and the cell wall or somatic (O) antigens arc determined by the sugar side-chains on the lipopolysaccharide molecules of the outer membrane. The flagellar (H) and fimbrial (F) antigens are proteins. Some of the well known fimbrial antigens, K88 (F4) and K99 (F5) are adhesins that allow palhogenic *E.coli* strains to adhere to intestinal cells and colonise the small intestine. The O, H and K antigens can be used to serotype strains of E.*coli;* each serotype is designated by the numbers of the antigens that it bears, for example O157:K85:H19.

E. coli has Three Antigens:

O - Somatic, Greek Ohne Hauch—without flagella;

H - Flagella; Greek Hauch—flagella and

K - Kapsular antigens.

K antigen is an envelope antigen, which encloses the O antigen, renders the strain inagglutinable by the O antiserum and contributes to virulence by inhibiting phagocytosis.

It may be of three types —L, A and B. Though L type is common, the B antigen is medically important as it is found on enteropathogenic E. coli.

F (Fimbrial) Antigen

The F antigen has no significance in antigenic classification of E. coli. Type I fimbriae mediates adhesion of bacterium to human and animal cells. Such adhesion enhances bacterial pathogenicity e.g. urinary tract infection in which type I fimbriae has some possible role to play.

Several fibrin structures resembling fimbriae have been demonstrated. They, most probably, play a very important role in pathogenesis of diarrhoeal diseases and urinary tract infection.

Toxin

Besides the endotoxin associated with O antigen, some strains produce two types of exotoxin—enterotoxin and haemolysin.

Enterotoxins responsible for diarrhoea are of two types—heat labile (LT); heat stable (ST). LT is similar to cholera enterotoxin antigenically and in its mechanism of action—by stimulating the adenyl cyclase - cyclic adenosine monophosphate (cAMP) system to produce fluid accumulation in the intestinal lumen. ST appears to stimulate fluid secretion into the gut through the mediation of cyclic guanosine monophosphate (cGMP) resulting into dehydration.

Haemolysin by Escherichia Coli

Three types of haemolysins produced by E.coli are not related to pathogenesis. E. coli forms a part of normal intestinal flora of man and animal and the commensal strains belong to several O groups. There are many strains of *E.coli* which include commensal strains as well as strains with virulence determinants that cause a wide variety of infections of all age groups of men and animals.

The virulent strains of E. coli are specific pathogens in the gut (enteritis) and of extra-intestinal sites (urinary tract infection, wound infection).

The worst type of *E.coli*, known as *E.coli O157:H7* causes bloody diarrhea and can sometimes cause kidney failure and even death. *E.coli* O157:H7 makes a toxin called Shiga toxin and is known as a Shiga toxin-producing *E.coli* (STEC).

Pathogenesis and Pathogenicity

The virulence factors of pathogenic strains of *E.coli* include capsules, endotoxin, and structures responsible forcolonization, enterotoxins and other secreted substances.Capsular polysaccharides, which are produced by some E. coli strains, interfere with the phagocytic uptake ofthese organisms. Capsular material, which is weaklyantigenic, also interferes with the antibacterial effectivenessof the complement system.

Endotoxin, a lipopolysaccharide (LPS) component of the cell wall of Gram-negative organisms, is released on death of the bacteria. It is composed of a lipid A moiety, core polysaccharide and specific side chains. The role of LPS in disease production includes pyrogenic activity, endothelial damage leading to disseminated intravascular coagulation, and endotoxic shock. These effects are of greatest significance in septicaemic disease.

Fimbrial adhesins which are present on many enterotoxigenic strains of *E.coli* allow attachment to mucosal surfaces in the small intestine and in the lower urinary tract.

Two types of enterotoxins, heat labile (LT) and heatstable (ST) have been identified. Each type of enterotoxin has two subgroups. Many strains of enterotoxigenic E.*coli* (ETEC) from pigs produce LT1 which induces hypersecretion of fluid into the intestine through stimulation of adenylate cyclise activity.

Verotoxins (VT) are similar structurally, functionally and antigenically to the Shiga toxin of *Shigelladysenteriae.* These toxins are heat-labile and lethal for cultured Vero cells. Verotoxigenic *E.coli* (VTEC) colonizing the intestines can damage enterocytes and, when verotoxin is absorbed into the bloodstream, it exerts a deleterious effect on endothelial cells in relatively defined anatomical locations such as the central nervous system in pigs. The verotoxin VT2e is implicated in oedema disease of pigs.

Clinical Infections

Clinical infections in young animals may be limited to the intestines (enteric colibacillosis, neonatal diarrhoea), or may manifest as septicaemia (colisepticaemia, systemic colibacillosis) or toxaemia (colibacillary toxaemia). In older pigs, post-weaning enteritis and oedema disease are manifestations of toxaemia. Non-enteric localized infections in adult animals, many due to opportunistic invasion, can involve the urinary tract, mammary glands and uterus.

In poultry, airsacculitis and pericarditis may develop following septicaemia. Coligranuloma (Hjarre's disease) is characterized by chronic inflammatory changes which are encountered at postmortem in laying hens and resemble tuberculous lesions.

Infection of the mammary glands of cows and sows by members of the *Enterobacteriaceae*,including *E.coli* occurs opportunistically. In dairy cows, the source of infection is faecal contamination of the skin of the mammary gland and relaxation of the teat sphincter following milking increases vulnerability to infection.

Cows with low somatic cell counts are particularly susceptible to infection. No specific serotypes of *E.coli* have been linked with this form of mastitis. The acute form of the disease is characterized by endotoxaemia and can be life-threatening. Peracute disease may be fatal in 24 to 48 hours. Affected animals are severely depressed with drooping ears and sunken eyes. Mammary secretions are watery and contain white flakes.

Diagnostic Procedures

The age and species of the affected animal, the clinical signs and the duration of illness may suggest the type of infection and the category of disease. The history, progress of the disease and the system or organ affected influence the selection of specimens, the laboratory procedures for diagnosis and appropriate treatment and control measures.

Suitable specimens include faecal samples from animals with cnteric disease, tissue specimens from cases of septicaemia, mastitic milk, samples of mid-stream urine and cervical swabs from suspected cases of pyometra or metritis.

Specimens cultured on blood and MacConkey agar are incubated aerobically at 37 – 42°C for 24 to 48 hours.

Identification criteria for isolates:

- On blood agar the colonics are greyish, round and shiny with a characterisitic smell. Colonies may be haemolytic or non-haemolytic.
- On MacConkey agar colonies are bright pink.
- IMViC tests can be used for confirmation
- The colonies of some E.*coii* strains *have a* metallic sheen on EMB agar.
- A full biochemical profile may be necessary to identify isolates from coliform mastitis or cystitis.
- Some serotypes are found in association with certain disease conditions. Slide agglutination tests for 'O' and 'H' antigens are employed for serotype identification.

Enterotoxins in the small intestine can be detected, using methods employing monoclonal antibodies. Some of thesc reagents are available commercially. For expression of fimbrial antigens, isolates should be subcultured on Minca medium. Fimbrial antigens can be identified using ELISA or latex agglutination. DNA probes specific for genes encoding heat-labile and heat-stablc enterotoxins may be used to identify cnterotoxigenic strains of *E.coli.*

Treatment

Milk feeding can be resumed gradually when clinical improvement is evident. Severely dehydrated calves require parenteral fluid replacement therapy. Calves with hypogammaglobulinaemia can be given bovine gammaglobulin intravenously. In most domestic species, enteric diseases may be treated by oral administration of antimicrobial compounds which are active in the gastrointestinal tract. Systemic and localized infections require parenteral administration of therapeutic agents. Treatment should be based on susceptibility testing of isolates.

Control

Newborn animals should receive ample amounts of colostrum shortly after birth. Colostral antibodies can prevent colonization of the intestine by pathogenic E.*coli.* Absorption of gammaglobulin from the intestine declines progressively after birth and is negligible by 36 hours.

Vaccination is of value for a limited number of the diseases caused by *E.coli.* Commercially available killed vaccines containing prevalent pathogenic *E.coli* serotypes can be given to pregnant sows. Alternatively, autogenous, killed vaccines prepared from strains of *E.coli* implicated in disease outbreaks on a farm can be used. Vaccination of pregnant cows with purified *E.coli* K99 fimbrial or whole-cell preparations, often combinedwith rotavirus antigen, can be used to enhancecolostral protection.

Diseases Caused by *Escherichia Coli*

Animals involved	Disease	Clinical signs and pathogenesis
Pigs		
Piglets less than 1 week old Pigs about 2 weeks after weaning Weaned pigs Sows after farrowing Gilts after farrowing	Neonatal diarrhoea (colibacillosis) Colisepticaemia Piglet meningitis Weanling enteritis (colibacillosis) Oedema disease Coliform mastitis Mastitis- metritis-agalactia (MMA) syndrome	Profuse watery diarrhoea and severe dehydration, mortality 90-100% Occasionally septicaemia with invasive strains and death of a piglet within 48 hours of birth Acute meningitis and fibrinous polyserositis in piglets has been reported Diarrhoea, anorexia and fever. Mortality lower than in neonatal pigs Enterotoxins involved (STb and LT) Often sudden death. Oedema of forehead, eyelids, stomach wall and larynx (hoarse squeal). Nervous signs such as ataxia, convulsions and paralysis may be seen. Verotoxin (variant of SLT-2 toxin) involved often associated with *E. coli* 0139 and 0141 One or more mammary glands affected Complex syndrome involving hysteria, hormonal imbalance and coliform infection (often *E. coli)*

Animals involved	Disease	Clinical signs and pathogenesis
Cattle		
Calves less than 1-week old Calves less than 1-week old Calves surviving a Colisepticaemia Dairy cows soon after parturition	'White scours' (coli bacillosis) Colisepticaemia Joint ill Coliform mastitis	The appetite is normal at first but decreases as the faeces become more fluid. White pasty faeces around rectum. Dehydration and emaciation occur. Death usually within 4- 5 days if untreated. Enterotoxins involved (Sta) and possibly aShiga-like toxin in some cases.
		Sudden death due to endotoxic shock or diarrhoea, depression, respiratory distress and death. Endotoxin mainly involved from invasive *E. coli* strains *E. coli* localised in joints and/or kidneys ('white spot'). Entry can be via the umbilicus Peracute disease: fever, anorexia, depression and sunken eyes. Death due to endotoxic shock. Most common in housed cows
Sheep		
Neonatal lambs Neonatal lambs Ewes	Colibacillosis and Colisepticaemia 'Watery mouth' Coliform mastitis	Syndromes similar to those that occur in calves but less common. Enterotoxigenic and enteropathogenic strains involved in colibacillosis and *E. coli* 078 is common in Colisepticaemia The lamb is dull and anorectic, saliva drools over the muzzle and there is abdominal tympany. There are splashing sounds within the abomasum and death within 6 - 24 hours. Associated with *E. coli* endotoxaemia Peracute and similar to the condition in cows. Most commonly seen in ewes housed during lambing
Dogs (CATS)		

Animals involved	Disease	Clinical signs and pathogenesis
Neonatal pups Bitches Adult dogs	Colisepticaemia Pyometra Urinary tract infection	Septicaemia, often fatal, associated with the progesterone-stimulated endometrium. A vaginal discharge may occur 4-8 weeks after oestrus. In a closed-cervix pyometra the bitch is toxaemic and ill. Endotoxin involved Cystitis (often in bitches) is most common but *E. coli* can ascend higher in the urinary tract. A bacteriuria occurs with greater than 10^5*E. coli* / mlurine
Poultry		
Young chicks All ages All ages	Omphalitis Colisepticaemia Coligranuloma	Infection of the vestigial yolk sac. The contents are dark, fluid and evil-smelling. Common name is 'mushy-yolk disease' Primary or secondary infection via the intestines or respiratory tract. Many body organs are affected with airsacculitis, peritonitis and ovarian infection Chronic condition, possibly following a colisepticaemia. There are nodular lesions in the liver and intestines
Other Animals		
Neonatal animals such as foals and rabbits	Colibacillosis and colisepticaemia	Diarrhoea and septicaemia. Less common than in calves and piglets

27

Klebsiella

Klebsiella pneumoniae (also known as *Friedlander's bacillus*) is a Gram-negative, non-motile, encapsulated, lactose fermenting, facultative anaerobic, rod shaped bacterium found in the normal flora of the mouth, skin, and intestines.

Klebsiella spp. are Gram-negative, nonmotile, usually encapsulated rod-shaped bacteria, belonging to the family Enterobacteriaceae. These bacteria produce lysine decarboxylase but not ornithine decarboxylase and are generally positive in the Voges-Proskauer test. Members of the Enterobacteriaceae family are generally facultative anaerobic, and range from 0.3 to 1.0 μm in width and 0.6 to 6.0 μm in length. *Klebsiella* spp. often occurs in mucoid colonies. The genus consists of 77 capsular antigens (K antigens), leading to different serogroups.

Morphology of Klebsiella pneumoniae (K. pneumoniae)

- Shape – *Klebsiella pneumoniae* is a short, plump, straight rod shape (bacillus) bacterium.
- Size – The size of *Klebsiella pneumoniae* is about 1–2 μm × 0.5–0.8 μm (micrometer).
- Arrangement Of Cells – *K. pneumoniae* is arranged singly, in pairs, or in short chains and sometimes in clusters.
- *Motility* – *Klebsiella pneumoniae* is a non-motile bacterium.
- *Flagella* – *K. pneumoniae* is a non-flagellated bacterium.
- *Spores* – The *Klebsiella pneumoniae* is a non–sporing bacterium.
- *Capsule* – Capsules are present in *Klebsiella pneumoniae* which can easily be demonstrated using India ink preparation, appear as a clear halo in a dark background.
- *Gram Staining Reaction* – *Klebsiella pneumoniae* is a Gram -ve (Negative) bacterium.

Culture requirements of *Klebsiella pneumoniae* (*K. pneumoniae*)

Special requirements – Klebsiella pneumoniae have no complex nutritional requirements and readily grow in an ordinary media like Nutrient Agar medium (NAM). Commonly the NAM & MacConkey Agar medium is used for the cultivation of *Klebsiella pneumoniae* in Laboratory.

There are various culture media used for the cultivation of *Klebsiella pneumoniae* in the laboratory and most commonly the Nutrient Agar Medium and MacConkey Agar Medium is used, the other media are as follows;

- Columbia Horse Blood Agar medium
- Sheep Blood Agar medium
- Trypticase Soy Agar medium
- Eosin Methylene Blue Agar (EMB Agar) Medium.
- The liquid medium (Nutrient Broth medium, TSB medium)
- The Eosin Methylene Blue Agar (EMB Agar) medium which is the Selective medium for *Klebsiella pneumonia.*

In *Liquid culture media* like Trypticase soy broth or Nutrient broth, the growth of the bacterium occurs as turbidity in the broth medium which is further analyzed for the morphology (under the microscope), gram reaction, biochemical tests, and *Klebsiella pneumoniae* specific tests.

In *Blood Agar medium*, the *Klebsiella pneumoniae* colonies are non-hemolytic i.e. shows Gamma Hemolysis (γ-hemolysis).

In *MacConkey Agar medium*, the colonies of *Klebsiella pneumoniae* are pink colored due to the lactose fermentation which is of great importance in differentiating *K. pneumoniae* from other Bacteria present in the specimen, especially from Gram-positive bacteria and Salmonella species which are non–lactose fermentors and gives colorless colonies on MacConkey agar medium.

In *Eosin Methylene Blue Agar (EMB) medium*, the colonies of *Klebsiella pneumoniae* are Pink to purple in color without green metallic sheen which is of great importance in differentiating *K. pneumoniae* from other bacteria normally found in Specimen and especially from E. coli which grows with Green metallic sheen.

Key biochemical reactions

Oxidase - negative

Catalase - positive

Indole - negative

DNase - negative

Voges-Proskauer - positive

Urease - positive

H_2S - negative

Lysine-decarboxylase - positive

Host Range

Humans, mammals (including horses, bovines, rhesus and squirrel monkeys, guinea pigs, muskrats, lemurs, and bats), aquatic animals (including elephant seals, California sea lions, and harbor seals), reptiles (including snakes, crocodiles, and American alligators), birds, insects, and plants (banana, rice sugar cane and maize).

Susceptibility to Disinfectants

Gram-negative bacteria are generally susceptible to a number of disinfectants, including phenolic compounds, hypochlorites (1% sodium hypochlorite), alcohols (70% ethanol), formaldehyde (18.5 g/L; 5% formalin in water), glutaraldehyde, and iodines (0.075 g/L).

Physical Inactivation

Reduction in the growth and metabolic activity of *K. pneumoniae* at temperatures >35 °C has been reported. Significant growth reduction has been demonstrated at 60 °C; however, the bacteria still show some metabolic activity (i.e. not completely inactivated). Bacteria are also sensitive to moist heat and dry heat.

Pathogenicity/Toxicity

Pathogenicity factors of *Klebsiella* spp. include adhesins, siderophores, capsular polysaccharides (CPLs), cell surface lipopolysaccharides (LPSs), and toxins, each of which plays a specific role in the pathogenesis of these species.

Depending on the type of infection and the mode of infectivity, cells of *Klebsiella* spp. may adhere and attack upper respiratory tract epithelial cells, cells in gastrointestinal tract, endothelial cells, or uroepithelial cells, followed by colonization of mucosal membranes. Common underlying conditions include chronic liver disease (cirrhosis), chronic renal failure, cancer, transplants, burns, and/or use of catheters.

Respiratory disease

- *K. pneumoniae* – a leading cause of community-acquired and nosocomial pneumonia and lung abscesses. Infection of the upper lobe is more common. Symptoms include: fevers, chills, and leukocytosis with red currant jelly-like sputum. Rare complications include lung infection involving necrosis and sloughing of the entire lobe.
- *K. ozaenae* – causes ozena, a primary atrophic rhinitis (AR) which involves chronic inflammation of the nose.
- *K. rhinoscleromatis* – causes rhinoscleroma (RS), a chronic granulomatous infection which predominantly affects the cavity of the nose.

Central nervous system (CNS) infections:

- *K. pneumoniae* and *K. oxytoca* – cause community-acquired meningitis and brain abscesses. Clinical symptoms include: fever, altered conciousness, seizures, and septic shock.
- *K. ozaenae* – associated with rare cases of cerebral abscess and meningitis.

Urinary tract infections (UTIs):

- *Klebsiella* spp. are a frequent cause of UTIs. Significant bacteriuria has been ascribed to *K. ozaenae.*

Hepatic disease

K. pneumoniae – an important causative pathogen for pyogenic liver abscesses with symptoms including fever, right-upper-quadrant pain, nausea, vomiting, diarrhea or abdominal pain, and leukocytosis. Abscesses occur predominantly in the right lobe and are solitary.

Other infections

K. granulomatis – causes donovanosis or granuloma, a chronic ulcerative disease that primarily affects the genitalia. Symptoms include development of small papule or ulcer at the site of inoculation that later develop into large red ulcers (lesions) that extend along the moist folds of the genitalia.

Peracute Bovine Mastitis

Klebsiella species are Gram-negative coliform bacteria that can cause mastitis, leading to significant economic losses on dairy farms. *K. oxytoca* and *K. pneumoniae* are the species that are responsible for causing mastitis.

Source and Transmission

Like other coliforms, *Klebsiella* are found in manure, and manure easily contaminates the environment of the dairy cow. Drinking water, feed, other cows, and bedding (especially wood by-products) are some sources of environmental *Klebsiella*. Infection through the teats can lead to spread of the pathogen during milking as milk from an infected cow contaminates the milking unit and transmits the infection to the next cow that is milked.

Pathogenesis

Commonly, these organisms are found in organic matter, including beddingand manure. *Klebsiella* spp. is naturally abundant on the forest floor; thus, their presence in sawdust is common. Therefore, cows bedded on fresh sawdust are at an increased risk for mastitis caused by *Klebsiella* spp.

Loads of sawdust containing high counts of *Klebsiella* spp. have been linked to herd outbreaks of *Klebsiella* mastitis. Poor udder cleanliness, inadequate stall management, and damaged teat ends are risk factors for *Klebsiella* infections in uninfected cows.

Symptoms

Of all *Klebsiella* cases, approximately a third are mild (abnormal milk), a third are moderate (abnormal milk and swollen udders), and a third are severe (systemic signs, including fever, off feed, decreased milk production, shock, or recumbency). *Klebsiella* invades deep into the secretory tissue of the udder, compromising the secretory capacity of the mammary gland. Consequently, some *Klebsiella* infections become chronic, and infected cows suffer a long-term reduction in milk production.

Diagnosis

Laboratory Identification of Klebsiella

K. pneumonia colonial morphology on blood agar is mucoid and 3 to 4mm in diameter. On MAC, K. pneumoniae colonies are pink (LF), mucoid (usually), and 3 to 4 mm in diameter. Colonies on Hektoen enteric agar and XLD are yellow.

(Large, mucoid, glistening pink colonies on a MacConkey agar plate are of typical colonies produced by many Klebsiella and Enterobacter spp.).

Laboratory Tests

Many commercial mini systems are available for identifying Klebsiella spp. and other members of the family Enterobacteriaceae. *K. pneumoniae* is positive for Glucose, Lactose, Sucrose, ONPG, Methyl red, Citrate, Urease, Malonate and Lysine decarboxylase. Serology Serological identification of Klebsiella species is based on their O (body or somatic) antigens and K (capsular) antigens reacting with pooled antisera. Klebsiella are differentiated into 72 serotypes based on the identification of the capsular (k) antigens. The capsular identification is performed by microscopical demonstration of capsule 'swelling' (Quellung reaction) in wet films with the type specific capsular antiserum.

Treatment

Once a cow is tested positive for *Klebsiella*, the somatic cell count (SCC) history of the cow should be reviewed to decide whether a treatment may be needed. One or more months of a SCC exceeding 2,00,000 cell/mL is an indication of a chronic infection and antimicrobial treatment would be warranted.

Prevention and Control

Identification of chronically infected cows is crucial in the control of transmission; an effective on-farm culturing program should be implemented for the early detection of infected cows. Cows with chronic mastitis should be segregated and milked last, and culled when possible. Proper milking practices, including pre- and post-milking teat disinfection, are important for good udder hygiene and minimizing spread of the infection during milking.

The use of a coliform mastitis vaccine (J5 bacterin) has been shown to reduce the severity of clinical Gram-negative mastitis, which includes mastitis caused by *Klebsiella* spp. It is important to remember, however, that these vaccines do not reduce the incidence of mastitis.

28

Salmonella

Salmonella consists of bacilli leading to Enteric fever, Gastroenteritis, Speticaemia etc. The important member of the genus is Salmonella typhi, which causes Typhoid fever. Salmonella are of two groups;

(i) Enteric fever group consisting of typhoid & Paratyphoid bacilli exclusively or primary human parasites

(ii) Food poisoning group, which are animal parasite but may infect humans causing gastrointestinal infections

Salmonella is oxidase negative, catalase positive, indole and Voges Proskauer (VP) negative, methyl red and Simmons citrate positive, H_2S producing and urea negative.

Characteristics

Cells are rod-shaped, non-spore-forming, and predominantly motile by means of peritrichous flagella with diameters of around 0.7-1.5μm and lengths of 2-5μm with a few exceptions. On blood agar, colonies are 2-3mm in diameter. Colonies are generally lactose non-fermenters. They obtain their energy from oxidation and reduction reactions using organic sources, and are facultative anaerobes. They produce acid from glucose usually with the production of gas, and are oxidase negative. Most produce hydrogen sulphide except Salmonella paratyphi A and Salmonella typhi, which is a weak producer. They are identified with a combination of serological and biochemical tests.

Salmonella species are classified and identified into serotypes according to the White-Kauffmann-Le Minor scheme; there are more than 2,500 Salmonella serotypes that have been described and reported.

Isolation of Salmonella

Some farm animals are infected with Salmonella without showing signs of the illness, i.e. they are subclinically infected. Faeces from these herds may contain Salmonella in low numbers. In food, Salmonella may also be present in low numbers in addition to a lot of other micro-organisms, and they may be injured. To diminish the risk of obtaining false negative results, a non-selective pre-enrichment of faeces or food sample, a combination of two selective enrichments and plating on two selective media are performed:

- Pre-enrichment in non-selective medium (buffered peptone water).
- Selective enrichment in Tetrathionate broth (Müller-Kauffmann) and Rappaport Vassiliadis soy peptone (RVS) broth.
- Subcultivation on Xylose Lysine Desoxycholate (XLD) agar and on Brilliant Green agar (BGA) (or another selective agar media).

Primary Isolation media

- Blood agar incubated in 5-10% CO2 at 35–37°C for 18-24hr.
- Xylose-lysine-desoxycholate agar (XLD) agar incubated in air at 35–37°C for 18-24hr.
- Desoxycholate citrate (DCA) agar incubated in air at 35–37°C for 18-24hr.
- Brilliant Green agar (BGA) incubated in air at 35–37°C for 18-24hr.

Cultural Characteristics

Salmonellae are aerobic and facultatively anaerobic bacteria growing readily on simple media over a range of pH 6-8 & temperature 15-41°C with optimum temperature of 37°C. Colonies are large, circular and smooth on MacConkey and Deoxycholate citrate media, colonies are colourless due to absence of lactose fermentation.

Colonial Appearance

- Blood agar - Colonies are moist and 2-3mm in diameter.
- CLED agar - Salmonella species are non-lactose fermenters (some serotypes eg Salmonella Arizonae and Salmonella Indiana may ferment lactose).
- XLD agar – Colonies are red, and usually with a black centre (some serotypes eg Salmonella paratyphi A and Salmonella typhi may not produce a black centre).
- DCA agar - Colonies are colourless, and usually with a black centre (some serotypes eg Salmonella paratyphi A and Salmonella typhi may not produce a black centre).

- BGA agar - Colonies appear as red-pink, 1-3mm in diameter, surrounded by brilliant red zones in the agar.

Biochemical Reaction

Salmonellae ferment glucose, mannitol and maltose forming acid and gas. Whereas S. typhi is an aerogenic i.e. it does not form fermentation of sugars like glucose etc. Lactose, Sucrose and Salicin are not fermented. Indole is not produced. They are MR positive, VP negative and citrate positive.

Resistance

The bacilli are killed at 55°C in one hour or at 60°C in 15 minutes. Boiling or chlorination of water and pasteurization of milk destroy the bacilli. In polluted water it may survive for weeks and in ice for months.

Antigenic Structure

Salmonellae possess the antigens and based on which they are classified as;

(i) Flagella antigen H,

(ii) Somatic antigen O and

(iii) Surface antigen Vi

H antigen

This antigen present on flagella is heat labile protein. It is destroyed by boiling or by treatment with alcohols but not by formaldehyde.

O antigen

O antigen is a Phospholipid-protein-polysaccharide complex which forms an integral part of the cell wall. It is identical with endotoxin. This is unaffected by boiling, alcohol or weak acids

Classification and Nomenclature

Classification within the genus is on antigenic characterisation based on Kauffman-White scheme and this depends on identification by agglutination of the O and H antigens of the strains. Salmonellae are classified into serological groups based on the presence of distinctive O antigen factors and designated as 1, 2, 3 etc.

Biochemically Kauffman proposed Salmonellae classification as;

- Subgenus I: Largest and medically most important group causing human and animal infections
- Subgenus II: Species isolated from reptiles.
- Subgenus III: Species isolated from reptiles and human beings
- Subgenus IV: These are rarely encountered.

Pathogenesis and Pathogenicity

Although many aspects of the pathogenesis of salmonellosis are poorly understood, particularly the relationship between salmonella toxins and cell damage, some of the general features associated with virulence are known. The virulence of salmonellae relates to their ability to invade host cells, replicate in them and resist both digestion by phagocytes and destruction by the complement components of the plasma.

Following adherence, probably through fimbrial attachment, to the surface of intestinal mucosal cells, the bacteria induce ruffling of cell membranes. The ruffles facilitate uptake of the bacteria in membrane-bound vesicles, which often coalesce. The organisms replicate in these vesicles and are eventually released from the cells, which sustain only mild or transient damage. The complex invasion process is mediated by the products of a number of chromosomal genes, whereas growth within host cells depends on the presence of virulence plasmids.

Clinical Infections

Salmonellosis is of common occurrence in domestic animals and the consequences of infection range from subclinical carrier status to acute fatal septicaemia. Some Salmonella serotypes such as Salmonella Pullorum and Salmonella gallinarum in poultry, Salmonella choleraesuis in pigs and Salmonella Dublin in cattle are relatively host-specific. In contrast, Salmonella typhimurium has a comparatively wide host range. It is recognised that healthy adult carnivores are innately resistant to salmonellosis.

Salmonellae often localize in the mucosae of the ileum, caecum and colon, and in the mesenteric lymph nodes of infected animals. Although most organisms are cleared from the tissues by host defense mechanisms, subclinical infection may persist with shedding of small numbers of salmonellae in the faeces. Salmonella Dublin causes a variety of clinical effects in cattle. Terminal dry gangrene and bone lesions are common manifestations in chronic infections with Salmonella dublin in calves.

Enteric Salmonellosis

Enterocolitis caused by salmonella organisms can affect most species of farm animals, irrespective of age. Acute disease is characterized by fever, depression, anorexia and profuse foul-smelling diarrhoea often containing blood, mucus and epithelial casts. Dehydration and weight loss follow and pregnant animals may abort.

Septicaemic Salmonellosis

In pigs with septicaemic Salmonella choleraesuis infection, there is a characteristic bluish discolouration of the ears and snout. Intercurrent viral infections often predispose to severe clinical forms of the disease. The close clinical and pathological relationships which have been recognized in animals infected with Salmonella choleraesuis ('hog-cholera bacillus') and classical swine fever virus, either jointly or separately, exemplify both the importance of intercurrent infections and the difficulty of clinically distinguishing the diseases caused by these agents.

Salmonellosis in Poultry

Salmonella pullorum, Salmonella gallinarum and Salmonella enteritidis can infect the ovaries of hens and be transmitted through eggs. The presence of Salmonella enteritidis in undercooked egg dishes may result in human food poisoning.

Pullorum disease or bacillary white diarrhoea (Salmonella pullorum) infects young chicks and turkey poults up to 2 to 3 weeks of age. The mortality rate is high and affected birds huddle under a heat source and are anorexic, depressed and have whitish faecal pasting around their vents. Characteristic lesions include whitish nodes throughout the lungs and focal necrosis of liver and spleen.

Fowl typhoid (Salmonella gallinarum) can produce lesions in young chicks and poults similar to those of pullorum disease. However, in countries where fowl typhoid is endemic, a septicaemic disease of adult birds occurs, often resulting in sudden deaths. Characteristic findings include an enlarged, friable, bile-stained liver and enlarged spleen. As Salmonella pullorum and Salmonella gallinarum possess similar somatic antigens. Both have been eradicated from many countries by a serological testing and slaughter policy for pullorum disease. Paratyphoid is a name given to infections of poultry by non-host-adapted salmonellae such as Salmonella enteritidis and Salmonella typhimurium. These infections are often subclinical in laying birds.

Diagnostic Procedures

Identification criteria for isolates

Specimens should be cultured directly onto BG and XLD agars and also added to selenite F, Rappaport or tetrathionate broth for enrichment and subsequent subculture. The plates and enrichment broth are incubated aerobically at 37°C for up to 48 hours. Subcultures are made from the enrichment broth at 24 and 48 hours.

On brilliant green agar, colonies and medium are red indicating alkalinity. On XLD agar, colonies are red (alkaline) with a black centre, indicating H_2S production.

Suspicious colonies, subcultured from the selective media into TSI agar and lysinc decarboxylase broth, should be examined after incubation for 18 hours at 37°C to establish their biochemical identity as salmonellae.

If reactions in TSI agar and lysine decarboxylase broth are inconclusive, a biochemical profile using a battery of biochemical tests may allow definitive identification.

The isolates from the TSI agar slant are confirmed as salmonellae using commercially available antisera for O and H antigens in a slide agglutination test. Serotypes with O antigens in common are assigned to a serogroup.

Serotypes which have flagellar (H) antigens in two phases, phase 1 (specific) and phase 2 (non-specific),are termed diphasic. Biotyping is required for serotypes which are antigenically indistinguishable such as Salmonella pullorum and Salmonella gallinarum.

Phage typing is used in epidemiological studies to identify isolates with specific characteristics such as multiple resistances to antibiotics and enhanced virulence. Examples of important phage types are Salmonella typhimurium DT (definitive type) 104 which exhibits multiple resistance to antibiotics and Salmonella enteritidis PT (phage type) 4 which is found in poultry products and is a common cause of food poisoning in humans.

Serological tests such as ELISA and agglutination techniques are of greatest value when used on a herd or flock basis. A rising antibody titre using paired serum samples is indicative of active infection. DNA probes can be used to screen large numbers of faecal samples for salmonellae.

Treatment

Antibiotic therapy should be based on results of susceptibility testing because R-plasmids coding for multiple resistance are comparatively common in salmonellae. Oral antimicrobial therapy should be used judiciously for treating enteric salmonellosis because it may disturb the normal intestinal flora, extend the duration of salmonella excretion and increase the probability of drug resistance developing. In the septicaemic form of the disease, intravenous antibiotic therapy must be used. Fluid and electrolyte replacement therapy is required to counteract dehydration and shock.

Control

Control is based on reducing the risk of exposure to infection. Intensively reared, food-producing animals are more likely to acquire infection and are also a major source of human infection.

Measures for excluding infection from a herd or flock free of salmonellosis:

- A closed-herd policy should be implemented when feasible.
- Animals should be purchased from reliable sources and remain isolated until negative for salmonellae on three consecutive samplings.
- Stem should be taken to prevent contamination of foodstuffs and water. In this context, rodent control is important.
- Protective clothing and footwear should be worn by personnel entering hatcheries and minimal disease pig units.
- Measures for reducing environmental contamination:
- Effective routine cleaning and disinfection of buildings and equipment is essential.
- Overstocking and overcrowding should be avoided.
- Slurry should be spread on arable land where possible. An interval of at least two months should elapse before grazing commences on pastures following the application of slurry.
- The continuous use of paddocks for susceptible animals should be avoided.

Diseases Caused by Selected Salmonella Serotypes

Host	Salmonella serotypes	Disease
Humans	*S. typhi* *S. paratyphi A* *S. schottmuelleri* *S. enteritidis,* *S. typhimurium* and Others	Typhoid fever Paratyphoid fever Paratyphoid fever Food poisoning
Cattle	*S.dublin* *S. typhimurium,* *S. bovismorbificans* and others	Subclinical excreters, latent carriers, enteritis, septicaemia, meningitis in calves, abortion (with or without other apparent clinical signs), osteomyelitis, joint ill, terminal dry gangrene in calves Enteritis or septicaemia
Pigs	*S. choleraesuis and* *S. choleraesuis* biotype Kunzendorf *S. typhisuis* *S. typhimurium* and Others	Severe outbreaks clinically similar to swine lever (hog cholera), and swine fever can be followed by a secondary infection with S.choleraesuis, earning it the name of 'hog cholera bacillus'. Rectal stricture is sometimes a sequel of the disease Chronic enteritis in young pigs. Far less virulent than S. Choleraesuis Enteritis or septicaemia
Sheep	*S. abortusovis* *S. montevideo* *S.dublin* *S. typhimurium,* *S. anatum* and others	Abortion in ewes. S. abortusovis is present in Britain, Europe and the Middle East Enteritis or septicaemia
Horses	*S. abortusequi* *S. typhimurium* and others	Abortion in mares. The serovar is present in Europe, South Africa and South America but is now rare in the USA Enteritis or septicaemia, especially in foa ls and stressed adults

Poultry and other birds	*S. pullorum*	Pullorum disease (bacillary white diarrhoea) in chicks. Transovarian transmission
	S. gallinarum	Fowl typhoid in all ages, mainly adults. Egg transmitted
	S. arizonae	Severe infections (enteritis and septicaemia) in chicks and turkey poults. Egg transmitted. Serovar associated with reptiles. Occasional infections in other animals
	S. enteritidis, *S. typhimurium* and many other serotypes	Collectively known as 'fowl paratyphoid'. Inapparent infections, enteritis and septicaemia. S. enteritidis may be egg transmitted. S. typhimurium can cause sudden deaths (septicaemia) in pigeon squabs, or if they survive, swollen wing joints

29

Yersinia

Yersinia species are non-lactose fermenters and, with the exception of *Y. pestis* are motile. Although there are more than 10 Yersinia species, only *Y. pestis, Y. enterocolitica* and *Y. pseudotuberculosis* are pathogenic for animals and man. Yersinia ruckeri causes perioral haemorrhagic inflammation in some species of fish. Growth of yersiniae tends to be less rapid than other members of the Enferobacteriaceae. They characteristically demonstrate bipolar staining in Giemsa-stained smears from animal tissues

Natural Habitat

Y. pestis, the cause of bubonic plague in man and a sylvatic cycle in animals, is transmitted mainly by fleas from tolerant rodents. Human infections through cuts, bites, scratches and aerosols can also occur. Cats are susceptible to Y.pestis and naturally infected cats can pose a health hazard for humans in endemic areas. *Y. pseudotuberculosis* persists in wild rodents and birds as well as in the environment. The intestinal tract of wild and domestic animals appears to be the reservoir for *Y. enterocolitica*. Pigs, particularly, are carriers of *Y. enterocolitica* strains pathogenic for humans.

Isolation

Yersinia species grow on nutrient, blood and MacConkey agars but the colonies, after 24 hours incubation, tend to be smaller than those of the other members of the Enterobacteriaceae. *Y. pestis* grows poorly on agars containing desoxycholate whereas *Y. enterocolitica* and *Y. pseudotuberculosis* grow well on these media. Yersinia selective medium (CIN agar) containing the antibiotic supplement cefsulodin (15mg/litre), irgasin (4 mg/litre) and novobiocin (2.5 mg/litre) is designed for the isolation of Y. enrerocolitica from faeces.

A cold-enrichment procedure may be necessary for the isolation of both *Y. enterocolitica* and *Y. pseudotuberculosis* from faecal specimens. A faecal specimen (approximately 5 per cent by volume) is placed in1/15M phosphate buffered saline (Oxoid) and held in the refrigerator (4 °C) for 3 weeks. Subcultures, at weekly intervals, can be made on MacConkey and Yersinia selective medium. Yersinia species usually grow faster at 37°C but prefer lower incubation temperatures, particularly on primary isolation. Sometimes additional culture plates, incubated at 22-25°C, can be useful for initial isolation.

Colonial Morphology

The Yersinia species are lactose-negative although lactose-positive strains of *Y. enterocolitica* occur and this property is thought to be plasmid-mediated. *Y. enterocolitica* will grow well on media, such as brilliant green and XLD agar intended for salmonella isolation but Y.pseudotuberculosis is less tolerant. Y.enerocolitica and *Y. pseudotuberculosis* are non-haemolytic on blood agar. The colonies of *Y. enterocolitica* on Yersinia selective medium (Oxoid) have dark red centres with a transparent periphery.

Biochemical Tests

Y. pestis, Y.pseudotuberculosis and *Y. enterocolitica* are nonmotile at 37°C but *Y. pseudotuberculosis* and *Y.enterocolitica* strains can be motile at 28°C. *Y. pseudotuberculosis* is almost always urease-positive as are most strains of *Y. enterocolitica*, but *Y. pestis* does not produce this enzyme. They are negative for the pyrazinamidase test, salicin fermentation and aesculin hydrolysis, whereas the non-pathogenic strains are positive to these tests. The pathogenic strains grow as small red colonies on congored-magnesium oxalate (CR-MOX) agar but non-pathogenic strains are unable to grow on this medium.

Antigens of Yersinia Species

Y. pseudotuberculosis can be divided into 6 serogroups based on the thermostable O antigens (I-VI) and five H (flagella) antigens (a-e). There are shared antigens between the closely related *Y. pestis* and *Y. pseudotuberculosis*. An antigenic relationship exists between *Y.pseudotuberculosis* and salmonella O antigens in serogroups B, D and E and also with some *E.coli O antigens*. *Y. enterocolitica* shares an antigen with Brucella species that can give rise to false brucella agglutination test results. *Y.enterocolitica* is divided into more than 37 serotypes, many of which do not appear to be pathogenic.

Pathogenesis and Pathogenicity

Pathogenic yersiniae are facultative intracellular organisms which possess plasmid and chromosomal encoded virulence factors, many of which are required for survival and multiplication in macrophages. Yersinia pseudotuberculosis and Y.enterocolitica are less virulent than Y.pestis and rarely produce generalized infections. The pathogenetic mechanisms in enteric disease caused by *Y.enterocolitica* and *Y. pseudotuberculosis* are incompletely understood. It is probable that both organisms gain entry to the mucosa through M cells of Peyer's patches. Adhesion to and subsequent invasion through these cells are facilitated by factors such as invasion and adhesion / invasion proteins which have an affinity for integrins on cell surfaces. Once in the mucosa, the bacteria are engulfed by macrophages in which they survive and are transported to the mesenteric lymph nodes. Replication in the nodes follows with the development of necrotic lesions and neutrophil infiltration. Survival of *Y.pseudotuberculosis* and *Y.enterocolitica* is enhanced by antiphagocytic proteins secreted by the organisms which interfere with the normal functioning of neutrophils in the host.

Yersinia pestis is more invasive than *Y.pseudotuberculosis* and *Y. enterocolitica* and possesses additional virulence factors. These include an antiphagocytic protein capsule (Fraction 1) and a plasminogen activator which aids systemic spread. Endotoxin, with properties similar to the endotoxin produced by other members of the Enterobacteriaceae, also contributes to the pathogenesis of disease.

Clinical Infections

Yersinia pseudotuberculosis causes enteric infections, in a wide variety of wild and domestic animals which are often subclinical. The septicaemic form of disease, known as pseudotuberculosis, can occur in laboratory rodents and aviary birds. Sporadic abortions caused by *Y.pseudotuberculosis* have been reported in cattle, sheep and goats.

Wild and domestic animals may act as reservoirs of *Yersinia enterocolitica* which is primarily a human enteric pathogen. The pig is the natural reservoir for *Y. enterocolitica* serotype 03 biotype 4, which is an important pathogen in humans. Rare cases of enteric disease, precipitated by stress, may be encountered in pigs, farmed deer, goats and lambs. Yersinia enterocolitica has been implicated in sporadic ovine abortion. *Yersinia pestis*, the cause of human bubonic plague ('black death'), can infect both dogs and cats in endemic areas. Cats, which are particularly susceptible, may be a source of infection for owners and attending veterinarians.

Enteric Yersiniosis

Enteritis caused by *Y.pseudoruberculosis* is relatively common in young farmed deer. Enteric disease has been reported in sheep, goats and cattle less than one year of age. Subclinical infection in many species is common and clinical disease may be precipitated in the winter months by stress factors such as poor nutrition, weaning, transportation and cold wet conditions. There may be prolonged survival of *Y. pseudotuberculosis* on pasture in cold wet weather, facilitating faecal-oral transmission.

Enteritis in young deer and lambs is characterized by profuse watery diarrhoea, sometimes blood-stained, which may be rapidly fatal if untreated. The luminal contents of the small and large intestine are watery and mucosal hyperaemia is evident at postmortem examination. Severely affected animals may show mucosal ulceration. The mesenteric lymph nodes are often enlarged and oedematous and scattered pale necrotic foci may be present in the liver. A clinically similar but less severe enterocolitis caused by Yersinia enterocolitica has described in young ruminants.

Diagnosis

The species and age group affected, especially during cold wet spells of weather, may suggest yersiniosis. Histological examination of intestinal lesions may reveal clusters of organisms in micro-abscesses within the mucosa. Confirmation requires isolation and identification of *Y.pseudotuberculosis* or, occasionally, *Y. enterocolitica*:

- Samples from tissues can be plated directly onto blood and MacConkey agars and incubated aerobically at 37°C for up to 72 hours.
- Faecal samples may be plated directly onto special selective media.
- A cold enrichment procedure may facilitate recovery of yersiniae from faeces especially if they are present in low mumbers. A 5% suspension of faeces in phosphate buffered saline, held at 4°C for three weeks, is subcultured weekly onto MacConkey agar.
- Serotyping may be necessary to establish whether the isolates belong to known pathogenic serotypes.

Treatment and control

Fluid replacement therapy together with broad spectrum antimicrobial treatment should be initiated promptly in young animals.

A formalin-killed Y.pseudotuberculosis vaccine composed of serotypes I, II and III, administered in two doses three weeks apart has been shown to decrease the occurrence of clinical disease in young deer,

Stressful conditions should, where practicable, be minimized.

Diseases Caused by Yersinia Species

Yersinia species	Host	Disease(s)
Y. pestis	Humans	Bubonic plague ('black death')
	Rodents	Sylvatic plague. Infection in rodents usually latent with occasional outbreaks of disease
	Cats	Cats in endemic areas may show mandibular lymphadenitis, fever, depression, anorexia, sneezing and occasionally nervous disturbances. Most infections are fatal
Y. pseudotuberculosis	Guinea-pigs, other rodents, rabbits, wild and captive birds	Pseudotubercu losis 1. Septicaemic syndrome. 2 Classical syndrome with nodules in internal organs Seen most commonly in guinea-pigs and canaries
	Farm animals	Latent infections, Occasional disease such as in captive deer
	Sheep	Orchitis and epididymitis reported
	Humans	Mesenteric lymphadenitis, acute terminal ileitis and rare cases of septicaemia. Mainly children and young adults
Y. enterocolitica	Farm animals	Latent infections with sporadic cases of enteritis or generalised infections. Captive deer particularly susceptible
	Humans	Food poisoning (enteritis), mesenteric lymphadenitis (pseudo-appendicitis) Most common in children

30

Proteus and Providencia

The genera *Proteus* and *Providencia* belong to the tribe *Proteae* of the family *Enterobacteriaeceae*. The members of both these genera are gram negative, motile bacilli, aerobes and facultative anaerobes and can grow on basic media. A characteristic feature which distinguishes tribe *Proteae* from other members of *Enterobactereaceae* is the presence of the enzyme phenylalanine-deaminase which converts phenylalanine to phenylpyruvic acid (PPA reaction).They also produces a powerful urease enzyme which rapidly hydrolyses urea to ammonia.

Genus Proteus

They are gram-negative bacilli, 1-3 μm long and 0.6 μm wide. They are noncapsulated and are actively motile by peritrichous flagella. The name 'Proteus' refers to their pleomorphism, after the Greek God Proteus who could assume any shape. Four species: *Proteus mirabilis, P.vulgaris, P.penneri* and *P.myxofaciens* are recognized. Proteus mirabilis, P.vulgaris are widely recognised as human pathogens.

Culture Characteristics

These can grow on ordinary media like nutrient agar with a characteristic fishy or seminal odour. On MacConkey and Teepol lactose agar, lactose non-fermenting pale colonies, around 2-3 mm in size are formed. On non-inhibitory solid media such as blood and nutrient agar *Proteus mirabilis* and *P vulgaris* show characteristic swarming growth in the form of a uniform film, which spreads over the whole surface of the plate. In young swarming cultures, many of the bacteria are long, curved and filamentous, sometimes reaching upto 80 μm in length. When two different strains of swarming proteus mirabilis encounter one another on an agar plate, swarming ceases and a visibile line of demarcation forms. This is known as the Dienes phenomenon.

In liquid medium (peptone water, nutrient broth), *Proteus* produces uniform turbidity with a slight powdery deposit and an ammoniacal odour.

Swarming inhibitory methods: Swarming of *Proteus* can be prevented by;

- Increasing the concentration of agar from 1-2% to 6%.
- Incorporation of sodium azide, boric acid, or chloral hydrate.
- Introducing growth inhibitors like sulphonamides.
- On Teepol Lactose agar by Teepol (surface active agent)
- On MacConkey agar or DCA by presence of bile salts.
- On CLED agar by the absence of electrolytes.

Biochemical Reactions

Like all other members of the family Enterobacteriaceae, all the species of genus *Proteus* are catalase positive, oxidase negative, reduce nitrates to nitrites and show fermentative reaction on Hugh Leifson's of media. All members of the tribe *Proteeae* are PPA positive and, hydrolyse urea to ammonia which differentiates them from other *Enterobacteriaeceae*.

Antigenic Structure

The bacilli possess thermostable 'O' antigen (somatic) and thermolabile 'H' (flagellar) antigen. Weil and Felix observed that certain non-motile strains of *P.vulgaris*, called the 'X' strains, were agglutinated by sera from patients with typhus fever. This heterophile agglutination due to the sharing of carbohydrate antigen by certain strains of *Proteus* and *Rickettsia* forms the basis of the Weil-Felix reaction and is used for diagnosing rickettsial infections. Three non-motile *Proteus* strains OX-2 and OX-19 of *P vulgaris* and OX-K of *P.mirabilis* are used in the agglutination test.

Typing Methods

- Serotyping
- Phage typing
- Dienes typing

Bacteriocin (proticin) typing

Dienes phenomenon:This method forms the basis of typing swarming strains of *Proteus* for local epidemiological studies. Different cultures are inoculated as discrete spots on the same plate and allowed to swarm towards one another. A line of complete or partially inhibited growth is formed where cultures of different strains meet; no line is formed between cultures of the same strain.

Pathogenecity

Proteus species are saprophytic and widely distributed in nature. They also occur as commensals in the intestine. They are opportunistic pathogens and may cause many types of infections such as:

Urinary tract infections (UTI) with predilection for upper UTI:

It produces urease which liberates ammonia from urea. The alkaline conditions lead to the precipitation of phosphates and the formation of calculi in the urinary tract.

Proteus species may also be associated with the following conditions;

- Pyogenic lesions
- Wound infections
- Otitis media
- Meningitis
- Septicaemia
- Osteomyelitis

Laboratory Diagnosis

(i) Specimen:
  - UTI- midstream urine
  - Wound/ abscesses, osteomyelitis, otitis media: pus
  - Meningitis: CSF
  - Septicaemia – blood culture

(ii) Culture: Clinical specimens should be cultured on MacConkey agar/ Teepol lactose agar, 6% blood agar and in case of urine on CLED (Cysteine lactose electrolyte deficient agar). Culture media are incubated at 37ºC for 18-24 hours. Pale coloured Non-Lactose Fermenting (NLF) colonies are seen on MacConkey agar. Identification is done by standard biochemical reactions.

(iii) Antibiotic susceptibility: *Proteus* are resistant to many of the common antibiotics, except *P mirabilis* which is sensitive to ampicillin and cephalosporins but nitrofurantoin is not effective.

Genus Providencia

Like *Proteus*, strains of *Providencia* are Non-lactase fermenting (NLF), methyl red and PPA positive bacilli which are motile by peritrichous flagella. However, they do not swarm on solid media. They can often be recognized by their 'fruity' smell. Three important pathogenic species include *Prov.alcalifaciens, Prov.rettgeri* and *Prov.stuartii*. It has been suggested that *Prov alcalifaciens* causes diarrhea, *Prov.rettgeri* and *Prov. stuartii* have been associated with urinary-tract, wound and other infections. *Providencia* are very resistant to antibiotics, particularly *Prov.stuartii* which is also resistant to disinfectants, making it a major pathogen in burn units.

31

Pseudomonas

History

Pseudomonas is comprised of two Greek words-Pseudo meaning false and Monas meaning unit, so literally the term means “false unit”. However Pseudomonas is a real bacteria so we really do not know the basis of the nomenclature by Walter Migula. A scientist by the name of Migula gave the Genus name of Pseudomonas in 1894 to these bacteria.

Classification

The classification of Pseudomonas is given below:

- Class : Gamma Proteobacteria
- Order : Pseudomonadales
- Family : Pseudomonadaceae
- Genus : Pseudomonas
- Eight groups : *P. aeruginosa* ; *P. chlororaphis; P. fluorescens; P.pertucinogena; P.putida; P. stutzeri; P. syringae; P. incertae sedis.*

The family has 191 valid species. These include bacteria which are saprophytic, free living, and human, animal and plant pathogens. The type species is *Pseudomonas aeruginosa.*

Usual Habitat

Pseudomonas species are environmental organisms which occur worldwide in water and soil, and on plants. Pseudomonas *aeruginasa* is also found on the skin, on mucous membranes and in faeces. *Burkholderia pseudomallei,* which is found in soils, occasionally infectsanimals and man. Wild rodents can act as reservoirs of this organism. It is widely distributed in some tropical and subtropical regions of Southeast Asia and Australia.Although *B.mallei* can survive in the environment for upto 6 weeks its reservoir is infected *Equidae.*

Clinical Infections

Burkhlderia mallei amajor pathogen of *Equidue* causes both acute and chronic disease. It manifests mainly as lesions in the skin and the respiratory tract, Infection with *B. pseudomallei* can cause chronic suppurative lcsions in the lungs and other organs of a wide range of species. In contrast, *P. aeruginosa* is an opportunistic pathogen which may occasionally cause acute systemic disease.

Morphology

Pseudomonas is rod shaped, slender (0.5 to 0.8 µm by 1.5 to 3.0 µm) Gram negative organism, occurs as single bacteria, in pairs, and occasionally in short chains. motile by polar flagella, sometimes more than two flagella may be present. Some strains of Pseudomonas particularly those isolated from cases of cystic fibrosis are very mucoid and have kind of pseudo capsule (glycocalyx) made of polysaccharides. The glycocalyx protects pseudomonas from host defense.

Pseudonlonas aeruginosa, Burkholderia mallei and *B.pseudomallei* are Gram-negative rods (0.5 to 1.0 x 1 to*5 µm*) which are obligate aerobes and oxidize carbohydrates. Most isolates are oxidase-positive and catalase positive. They are motile by one or more polar flagella, with the exception of *B.mallei* which is non-motile. The majority of these organisms have no special growth requirements and grow well on MacConkey agar. *Burkholderia mallei* require 1% glycerol in media for optimal growth. *Pseudomonas aeruginosa,* characterized *by* the production of diffusible pigments, causes a variety of opportunistic infections in a wide range of animals.

A number of other *Pseudomonas* species may *be* isolated from clinical specimens. *Pseudomonass fluorescens* and *P.putida* occasionally infect freshwater fish. *Burkholderia* species, previously classified in the genus *Pseudomnonas,* include B.*mallei,* the cause of glanders and *B.psoudomallei,* the cause of melioidosis. Both diseases are zoonoses.

Culture

The *Pseudomonas* species are non-fastidious and will grow on trypticase soy agar, blood agar and on less complex media. The growth of *P. mallei* is enhanced by 1 per cent glycerol. A selective medium for *P.mallei* can be made by adding 1000 units polymyxin E, 1250 units bacitracin and 0.25 mg actidione to 100 ml of trypticase soy agar. Commercial selective media are available for *P. aeruginosa* and usually contain 0.03 per cent cetrimide (cetyl trimethyl ammonium bromide).

P.aeruginosa will also grow on many of the selective media intended for the *Enterobacteriaceae* such as MacConkey, brilliant green and XLD agars.

The cultures for *P.aeruginosa, P.pseudomallei* and *P.mallei* are incubated aerobically at 37°C for 24-48 hours. Some of the saprophytic pseudomonads, such as *P.fluorescens,* grow extremely poorly, or not at all, at 37°C as 30°C is often the upper temperature limit of their growth range.

P. aeruginosa is an obligate aerobe that grows readily on many types of culture media, sometimes producing a sweet or grapelike or corn taco–like odor. Some strains hemolyze blood. *P aeruginosa* forms smooth round colonies with a fluorescent greenish color. It oft en produces the nonfl uorescent bluish pigment pyocyanin, which diff uses into the agar. Other *Pseudomonas* species do not produce pyocyanin.

Many strains of *P.aeruginosa* also produce the fluorescent pigment pyoverdin, which gives a greenish color to the agar. Some strains produce the dark red pigment pyorubin or the black pigment pyomelanin.

P.aeruginosa in a culture can produce multiple colony types. *P aeruginosa* from different colony types may also have different biochemical and enzymatic activities and different antimicrobial susceptibility patterns. Sometimes it is not clear if the colony types represent different strains of *P. aeruginosa* or are variants of the same strain.

P.aeruginosa grows well at 37–42°C; its growth at 42°C helps differentiate it from other *Pseudomonas* species in the fluorescent group. It is oxidase positive. It does not ferment carbohydrates, but many strains oxidize glucose. Identification is usually based on colonial morphology, oxidase positivity, the presence of characteristic pigments, and growth at 42°C.

Pseudomonas is a strict (obligate) aerobe, but sometimes it can grow anaerobically if nitrates (NO_3 act as respiratory electron acceptor) are present in the medium.

Pseudomonas can grow at wide ranges of temperature; the optimum temperature is 37°C. It can grow on ordinary media like nutrient agar and grows almost on all the culture media used routinely in the bacteriology lab. Pseudomonas has been seen to grow in distilled water also.

Pseudomonas produces large, opaque, flat colonies with irregular margins and distinctively fruity odour colonies. The colour of growth will depend upon the type of pigments (enumerated below) produced by the organism. The isolates from water and soil produce small round colonies. The isolates from clinical specimens like respiratory, urine, etc. may produce mucoid colonies. The bacteria which form mucoid colonies are more virulent compared to others.

The pigments produced by Pseudomonas are:

- The fluorescent pigment pyoverdin (greenish yellow)
- The blue pigment pyocyanin (bluish green)
- Pyorubin (red)
- Pyomelanin (brown)

Colonial Morphology

P.aeruginosa: The colonies are large (3-4 mm), flat, greyish-blue with a characteristic fruity, grapelike odour of aminoacetophenone. Most strains give a clear zone of haemolysis on blood agar. Pyocyanin, a bluish pigment unique to *P.aeruginosa,* gives the blue colour associated with many cultures. Colonial variation includes S-forms (soft and shiny), R-forms (dry and granular) that are not unlike (he colonies of some *Bacillus* species, and mucoid M-forms that are frequently biochemically atypical. Some strains have colonies with a distinctive metallic sheen. *P. aeruginosa* produces large, pale colonies on MacConkey agar (unable to utilise lactose)with greenish-blue pigment superimposed. Red colonies and medium, indicative of an alkaline reaction, are seen on brilliant green and XLD agars. No H_2S is produced on XLD medium.

P. pseudomallei: Colonial growth varies from smooth and mucoid to rough with a dull, wrinkled, corrugated surface. In the smooth form the colonies are round, low-convex, entire, shiny and greyish-yellow. After several days the colonies become opaque, yellowish-brown and umbonate. The growth has a characteristic earthy or musty odour. Partial and, later, complete haemolysis occurs on sheep blood agar. *P. pseudomallei* grow on MacConkey agar, utilising lactose, but there is no growth on deoxycholateor Salmonella-Shigella (SS) agars.

P. mallei: growth is slower than that of *P.aeruginosa* and *P.pseudomallei* but in 24-48 hours thecolonies are 1- 2 mm in diameter, smooth and white tocream. As they age, they become granular and yellow is hor brown in colour. *P.mallei* is unable to grow onMacConkey agar.

The pigments of *P. aeruginosa;*

Strains of *P. aeruginosa* produce the diffusible pigments pyocyanin (blue). pyoverdin (yellow), pyorubin (red) and pyomelanin (dark brown) in varying combinations and amounts. Some strains produce all four pigments. Pyorubin and pyomelanin are less comm only produced, develop slowly and are seen best by growing the strains on nutrient agar slants at room temperature for up to 2 weeks. As pyocyanin is unique to*P. aeruginosa* this is an important diagnostic characteristical though strains vary in the amount of the pigment they produce. Media such as Pseudomonas agar P (Difco) will enhance pyocyanin production and Pseudomonas agar F (Difco) enhances pyoverdin production. Pyoverdin, once called ' fluorescein',will fluoresce under ultra- violet light. Some strains of*P. aeruginosa* do not produce pyocyanin and thesemay also be atypical in certain biochemical reactions,making them difficult to identify.

Biochemical Characteristics

- Pseudomonas has oxidative metabolism. Since organism is non fermentative the acid is not produced from peptone water sugars.
- The important biochemical characteristics of Pseudomonas include:
- Oxidase test positive;
- Catalase test positive
- Nitrates are reduced to nitrites;
- Arginine dihydrolase test positive;
- Glucose is utilized oxidatively □ Oxidative reaction in of media
- Indole, Methyl red (MR), Vogues Prauskar (VP) and H_2S production test are negative.
- Commonest screening diagnostic biochemical test used in lab is the oxidase test.

Antigenic Structure and Toxins

Pili (fimbriae) extend from the cell surface and promote attachment to host epithelial cells. The exopolysaccharide is responsible for the mucoid colonies. The lipopolysaccharide, which exists in multiple immunotypes, is responsible for many of the endotoxicproperties of the organism. *P.aeruginosa* can be typedby lipopolysaccharide immunotype and by pyocin (bacteriocin) susceptibility.

Most *P.aeruginosa* isolates from clinical infections produce extracellular enzymes, including elastases,proteases, and two hemolysins (a heat-labile phospholipase C and a heat-stable glycolipid).

Many strains of *P.aeruginosa* produce exotoxin A, which causes tissue necrosis and is lethal for animals when injected in purified form. The toxin blocks protein synthesis by a mechanism of action identical to that of diphtheria toxin, although the structures of the two toxins are not identical.

Virulence and Pathogenicity

Pseudomonas can infect any tissue, any organ system in an immune-compromised host. *P. aeruginosa* produces exotoxin A which is a virulence factor. This exotoxin inactivates ADP ribosylate eukaryotic elongation factor 2 (EF2) and thus interferes with the synthesis of protein resulting in death of the cell. *P.aeruginosa* also produces an exoenzyme "Exo A" which damages the cell membrane leading to lysis of the membrane and cell death.

The most important risk factor for Pseudomonas infections is break down of host defense due to disease or other factors. Pseudomonas is both invasive and toxinogenic. Infection involves the following three steps:

- Bacterial attachment and colonization;
- Local invasion;
- Disseminated systemic disease.

Pathogenesis

P. aeruginosa is pathogenic only when introduced into areas devoid of normal defenses, such as when mucous membranes and skin are disrupted by direct tissue damage as in the case of burn wounds.

The bacterium attaches to and colonizes the mucous membranes or skin, invades locally, and produces systemic disease. These processes are promoted by the pili, enzymes, and toxins described earlier. Lipopolysaccharide plays a direct role in causing fever, shock, oliguria, leukocytosis and leukopenia, disseminated intravascular coagulation, and respiratory distress syndrome.

P.aeruginosa and other pseudomonads are resistant to many antimicrobial agents and therefore become dominant and important when more susceptible bacteria of the normal microbiota are suppressed.

Bacterial attachment and Colonization

Pseudomonas infection may be endogenous (may be from intestines) or may be acquired from outside (exogenous). The adhesins are the pili of *P aeruginosa* with which bacteria adhere to specific galactose or mannose or sialic acid receptors on the mucosal epithelial cells of the upper respiratory tract and others. Production of protease enzyme by bacteria breaks down the fibronectin and exposes the pilus specific receptors on the epithelial cell surface.

Tissue injury caused by viral infection and other phenomenon facilitates colonization by Pseudomonas (Opportunistic colonization). Pseudomonas can also colonize by formation of biofilm which we will discuss later. Pseudomonas pili, mucoid polysaccharide, probably surface-bound exoenzyme S and possibly other cell surface adhesins help *Pseudomonas* to colonize.

Pseudomonas can invade the blood stream from initial site of infection and through blood is disseminated to different organs. The factors which help bacteria to invade as described above help to invade the organs, tissues wherever the bacteria reach. Bacterial endotoxin during septicaemia, may cause fever, hypotension, and intravascular coagulation.

The virulence/pathogenicity armamentarium of Pseudomonas includes:

- Adhesins
- Pili (N-methyl-phenylalanine pili)
- Polysaccharide capsule (glycocalyx)
- Slime
- Invasins
- Elastase
- Alkaline protease
- Hemolysins (phospholipase and lecithinase)
- Cytotoxin (leukocidin)
- Pyocyanin
- Toxins
- Exoenzyme S
- Exotoxin A
- Lipopolysaccharide (LPS)
- Antiphagocytic elements
- False capsules, slime layer
- Lipo polysaccharide
- Biofilm formation

Pathogenesis and Pathogenicity

Pathogenic strains of *P.aeruginosa* produce a variety of toxins and enzymes which promote tissue invasion and damage. Attachment to host cells is mediated by fimbriae. Colonization and replication are aided by antiphagocytic properties of exoenzyme S, extracellular slime and outer membrane lipopolysaccharides. Resistance to complement-mediated damage and the abiIity to obtain iron from host tissues are additional virulence factors.

Tissue *damage* is caused by toxins such as exvtoxin A, phospholipase C and proteases. Exotoxin A is a bipartitetoxin with binding and active components. The activecomponent, once internalized in a cell blocks protein synthesis by ADP-ribosy lation and elongation of Factor 2 with resultant cell death. The cytoplasmic membranes of neutrophils are damaged by a leukocidin. Dissemination is aided by exoenzyme S and systemic toxicity is attributed to exotoxin A and endotoxin. The host defence mechanisms against P.*aeruginosa* include opsonizing antibodies and phagocytosis by macrophages.

Diagnostic Procedures

- Specimens for laboratory examination include pus, respiratory aspirates, mid-stream urine, mastitic milk and ear swabs.
- Blood agar and MacConkey agar plates, inoculated with suspected material are incubated aerobically at 37°C for 24 to 48 hours.

Identification criteria for isolates:

- Colonial morphology are the characteristic fruity, grape-like odour
- Pyocyanin production
- Lactose-negative, pale colonies on MacConkey agar
- Oxidase-positive
- Triple sugar iron agar unchanged
- Biochemical tests

Treatment and Control

Predisposing causes and sources of infection should be identified and, where possible, eliminated. *Pseudomunus aeruginosa* is extremely resistant to many antibiotics and susceptibility testing should be carried out on isolates. A combination of either gentamicin or tobrarnycin with either carbenicillin or ticaricillin may be effective. Vaccines may be required for farmed mink and chinchillas. As there arc antigenic differences between strains, polyvalent or autogenous formalin-killed bacterins should he employed. Humoral antibody induced by a polyvalent exotoxin A-polysaccharide vaccine appears to be protective.

Glanders

Glanders, caused by B. *mallei,* is a contagious disease of *Equidae* characterized by the formation of nodules and ulcers in the respiratory tract or on the skin. Humans and carnivores are also susceptible to infection.

Pathogenesis

Glanders in the horse is usually a chronic, disseminated, debilitating disease but the mechanisms of pathogenicity are not known. The presence of B.*mallei* in the host gives rise to a hypersensitivity reaction, the basis of the mallein test.

Transmission follows ingestion of food or water contaminated by nasal discharges of infected *Equidae.* Less commonly, infection may be acquired by inhalation or through skin abrasions. An acute septicaemic form of the disease is characterized by fever, mucopurulent nasal discharge and respiratory signs. Death usually follows within a few weeks. Chronic disease is more common and presents as nasal, pulmonary and cutaneous forms, all of which may be observed in an affected animal.

In the nasal form, ulcerative nodules develop on the mucosa of the nasal septum and lowcr part of the turbinates. A purulent, blood-stained nasal discharge and regional lymphadenopathy are usually present. The ulcers eventually heal leaving star-shaped scars. The respiratory form i s characterized by respiratory distress and the development of tubercle-like lesions throughout the lungs. The cutaneous form, termed farcy, is a lymphangitis in which nodules occur along the course of the lymphatic vessels of the limbs. Ulcers develop and discharge yellowish pus.

Chronically affected animals may die after several months or may recover and continue to shed organisms from the respiratory tract or skin. Carnivores may contract the disease by eating infected carcases.

Diagnostic Procedures

In regions where the disease is endemic, clinical signs may be diagnostic. Specimens for laboratory diagnosis should include discharges from lesions and blood for serology. Specimens must be processed in a biohazard cabinet.

Burkholderia mallei grow on media containing 1% glycerol and most strains will grow on MacConkey agar. Plates are incubated aerobically at *37'C* for 2 to 3 days.

Identification criteria for isolates:

- Colonial characteristics
- Majority of strains grow on MacConkey agar without utilizing lactose
- Comparatively unreactive biochemically and non-motile

Immunological Tests

Suitable serological tests include the complement fixation test and agglutination techniques.

Mallein Test

The mallein test is an efficient field test both for confirmation and for screening in-contact animals. Mallein, a glycoprotein extract of *P.mallei,* is injected intradermally (0.1 ml) just below the lower eyelid. A positive reaction is indicated by local swelling and mucopurulent ocular discharge after 24 hours.

Mallein tests are used to demonstrate the hypersensitivity developed after infection with *P. mallei.* Mallein is a glycoprotein extracted from the bacterium. In infected animals, subcutaneous inoculation of mallein (subcutaneous test) results in swelling at the injection site and fever. Instillation of mallein into the conjunctival sac (ophthalmic test) is followed in 6-12 hours by an inflammatory and purulent reaction in the eye, whereas inoculation of a small amount of mallein into the skin of the lower eyelid (intrapalpebral test) gives a localised, oedematous swelling and purulent conjunctivitis. Healed lesions of the nasal mucosa are activated in glanderous animals after mallein tests and this can be a useful diagnostic feature.

Animal Inoculation

The Straus reaction is seen in male guinea-pigs inoculated intraperitoneally with infective material containing either *P.pseudomallei* or *P.mallei.* A localised peritonitis and a purulent inflammation of the testicular tunica vaginalis develops in 2-3 days.

Antibiotic susceptibility tests

This need to be carried out with *P.aeruginosa* isolates as multiple drug resistance, associated with R factors, is frequently encountered with this bacterium.

Treatment and Control

A test and slaughter policy is enforced in countries where the disease is exotic. In endemic areas, antibiotic therapy is inappropriate as treated animals often become subclinical carriers. Effective cleaning and disinfection of all contaminated carcass must be carried out. Formalin (1.5%) or an iodophor (2.0%) can be used, with a contact time of 6 hours.

Melioidosis

Melioidosis, caused by B.*pseudomellei,* is endemic in tropical and subtropical regions of southeastern Asia and Australia where rhe organism is widely distributed in soil and water. Infection may follow ingestion, inhalation or skin contamination from environmental sources. The bacterium is an opportunistic pathogen and stress factors or immunosuppression may predispose to clinical disease.

Many animal species, including humans, are susceptible and subclinical infections may occur. Because infection is usually disseminated, abscesses develop in many organs including lungs, spleen, liver, joints and central nervous system. Melioidosis is a chronic, debilitating, progressive disease, often with a long incubation period. Clinical signs, which are variable, relate to lesion severity and distribution. In horses melioidosis, which can mimic glanders, is often **referred to as pseudoglanders**.

Pathogenesis and Pathogenicity

The pathogenesis of melioidosis is poorly understood. Extracellular products of *B. psendomallei* such as an exotoxin, a dermonecrotic protease and a lecithinase have been implicated in disease production. Both strain virulence and host immunosuppression may influence the establishment and outcome of infection.

Diagnostic Procedures

- In regions where the disease is encountered, gross pathological findings may aid diagnosis.
- Specimens for laboratory diagnosis should include pus from abscesses, affected tissues and blood for serology.
- A biohazard cabinet must be used for processing specimens.
- A fluorescent antibody technique for demonstrating the organism in tissue smears is available in some reference laboratories.

- Blood agar and MacConkey agar plates, inoculated with suspected material, are incubated aerobically at 37°C for 24-48 hours.
- Identification criteria for isolates:
  - Colonial morphology and characteristic musty odour
  - Lactose utilized in MacConkey agar
  - Biochemical characteristics
- Slide agglutination test using specific antiserum
- ELISA, complement fixation and indirect haemagglutination tests can be used for detecting serum antibodies.

Treatment and Control

- Confirmation of infection followed by slaughter of infected animals is mandatory in countries where the disease is exotic.
- Treatment is expensive and unreliable. Relapses can occur after antibiotic therapy is discontinued.
- Vaccines are being developed in some countries.

Diseases and Main Hosts of the Pathogenic *Pseudomonas* Species

Species	Host(s)	Disease
P. *aeruginosa*	Cattle	Mastitis, uterine infections, skin infections, abscesses, enteritis and arthritis
	Sheep and goats	Mastitis, pneumonia, lung abscesses and 'green wool' (a skin infection in sheep)
	Pigs	Enteritis, respiratory infections and otitis
	Horses	Metritis, lung abscesses and eye infections
	Dogs and cats	Otitis externa, cystitis, endocarditis, dermatitis, wound infections and conjunctivitis
	Mink	Septicaemia and pneumonia
	Chinchilla	Generalised infection with conjunctivitis, otitis, pneumonia, enteritis and infection of the genital organs
	Reptiles	Necrotic stomatitis and other necrotic lesions, especially in captive snakes
	Many animal species	Infection of burns and other wounds, diarrhoea, genital and nosocomial infections

P.pseudomallei	Many animal species Horses Cattle Sheep Goats Pigs Dogs	Melioidosis (pseudoglanders) The disease can mimic glanders Acute and chronic forms with localisation of lesions in lungs, joints and uterus Arthritis and lymphangitis predominate Loss of condition, respiratory and central nervous disturbances, arthritis and mastitis As for goats but in addition diarrhoea and abortion Febrile disease with localising suppurative foci
P.mallei	Horses and other equids Humans, cats and other animals	Glanders: acute form with high fever, mucopurulent nasal discharge, respiratory signs, septicaemia and death within 2 weeks Chronic forms of glanders • Pulmonary: small nodules in lungs that break down and discharge P.*mallei* into the bronchioles • Cutaneous form: Farcy, which is a lymphangitis with ulcers along lymphatic vessels of the limbs and chest. The ulcers eventually heal leaving 'star shaped' scars. Acute, septicaemic disease

32

Pasteurella and Mannheimia

Pasteurella and Mannheimia species are small (0.2 x 1-2 µm) non-motile, Gram-negative rods or coccobacilli. They are oxidase-positive facultative anaerobes, and most species are catalase-positive. Although non-enriched media will support their growth, these organisms grow best on media supplemented with blood or serum. They usually remain viable for only a few days on culture plates.

Some species, such as Mannheimia haemolytica, Pasteurella trehalosi and P. aerogenes can tolerate the bile salts in MacConkey agar. In smears from infected tissues stained by the Giemsa method, pasteurellae exhibit bipolar staining.

The family Pasteurellaceae comprises five genera, Actinobacillus, Haemophilus, Mannheimia, Pasteurella and Lonepinella. These genera share a number of common features and some organisms have been reclassified within these genera following deoxyribonucleic acid hybridization studies and 16s rRNA sequencing.

The pasteurellae are small (0. 2 µm by up to 2.0 µm), Gram-negative rods or coccobacilli. They are nonmotile,non-sporing, facultatively anaerobic, fermentative(except for *P. anatipesrijer),* oxidase-positive andcatalase - positive (except for *P. caballi).* Althoughunenriched media support their growth, they grow beston media supplemented with serum or blood.

Recent Changes in Nomenclature

Previous name	Present name
Pasteurella ureae	*Actinobacillus ureae*
Haemophilus avium	*Pasteurella avium*
Pasteurella pneumotropica (Henriksen biotype)	*Pasteurella dogmatis*

Usual Habitat

Pasteurella species are worldwide in distribution with wide spectrum of hosts. Most are commensals on the mucous membranes of the upper respiratory and intestinal tracts of animals. The carrier rate for different species varies greatly. Most Pasteurella and Mannheimia species are commensals on the mucosae of the upper respiratory tract of animals. Their survival in the environment is relatively short.

Isolation

The routine medium for the isolat ion of *Pasteurella* spp. is ox or sheep blood agar. Clinical materials should be inoculated on both blood and MacConkey agars. Selective medium con taining clindamycin (2 µg / ml) should be used for the isolation of *P. mulrocida* from porcine nasal swabs. The plates are in cubated aerobically at 37 °C for 24 48 hours. *P. anatipestifer* grows best on blood or serum agar under 5-10 per cent CO_2 (a candle-jar is satisfactory).

Colonial Morphology

The colonies of all spec ies are usually evident in 24 hours. They are of moderate size, round and greyish.*P.anatipestifer* produces small "dewdrop' colonies within 48 hours. The colonies of *P. pneumootropica* are non-haemolytic and somewhat similar tothose of *P. multocida*. Type A strainsof *P.multocida* often produce relatively large, mucoidcolonies due to their large capsules of hyaluronic acid. *P. multocida* has a characteristic 'sweetish'odour, is non-haemolytic, does not grow on MacConkeyagar and is a good indole producer.

P.haemolytica is beta haemolytic and usually tolerates the bile salts in MacConkey agar to grow aspinpoint red colonies. It has no odour and doesnot produce indole. *P. testudinis* and *P. granulomatis* are also haemolytic on blood agar.

Biochemical Reactions

Characteristic colonies yielding small, Gram-negative rods or coccobacilli that are oxidase-positive (the enterobacteria are oxidase-negati ve) and catalase-positive (except for *P. cabalii)* are inoculated into TSI slopes. The usual reaction is a yellow slant and buttwith no gas or H2S production. *P. anatipestifer* is a non-fermenter and may eventually be reclassified in anothergenus. It is non-hae molytic, does not g row onMacConkey agar and is indole and urease-negative.

Pathogenesis and Pathogenicity

Many P.multocida infections are endogenous. The organisms, which are normally commensals of the upper respiratory tract, may invade the tissues of immunosuppressed animals. Exogenous transmission can also occur either by direct contact or through aerosols. Factors of importance in the development of disease include adhesion of the pasteurellae to the mucosa and the avoidance of phagocytosis. Fimbriae may enhance mucosal attachment and the capsule, particularly in type A strains, has a major antiphagocytic role. In septicaemic pasteurellosis, severe endotoxaemia and disseminated intravascular coagulation cause serious illness which can prove fatal.

Four main virulence factors have been identified in strains of M. haemolytica and P.trehalosi; fimbriae which may enhance colonization; a capsule that inhibits complement-mediated destruction of the organisms in serum; endotoxin which can alter bovine leukocyte functions and is directly toxic to bovine endothelial cells; leukotoxin, a pore-forming cytolysin that affects leukocyte and platelet functions when present at low concentrations and causes cytolysis at high concentrations.

The subsequent release from damaged cells of lysosomal enzymes and inflammatory mediators, such as tumour necrosis factor-a and eicosanoids, contributes to severe tissue damage in these infections.

Diagnostic Procedures

There may be a history of exposure to stress arising from transportation or overcrowding. Suitable specimens for laboratory examination from live animals include tracheobronchial aspirates, nasal swabs or mastitic milk. Tissue or blood smears from septicaemic cases, stained by Giemsa or Leishman methods, may reveal large numbers of bipolar-staining organisms.

Specimens should be cultured on blood agar and MacConkey agar. Plates are incubated aerobically at 37°C for 24 to 48 hours. Blood agar, supplemented with neomycin, bacitracin and actidione, can be used for the isolation of P.multocida from heavily contaminated specimens.

Identification criteria for isolates:

- Colonial characteristics
- Growth on MacConkey agar
- Positive oxidase test
- Biochemical profile
- Serological tests are generally of little diagnostic value in the majority of the diseases caused by pasteurellae and Mannheimia species.

Clinical Infections

Clinical infections caused by pasteurellae and Mannheimia species in domestic animals are mainly attributable to *P. muliocida, M.haemolytica* and *P. trehalosi. Pasteurella multocida* has a wide host range whereas M.haemolytica is largely restricted to ruminants and P.trehalosi to sheep. The diseases associated with *P.multocida* infection include haemorrhagic septicaemia in ruminants and occasionally in other domestic species, porcine atrophic rhinitis, fowl cholera and bovine pneumonic pasteurellosis. However, the main aetiological agent of bovine pneumonic pasteurellosis is *M. haemolytica*, and this organism is also responsible for pneumonia in sheep and septicaemia in young lambs. Infection with *P.trehalosi* frequently results in septicaemia in older lambs.

Mannheimia haemolytica can cause a severe necrotizing mastitis in ewes and both P.multocida and M.haemolytica has been isolated occasionally from cases of bovine mastitis. Both organisms have also been implicated aetiologically in the enzootic pneumonia complex of calves.

Haemorrhagic Septicaemia (HS)

Haemorrhagic septicaemia or barbone is an acute, potentially fatal septicaemia mainly affecting buffaloes and cattle. Predisposing factors such as overwork, poor body

Buffaloes tend to be more susceptible to the disease than cattle. All ages can be affected but in endemic areas the disease is most common in animals between 6 and 24 months of age. Older animals may have a degree of immunity from previous exposure.

Clinical Signs

The incubation period of the disease is 2 to 4 days and the course ranges from 2 to 5 days. Death, without prior signs of illness, may occur within 24 hours of infection. Sudden onset of high fever, respiratory distress and a characteristic oedema of the laryngeal region are features of the disease. The oedema may extend to the throat and parotid regions and to the brisket. Recumbency is followed by death from endotoxaemia. Mortality rates are usually over 50% and can approach 100%.

Diagnosis

A history of acute disease with high mortality in areas where haemorrhagic septicaemia is endemic may suggest a presumptive diagnosis of the condition.

Gross pathological changes may include widespread petechial haemorrhages, enlarged haemorrhagic lymph nodes and blood-tinged fluid in the pleural cavity and the pericardial sac. Giemsa-stained blood smears from a recently-dead animal often reveal large numbers of bipolar-staining organisms.

Isolation, identification and serotyping the P.multocida isolate is confirmatory. Serotypes B:2 and E:2 are the specific strains associated with the disease. An antibody titre of 1:160 or above in an indirect haemagglutination test is indicative of recent exposure to the pathogen.

Treatment and Control

Antibiotic therapy in the early febrile stage is usually effective. Although the organism is susceptible to penicillin, tetracyclines are more often used. A slaughter policy for affected and in-contact animals is usually pursued in countries where the disease is exotic. Vaccines available for control of the disease include bacterins and a modified live vaccine. Latent carriers can be detected using immunohistochemical techniques on samples of tonsillar tissue.

Bovine Pneumonic Pasteurellosis (Shipping Fever)

Shipping fever, characterized by severe bronchopneumonia and pieurisy occurs most commonly in young cattle within weeks of being subjected to severe stress, such as transportation, assembly in feedlots and close confinement. The condition is commonly associated with M.haemolytica, although P.multocida has also been isolated from lungs of affected cattle.

Clinical Signs

Clinical features include sudden onset of fever, depression, anorexia, tachypnoea and serous nasal discharge. In mixed infections, there is usually a marked cough and ocular discharge. Morbidity rates can reach 50% and mortality rates range from 1-10%.

Diagnosis

There may be a history of exposure to stress factors and of sudden onset of respiratory disease. Gross pathological findings are of diagnostic value. Cytospin preparations from bronchoalveolar lavage usually reveal large numbers of neutrophils. Isolation of M.haemolytica, often in association with other pathogens, from bronchoalveolar lavage or affected lung tissue is confirmatory.

Treatment and Control

Affected animals must be isolated and treated early in the course of the disease. Treatment with oxytetracycline, potentiated sulphonamides and ampicillin is usually effective.

Stress factors must be kept to a minimum. Procedures such as castration, dehorning, branding and anthelmintic therapy should be carried out several weeks before young cattle are transported. Vaccination regimes for respiratory pathogens should be completed at least 3 weeks before transportation. Vaccines for M. haemolytica which incorporate modified leukotoxin and surface antigens may induce protection. Recently, a vaccine containing serotypespecific antigens of both M. haemolytica A1 and M. haemolytica A6 has been developed.

Pasteurellosis in Sheep

Outbreaks of ovine pneumanic pasteurellosis are usually caused by M. haemolytica whereas P.multocida tends to produce sporadic cases of the disease. Manheimia haemolytica is a commensal of the upper respiratory tract in a proportion of healthy sheep.

Septicaernic pasteurellosis in lambs less than 3 months of age is caused by M. haemolytica. In older animals between 5 and 12 months-of-age, septicaemic pasteurellosis is usually associated with P.trehalosi infection. Pasteurella trehalosi is found in the tonsillar tissues of carrier sheep. As with most other pasteurella infections, clinical disease may be precipitated by a range of predisposing factors including transportation. Affected sheep which tend to be in good bodily condition may die suddenly and the mortality rate may approach 5%.

Atrophic Rhinitis of Pigs

Toxigenic strains of P.multocida type D or A cause a severe, progressive form of atrophic rhinitis. These toxigenic P.multocida isolates are designated AR+ (atrophic rhinitis-positive) strains. Infection with Bordetella bronchiseptica may cause mild, non-progressive turbinate atrophy without significant distortion of the snout.

Clinical Signs

Early signs, usually encountered in pigs between 3 and 8 weeks of age, include excessive lacrimation, sneezing and, occasionally, epistaxis. The snout gradually becomes shortened and wrinkled. As the disease progresses, a distinct lateral deviation of the snout may develop. Atrophic rhinitis is rarely fatal. Affected pigs are usually underweight and damage to the turbinate bones may predispose to secondary bacterial infections of the lower respiratory tract.

Diagnosis

In severely affected pigs characteristic facial deformities are diagnostic. Visual assessment of the extent of turbinate atrophy can be made following slaughter by transverse section of snouts between the first and second premolar teeth. Isolation and identification of P.multocida should be followed by tests to confirm that the isolate is a toxigenic strain. Suitable tests include demonstration of toxicity for tissue culture cells, an ELISA test for toxin detection and the detection of the toxin gene by a polymerase chain reaction technique.

Control

Chemoprophylaxis with sulphonamides, trimethoprim, tylosin or tetracyclines in weaner, grower and sow rations could be considered. Improvement in husbandry must be instituted to minimize the influence of predisposing factors. Vaccination with a combined B.brunchiseptica bacterin and P. multocida toxoid may reduce the severity of the disease and improve growth rates. Sows should be vaccinated at 4 and 2 weeks before farrowing and young piglets at 1 week and 4 weeks of age.

Fowl Cholera

Fowl cholera is a primary avian pasteurellosis caused by P.multocida capsular type A. It is highly contagious and affects both domestic and wild birds. The disease usually presents as an acute septicaemia which is often fatal. Turkeys tend to be more susceptible than chickens.

Post mortem lesions include haemorrhages on serous surfaces and accumulation of fluid in body cavities. In sporadic chronic cases of the disease, the signs and lesions are often related to localized infections. The wattles, sternal bursae and joints are usually swollen due to the accumulation of fibrinopurulent exudates.

In the acute septicaemic form of the disease, numerous characteristic bipolar-staining organisms can be detected in blood smears and P.multocida can be isolated from blood, bone marrow, liver or spleen. The bacterium may be difficult to isolate from chronic lesions.

Medication of the feed or water with sulphonamides or tetracyclines early in an outbreak of acute disease may decrease the mortality rate. Polyvalent adjuvant bacterins are widely used. Autogenous vaccines may be required if the commercial vaccines are ineffective. Modified live vaccines are available in some countries.

Snuffles in Rabbits

Snuffles is a common, recurring, purulent rhinitis in rabbits which is caused by type A strains of P.multocida. Bordetella bronchiseptica infection may sometimes cause similar clinical signs. Pasteurella multocida is a commensal in the upper respiratory tract of healthy carrier rabbits.

Clinical disease is often precipitated by stress factors such as overcrowding, chilling, transportation, concurrent infections and poor ventilation resulting in high levels of atmospheric ammonia. There is a purulent nasal discharge which cakes on the forelegs because affected rabbits paw their noses. Sneezing and coughing may be observed. Sequelae include conjunctivitis, otitis media and subcutaneous abscessation. Bronchopneumonia may develop in young rabbits.

Treatment or prophylactic therapy with antibiotics may be of value. Predisposingstress factors must he eliminated. No vaccine is available.

Principal hosts and Diseases of the *Pasturella* species

Species	Principal host(s)	Disease
P. multocida		
Type A	Cattle	Part of 'shipping fever' complex Part of the 'enzootic pneumonia' complex in calves
	Sheep	Occasional but severe mastitis
	Pigs	Pleuropneumonia
	Rabbits	Mastitis
	Poultry	Pneumonia (often secondary)
	Many domestic and wild animals	One cause of 'snuffles', pleuropneumonia, abscesses, otitis media, conjunctivitis and genital infections Fowl cholera (primary infection) Pneumonia and other infections in stressed animals Commensals in respiratory and digestive tract
Type B	Cattle, water buffalo, bison , yak and other ruminants	Epizootic haemorrhagic septicaemia (primary infection) Nasopharynx of carrier animals South-East Asia and other countries
Type D	Pigs	Atrophic rhinitis (with or without *Bordetella bronchiseptica*)
	Pigs, less commonly other domestic animals and poultry	Pneumonia (usually secondary

Species	Principal host(s)	Disease
Type E	Cattle and water buffalo	Epizootic haemorrhagic septicaemia (primary infection) Africa only
Type F	Turkeys mainly	Role in disease not clear
P. haemolytica		
Biotype A	Cattle Sheep	Part of 'shipping fever' complex Pneumonia (primary or secondary) Enzootic pneumonia (primary or secondary). Septicaemia in lambs under 3-months-old. Gangrenous mastitis ('blue bag')
Biotype T	Sheep	Septicaemia in lambs 5-12 months-old
P. *pneumotropica*	Rodents Dogs and cats	Pneumonia (secondary) and abscesses (possibly from bites) Commensal in nasopharynx Present in the nasopharynx of some animals. Not a significant pathogen
P.canis	Dogs (man)	Recovered from the oral cavity of dogs and from dog bites in humans
P.*dagmatis*	Dogs, cats (man)	Present in the oral cavity and intestinal tract of dogs and cats and recovered from dog and cat bites in humans
P.*stomatis*	Dogs, cats	Isolated from the respiratory tract but not usually pathogenic
P. caballi	Horses	Respiratory infections including pneumonia
P. *aerogenes*	Pigs (man)	Commensal in intestine of pigs. Rarely pathogenic but one isolate associated with abortion. Abscesses from pig bites in humans
P. anatipestifer	Ducklings mainly but also chickens, turkeys, pheasants, water fowl	'New duck disease': severe fibrinous polyserositis in ducklings 1-8 weeks old
P. *anatis*	Ducks	Recovered from intestines. Pathogenicity not demonstrated
P. gallinarum	Poultry	Commensal in respiratory mucosa. Occasional low grade infections
P. *avium* P. *fangaa* *P. vofantium*	Chickens	Isolated from the respiratory tract of healthy birds Pathogenicity not demonstrated for any of the three species
P. testudinis	Turtles, tortoises	Associated with abscesses. Possibly opportunistic
P. *granufomatis*	Cattle (Brazil)	Fibrogranulomatous disease

33

Actinobacillus

Actinobacillus species are non-motile, Gram-negative rods (0.3 to 0.5 x 0.6 to 1.4 µm) which occasionally have a coccobacillary appearance. They are non-motile, non-sporeformingand non-acid-fast. A surface slime is presentin the three major species (A. ligllieresii. A. equuli andA. suis) and may be related to the stickiness of theircolonies on agar. Surface cultures have low viabilityand die in 5-7 days. The actinobacilli ferment carbohydrates, without the production of gas, within 24 hours and most species produce urease and grow on MacConkey agar. Most species are urease and oxidase positive. The reactions in the oxidase andcatalase tests are variable. The genus is still in a stateof flux, mainly because of the close similaritiesbetween actinobacilli and species in the genera Pasteurella and Haemophilus.

These facultative anaerobes ferment carbohydrates producing acid but not gas. Actinobacilli exhibit some host specificity and are mainly pathogens of farm animals.

Usual Habitat

Actinobacilli are commensals on mucous membranes of animals particularly in the upper respiratory tract and oral cavity. As actinobacilli cannot survive for long in the environment, carrier animals play a major role in transmission.

Colonial Appearance

- A. lignieresii: small, glistening colonies develop in 24 hours. They are usually slightly sticky (viscid) on primary isolation but lose th is characteristic on subculture. The colonies are non-haemolytic and develop to about 2 mm in diameter in 48 hours. The organism grows well on MacConkey agar, the colonies are at first pale but become pinkish as A. lignieresii is a late lactose-fermemer.
- A. equuli: some strains are haemolytic and the colonies are sticky with this feature remaining on subculture. I t is a lactose-fermenter onMacConkey agar.

- A. suis: all strain s are haemolytic with colonies similar to those of A. lignieresii but more sticky. It grows well on MacConkey agar.
- A.pleuropneumoniae: small (1 mm) colonies surrounded by a zone of beta-haemolysis, which somewhat resemble those of a beta-haemolytic streptococcus. Two distinct colony forms are possible, one waxy and the other a soft glisteningtype. No growth occurs on MacConkey agar.
- A. capsulatus: the cells are capsulated and very sticky colonies are produced on blood agar. It grows well on MacConkey agar.
- A. actinomycetemcomitans: the colonies are small (1 mm after 2-3 days), adherent to the medium anddifficult to break-up. Colonies are described as 'star-like'. No growth occurs on MacConkey agar.
- A. seminis: small, round, pinpoint colonies that are greyish-white, non-haemolytic and appear after 24 48 hours on blood agar. No growth occurs on MacConkey agar.
- A. ureae: colonies on blood agar are mucoid and are usually accompanied by some greening of the medium.

Biochemical Reactions

A. pleuropneumoniae enhances the haemolysis caused by the beta-haemolysin of Staphylococclls aureus in a CAMP test. A. seminis is usually catalase-positive but oxidase- negative, and almost completely unreactive in conventionally performedbiochemical tests. A few strains show slight acid production, after incubation periods of up to 28 days, in arabinose, fructose, mannose, trehalose, glucose, maltose, mannitol and xylose.

Most strains are reported as being nitrate, H_2S, MR, VP, indole, urease, Phosphatase, gelatinase and citrate negative.

Pathogenesis and Pathogenicity

The virulence factors possessed by the actinobacilli are poorly defined with the exception of those associated with A. pleuropneumuniae, the cause of pleuropneumonia in pigs.

Clinical Infections

Actinobacilli can cause a variety of infections in farm animals including 'timber (wooden) tongue' in cattle, pleuropneumonia in pigs and systemic disease in foals and piglets.

Actinobacillosis in Cattle

Actinobacillosis, a chronic pyogranulomatous inflammation of soft tissues, is most often manifest clinically in cattle as induration of the tongue, referred to as timber tongue. Potentially important lesions occur in the oesophageal groove and the retropharyngeal lymph nodes.

The aetiological agent, Acrinobacillus lignieresii is a commensal of the oral cavity and the intestinal tract. It can survive for up to 5 days in hay or straw. The organisms enter tissues through erosions or lacerations in the mucosa and skin. A localized pyogranulomatous response is associated with club colonies containing the bacteria. In addition, spread through the lymphatics to the regional lymph nodes may induce pyogranulomatous lymphadenitis.

Bovine actinohacillosis is usually a sporadic disease, although herd outbreaks of limited extent can occur. Animals with timber tongue have difficulty in eating and drooling of saliva. Involvement of the tissues of the oesophageal groove can lead to intermittent tympany and enlargement of the retropharyngeal lymph nodes can cause difficulty in swallowing and stertorous breathing. Lesions of cutaneous actinobacillosis may be found on the head, thorax, flanks and upper limbs.

Animals with ulcerated discharging lesions can contaminate the environment. Localized pyogranulomatous lesions in the retropharyngeal lymph nodes are often found at slaughter.

Diagnosis

Induration of the tongue is characteristic of the disease and there may be a history of grazing rough pasture. Spccimens for laboratory examination include pus, biopsy material and tissues from lesions at post-mortem. Gram-negative rods are demonstrable in smears from exudates.

Pyogranulomatous foci containing club colonies may be evident in tissue sections.Cultures on blood agar and MacConkey agar are incubated aerobically at 37°C for 24 to 72 hours.

Isolation

Most of the actinobacilli can be cultured on sheep or ox blood agar and some will grow on MacConkey agar.However, A. pleuropleuropneumoniae benefits from factor V that can be provided by chocolate agar or a staphylococcalstreak on blood agar. The growth of all the actinobacilliand particularly A. pleuropleuropneumoniae isimproved by 5-10 per cent CO_2 (a candle-jar is satisfactory).The inoculated plates are incubated at 37°C for24-72 hours.

Identification criteria for isolates

- Small, sticky, non-haemolytic colonies on blood agar
- Slow lactose fermentation on MacConkey agar
- Biochemical profile

Treatment and Control

Animals with discharging lesions should be isolated. Sodium iodide parenterally or potassium iodide orally is effective. Potentiated sulphonamides or a combination of penicillin and streptomycin are usually effective. Oral isoniazid for 30 days has been used in animals with refractory lesions. Rough feed or pasture which may damage the oral mucosa should he avoided.

Infections in other Animals Caused by A. *Lignieresii*

Cutaneous actinobacillosis of sheep presents as granulomatous lesions mainly on the head without tongue involvement. Granulomatous mastitis in sows, bite wounds in dogs and glossitis in a horse have been attributed to infection with A. lignieresii.

Pleuropneumonia of Pigs

Pleuropneumonia, caused by A. pleuropneumoniae, can affect susceptible pigs of all ages and occurs in major pig rearing regions worldwide. This highly contagious disease, primarily in pigs under 6 months of age, appears to be increasing in prevalence as a consequence of intensive rearing practices.

Pathogenesis and Pathogenicity

Virulence factors associated with A.pleuropneumoniae have been partially elucidated. Virulent strains possess capsules which are both antiphagocytic and immunogenic, whereas non-encapsulated strains are avirulent. Fimbriae and other adhesins allow the organisms to attach to cells of the respiratory tract. Actinobacillus pleuropneumoniae produces three related cytotoxins which belong to the repeats-in structural- toxin (RTX) cytolysin family. These toxins act by producing pores in cell membranes. Neutrophils chemically attracted to infected lung tissue are damaged and release lytic enzymes. The sustained inflammatory response is considered to be a major factor in causing rapid tissue necrosis.

Clinical Signs and Epidemiology

Subclinical carrier pigs, which are encountered in unaffected populations, harbour the organisms in the respiratory tract and tonsillar tissues. Poor ventilation and sudden drops in ambient temperature seem to precipitate disease outbreaks. Aerosol transmission occurs in confined groups. In outbreaks of acute disease, some pigs may be found dead and others show dyspnoea, pyrexia, anorexia and a disinclination to move. Blood-stained froth may be present around the nose and mouth, and many pigs show cyanosis.

Pregnant sows may abort. Morbidity rates can range from 30-50% and case fatality rates may reach 50%. Concurrent infections with Pasteurella multocida and mycoplasma may exacerbate the condition. At post-mortem areas of consolidation and necrosis are found in the lungs along with fibrinous pleurisy. Blood-stained froth may be found in the trachea and bronchi.

Diagnosis

There may be a history of ventilation failure or environmental temperature decrease prior to an outbreak of pulmonary disease. Specimens for laboratory examination should include tracheal washings or affected portions of lung tissue. Areas of haemorrhagic consolidation close to the main bronchi and severe fibrinous pleuritis may suggest this condition.

Specimens, cultured on chocolate agar and blood agar, are incubated in an atmosphere of 5-10% CO_2 at 37°C for 2 to 3 days.

Identification Criteria for Isolates

- Small colonics surrounded by clear haemolysis
- No growth on MacConkey agar
- Positive CAMP test with Staphylococcus aureus
- Biochemical profile
- Twelve serotypes and two biotypes are recognised and isolates belonging to biotype 1 require V factor (NAD) for growth whereas those belonging to biotype 2 are NAD-independent.
- The serotypes prevalent in a particular region should be identified prior to the implementation of vaccination programmes. Serological techniques are also used for epidemiological studies.
- Immunofluorescent techniques or PCR-based techniques may be used to demonstrate the organism in tissues.

Treatment

As antibiotic resistance is encountered in some strains, chemotherapy should be based on the results of antibiotic susceptibility testing. Prophylactic administration of antibiotics to in-contact pigs may limit the severity of clinical disease.

Control

Polyvalent bacterins may induce protective immunity but fail to prevent transmission or the development of a carrier state. A subunit vaccine containing toxoids of the three A. pleuropneumoniae toxins and capsular antigen has been developed. Predisposing factors such as poor ventilation, chilling and overcrowding should be avoided.

Sleepy Foal Disease

Sleepy foal disease is an acute, potentially fatal septicaemia of newborn foals caused by Actinobacillus equuli. Although primarily a pathogen of foals, A.equuli occasionally produces disease conditions, such as abortion, septicaemia and peritonitis, in adult horses. The organism is found in the reproductive and intestinal tracts of mares. Foals can be infected inutero and after birth via the umbilicus. Affected foals are febrile and recumbent. Death usually occurs in 1 to 2 days. Foals which recover from the acute septicaemic phase may develop polyarthritis, nephritis, enteritis or pneumonia. Foals dying within 24 hours of birth have petechiation on serosal surfaces and enteritis. Meningoencephalitis can be detectable histologically. Foals which survive for 1 to 3 days have typical pin-point suppurative foci in the kidneys.

Diagnosis

- History of the disease occurring on the premises in previous seasons.
- The clinical signs in a neonatal foal may suggest the disease.
- Specimens should be cultured on blood agar and MacConkey agar and incubated aerobically at 37°C for 1 to 3 days.
- Identification criteria for isolates:
  - Sticky colonies with variable haemolysis on blood agar
  - Lactose-fermenting colonies on MacConkey agar
  - Biochemical profile

Treatment and Control

Unless the disease is detected early, antimicrobial therapy is of little benefit. The organism is usually susceptible to streptomycin, tetracyclines and ampicillin. Supportive treatment includes blood transfusion arid bottle-feeding with colostrum. Mares which have had affected foals should be monitored closely at subsequent foalings. Good hygiene should be observed. Prophylactic antibiotic therapy may be considered fur newborn foals. No commercial vaccines are available.

Diseases Caused by the Pathogenic Actinobacilli

Species	Host(s)	Disease
A. lignieresii	Cattle Sheep Pigs	Bovine actinobacillosis: 'wooden (timber) tongue' and pyogranulomatous lesions around the head, neck and limbs and occasionally in the internal organs Pyogranulomatous lesions usually in head and neck region Rare granulomatous abscesses in mammary glands
A. equuli	Neonatal foals Postnatal foals Mares Pigs Calves, dogs, rats and rabbits	Sleepy foal disease (shigellosis or viscosum infection): septicaemia and those foals that survive for a few days develop purulent nephritis, pneumonia and arthritis Purulent arthritis, enteritis and infection of aneurysms Occasional abortion and/or septicaemia Arthritis in piglets and endocarditis in older pigs Occasional infections
A. suis	Pigs under 3 months-old Older pigs Neonatal and postnatal foals	Septicaemia Arthritis, pneumonia and pericarditis Some strains cause syndromes similar to those in foals associated with A equuli
A. pleuropneumoniae	Pigs	Acute: severe fibrinous pleuropneumonia. Chronic: pleuritis and pulmonary sequestration and abscessation. Morbidity and mortality high in non-immune pigs
A. capsulatus	Rabbits	Arthritis
A. actinomycetem comitans	Humans Rams	Role in periodontal disease and endocarditis Associated with epididymitis
A. seminis	Rams	Epididymitis
A.ureae	Humans Sows	Upper respiratory tract infections Reported as a cause of abortion

34

Haemophilus

Introduction

The genus Haemophilus contains small, nonmotile, nonsporing, oxidase positive, pleomorphic, gram negative bacilli that are parasitic on human beings or animals. Haemophilus means blood loving organisms.

Haemophilus species are small (less than 1 pm x 1 to 3 µm), Gram-negative rods, which often appear coccobacillary and may occasionally form short filaments. These motile organisms, which are facultative anaerobes with variable reactions in catalase and oxidase tests, do not grow on MacConkey agar. They are fastidious bacteria requiring one or both of the growth factors X (haemin) and V (nicotinamide adenine dinucleotide, NAD). Optimal growth occurs in an atmosphere of 5-10% CO_2 on chocolate agar which supplies both X and V factors. Small, transparent, dewdrop-like colonies are formed by most *Haemophilus* species after incubation for 48 hours.

Colonies of *H.somnus* have a yellowish hue and some isolates are haemolytic on sheep blood agar. The main pathogens in the genus are *H.somnus* in cattle and sheep, *H.parasuis* in pigs and *H.paragallinarum* which is responsible for infectious coryza of chickens.

The haemophili are motile, facultative anaerobes, produce acid from glucose, reduce nitrates and are variable in the oxidase and catalase tests. They are nutritionally fastidious, will not grow on MacConkey agar and grow best on chocolate agar (supplying the X and V factors) under 5- 10 per cent CO_2 at 37°C.

The growth of many of the *Haemaphilus* species is enhanced by 10 per cent CO_2, As this is not inhibitory for any of them, CO_2 should be used for routine isolation.The inoculated chocolate agar plates are incubated under 10 per cent CO_2 at 35-37°C for 3-4 days, although some growth may be seen after 24 hours.

Colonial Morphology

Small dewdrop-like colonies may appear after 24-48 hours' incubation and none are consistently haemolytic. A few strains of *H. somnus* may show afrank clearing around the colonies especially onColumbia-base sheep blood agar. *H. somnus* colonies may appear yellowish especially in a loopful of growth or on a confluent lawn.

Haemophilus piscium: H. piscium, responsible for 'ulcer disease' in trout, isprobably not a true *Haemophilus* species as it does notrequire either the X or V factors for growth. It is aGram-negative rod (0.5-0.7 x 2-3 μm), non-motileand will not grow at 37°C, the optimum temperaturefor growth being 20- 25°C. Media for isolation shouldbe supplemented with a peptic digest of fish tissue.Colonies develop to 1-3 mm in diameter and are circular,convex and cream-coloured. On blood agarcomplete haemolysis is produced.

Usual Habitat

Haemophilus species are commensals on the mucous membranes of the upper respiratory tract. They are susceptible to desiccation and do not survive for long periods away from their hosts.

Pathogenesis and Pathogenicity

Young or previously unexposed animals are particularly susceptible to infections by *Haemophilus* species. Specific-pathogen-free pigs, which do not harbour *H.parasuis* as a commensal, often develop signs of disease on primary exposure to the pathogen. Environmental stress factors appear to contribute to the development of *Haemophilus* infections. Although virulence factors have not been fully identified, endotoxin is thought to play a role in the pathogenesis of infections.

Haemophilus somnus can adhere firmly to several host cell types, including endothelial and vaginal epithelial cells. The organism is reported to cause degeneration of macrophages and it suppresses neutrophil function.

Degeneration of vascular endothelial cells and transmural neutrophil infiltration are prominent findings in thrombotic meningoencephalitis. Certain outer membrane proteins which confer virulence allow widespread dissemination of the bacteria in the host. Immunity to *H.somnus* appears to be predominantly humoral in nature. However, phase variation in the lipopolysaccharide antigen types may influence survival and persistence in host animals.

Diagnostic Procedures

Specimens for laboratory examination depend on the clinical condition and type of lesions. *Haemophilus* species are fragile and neither refrigeration nor transport media maintain viability. Ideally, clinical specimens should he frozen in dry ice and delivered to a laboratory within 24 hours of collection. Either chocolate agar or blood agar inoculated with a streak of S.*aureus* incubated under 5 - 10% CO_2 at 37°C for 2 to 3 days in a moist atmosphere is used for isolation.

Identification criteria for isolates:

- Small, dewdrop-like colonies after 1 to 2 days
- Enhancement of growth by CO_2
- Requirement for X and V growth factors
- Biochemical profile
- Although serological tests have been developed for epidemiological purposes, these tests are of little diagnostic value because *Haemophilus* species are widely distributed in animal populations.

Clinical Infections

The *Haemophilus* species which are pathogenic for animals tend to be host-specific. Some *Haemophilus* species of uncertain pathogenicity which are occasionally isolated from domestic animals.

Infections Caused by *H. Somnus* in Cattle

Haemophilus somnus is part of the normal bacterial flora of the male and female bovine genital tracts. The organism can also colonize the upper respiratory tract. Environmental stress factors contribute to the development of clinical disease. *Haemophilus somnus* is more resistant in the environment than most other *Haemophilus* species. It can survive in nasal discharges and blood forup to 70 days at ambient temperatures and for up to 5 daysin vaginal discharges. Transmission is by direct contact orby aerosols. Serological surveys indicate that at least 25%of cattle have antibodies to H.*somnus*.

Clinical Signs

Because septicaemia is commonly associated with *H.somnus* infection, many organ systems may be involved and the resulting clinical presentation can be variable. Thrombotic meningoencephalitis (TME), a common consequence of septicaemia, is encountered sporadically in young cattle recently introduced to feedlots. Some animals may be found dead and others may present with high fever and depression, sometimes accompanied by blindness, lameness and ataxia. Sudden death due to myocarditis has also been described. Arthritis often develops in animals which survive the acute phase of the disease.

Haemophilus somnus is one of the bacterial pathogens commonly isolated from the enzootic calf pneumonia complex. Sporadic cases of abortion, endometritis, otitis and mastitis caused by H.*somnus* have been recorded.

Diagnosis

Severe neurological signs in young feedlot cattle may be indicative of TME. Multiple foci of haemorrhagic necrosis, detectable grossly in affected brains at postmortem are consistent with TME. Vasculitis, thrombosis and haemorrhage are detectable histologically in brain, heart and other of the male and female bovine genital tracts and other parenchymatous organs.

Treatment and Control

Animals with clinical signs of septicaemia should be isolated and those at risk should be monitored closely to detect early signs of the disease. Although oxytetracycline is usually used for therapy, penicillin, erythromycin and potentiated sulphonamides are also effective. Commercially-available bacterins may reduce morbidity and mortality rates if administered one month before outbreaks of disease is anticipated.

Infections Caused by *Haemophilus Somnus* in Sheep

Healthy sheep may carry H.*somnus* in the prepuce or vagina. Epididymitis in young rams, caused by H.*somnus* has been recorded. Vulvitis, mastitis and reduced reproductive performance in ewes have been attributed to infection with *H.somnus.* The organism has also been associated with septicaemia, arthritis, meningitis and penumonia in lambs.

Infections Caused by *Haemophilus Somnus* in Pigs

Glasser's Disease

Glasser's disease, caused by H.*parasuis,* manifested as polyserositis and leptomeningitis usually affecting pigs from weaning up to 12 weeks of age. Some cases present as polyarthritis.

Haemophilus parasuis is part of the normal flora of the upper respiratory tract of pigs. Piglets acquire the organism from sows shortly after birth either by direct contact or through aerosols. The presence of maternally derived antibodies prevents the development of clinical signs. However, Glasser's disease may occur sporadically in 2 to 4 week old piglets subjected to stressful environmental conditions. Active immunity to *H. parasuis* is usually established by 7 to 8 weeks of age.

Clinical signs

The incubation period is 1 to 5 days. Clinical signs usually develop in conventionally-reared pigs 2 to 7 days following exposure to stress factors such as weaning or transportation. Anorexia, pyrexia, lameness, recumbency and convulsions are features of the disease. Cyanosis and thickening of the pinnae are often encountered. Pigs may die suddenly without showing signs of illness.

Diagnosis

Because organisms such as *Streptococcus suis* and *Mycoplasma hyorhinis* produce clinicopathological changes similar to those of Glasser's disease, diagnosis requires isolation and identification of *H.parasuis.* Postmortem findings in Glasser's disease may include fibrinous polyserositis, polyarthritis and meningitis.

Isolation and identification of H. *parasuis* from joint fluid, heart blood, cerebrospinal fluid or post-mortem tissues of a recently-dead pig is confirmatory.

Treatment and Control

Antimicrobial drugs such as tetracyclines, penicillins or potentiated sulphonamides, administered early in the course of the disease, are usually effective. Predisposing stress factors should be identified and, where possible, eliminated. Commercially-available bacterins or autogenous bacterins may stimulate protective immunity. Immunity is serotype specific.

Infectious Coryza of Chickens

Infectious coryza, caused by *H.paragallinarum,* affects the upper respiratory tract and paranasal sinuses of chickens. Its economic importance relates to loss of condition in broilers and reduced egg production in laying birds. Chronically ill and, occasionally, clinically normal carrier birds act as reservoirs of infection. Transmission occurs by direct contact, by aerosols or from contaminated drinking water. Chickens become susceptible at about 4 weeks after hatching and susceptibility increases with age.

Clinical signs

The mild form of disease manifests as depression, serous nasal discharge and slight facial swelling. In severe disease, swelling of one or both infraorbital sinuses is marked and oedema of the surrounding tissues may extend to the wattles. In laying birds, egg production may be severely affected. A copious, tenacious exudate may be evident at post-mortem in the infraorbital sinuses and tracheitis, bronchitis and airsacculitis may be present.

Diagnosis

Facial swelling is a characteristic finding. Isolation and identification of H. *paragallinarum* from the infraorbital sinuses of several affected birds is confirmatory. Immunoperoxidase staining can be used to demonstrate *H. paragallinarum* in the tissues of the nasal passages and sinuses. Serological tests such as agglutination tests, ELISA or agar gel immunodiffusion tests are used to demonstrate antibodies about 2 to 3 weeks after infection and to confirm the presence of *H. paragallinarum* in a flock.

Treatment and Control

Medication of water and feed with oxytetracycline or erythromycin should be initiated early in an outbreak of disease. An all-in-all-out management policy should be implemented and replacement birds should be obtained from coryza free stock. Good management of poultry units minimizes the risk of infection. Bacterins may be of value in units where the disease recurs. Vaccines should be administered about 3 weeks before outbreaks of coryza are anticipated.

Diseases of the *Haemophilus* Species

Species	Host(s)	Disease or significance
H. influenzae	Humans	Variety of diseases ranging from respiratory infections to meningitis
H. parainfluenzae	Humans	Normal flora of upper respiratory tract and implicated in urethritis
H. somnus	Cattle	• Infectious thromboembolic meningoencephalitis (TEME): septicaemia with infarcts in cerebellum • Respiratory disease: pneumonia and pleurisy, often in mixed infections with other agents • Genital infections such as endometritis and abortion. It can occur in semen and genital tract of bulls
	Sheep	Commensal of genital tract of sheep and reported as a cause of epididymitis and orchitis in rams. May also cause pneumonia, mastitis, polyarthritis, meningitis and septicaemia

Species	Host(s)	Disease or significance
H.parasuis	Pigs	• Primary agent of Glasser's disease: polyserositis and meningitis in young pigs • Arthritis and pneumonia in older pigs • Secondary invader in swine influenza or in enzootic pneumonia *(Mycoplasma hyopneumoniae)*
H.paragallinarum	Poultry Japanese quail	Infectious coryza of chicken: nasal discharge, sneezing and oedema of the face. reduction in egg production Highly susceptible to the infection
H.haemoglobinophilus	Dogs	Commensal of lower genital tract and sometimes causes cystitis and neonatal infections. Isolated from balanoposthitis and vaginitis but the role of the bacterium in these conditions is uncertain
H. aphrophilus	Humans Dogs	Part of the oral flora, found in dental plaque, and can be an occasional cause of endocarditis Recovered from the pharynx of dogs
H.ovis	Sheep	Rare reports of bronchopneumonia
H. paracuniculus	Rabbits	Significance not known. Isolated from the intestine of rabbits with mucoid enteritis
H.influenzaemurium	Mice	Respiratory infections and conjunctivitis
H. piscium	Trout	Ulcer disease: ulcerating inflammation of gills and mouth

35

Brucella

Brucella species are small (0.6 x 0.6 to 1.5 pm), nonmotile, coccobacillary, Gram-negative bacteria. As they are not decolourized by 0.5% acetic acid in the modified Ziehl-Neelsen (MZN) staining technique, they are classed as MZN-positive. In MZN-stained smears of body fluids or tissues, they characteristically appear as clusters of red coccobacilli. For taxonomic purposes, all *Brucella* species should be classified as *Brucella melitensis* as DNA hybridization studies have shown that the genus contains only one species.

Brucella species are aerobic, capnophilic and catalase-positive. Apart from *B.ovis* and *B.neotomae,* they are oxidase-positive. All *Brucella* species are urease-positive except *B.ovis. Brucella ovis* and some biotypes of *B.abortus* require 5 to 10% CO_2 for primary isolation. Moreover, the growth of other *Brucella* species is enhanced in an atmosphere of CO_2. Media enriched with blood or serum is required for culturing *B.abortus* biotype 2 and *B.ovis.* Recently, brucellae have been detected in sea-mammals.

Smears are made from specimens and stained by the modified Ziehl- Neelsen (MZN) stain. Brucellae appear as small, red-staining coccobacilli in clumps because of their intracellular growth.

Isolation

The brucellae grow well on 5-10 per cent blood agar. However, other than foetal abomasal contents and colostrum, the specimens are likely to contain many contaminating bacteria and fungi, so selective media are usually required. The selective media contain a nutritive blood agar base with 5 per cent sterile sero-negative equine or bovine serum and an antibiotic supplement. The antibiotic supplement used in selective media for *B.ovis* usually differs from that for *B. abortus.* Skirrow agar is a satisfactorymedium for both the *Campylobacter fetus* subspecies and for brucellae, including the most fastidious species such as *B. abortus* biotype 2, *B.canis* and *B.ovis.*

Colonial Morphology

After 3-5 days' incubation on selective serum agar, pinpoint, smooth, glistening, bluish, translucent colonies appear. As they age the colonies becomeopaque and about 2-3 mm in diameter. Strains of *B. abortus, B. suis, B. melitensis* and *B. neotomae* are usually in the smooth form when first isolated. Colonies of rough morphology occur in each of these species on subculture. *B. ovis* and *B. canis* are always in the rough form. The rough colonies are dull, yellowish, opaque and when touched with an inoculation loop are found to be friable. Brucellae are nonhaemolyticon blood agar.

Biochemical Tests

For routine identification, a few biochemical tests, together with colonial morphology and staining properties will presumptively identify the isolate as a *Brucella* sp. In summary, the brucellae are non-motile, catalase-positive, oxidase-positive (except *B. ovis* and *B. neotomae),* give rapid urease activity (except *B.ovis* and some strains of *B. melitensis),* reduce nitrate and are indole-negative. Using known positive antisera a rapid slide agglutination test can be used for *B.abortus.*

Usual Habitat

As a general rule, brucellae have a predilection for both female and male reproductive organs in sexually mature animals and each *Brucella* species tends to infect a particular animal species. Infected animals serve as reservoirs of infection which often persists indefinitely. Organisms, shed by infected animals, can remain viable in a moist environment for many months.

Pathogenesis and Pathogenicity

The establishment and outcome of infection with brucellae depend on the number of infecting organisms and their virulence and also on host susceptibility. Brucellae, which lack the major outer-membrane lipopolysaccharide, produce rough colonies and are less virulent than those derived from smooth colonies. Although smooth and rough organisms can enter host cells. Rough forms are usuallv eliminated unlike smooth forms which may persist and multiply. Virulent brucellae, when engulfed by phagocytes on mucous membranes, are transported to regional lymph nodes. Brucellae persist within macrophages but not within neutrophils. Inhibition of phagosome-lysosome function is a major mechanism for intracellular survival and an important determinant of bacterial virulence.

However, many of the mechanisms used by brucellae to survive within macrophages are not fully elucidated. Various stress proteins are thought to allow the organisms to adapt to harsh conditions encountered within macrophages. In addition, superoxide dismutase and catalase production may play a role in resistance to oxidative killing. Intermittent bacteraemia results in spread and localization in the reproductive organs and associated glands in sexually mature animals. Erythritol, a polyhydric alcohol which acts as a growth factor for brucellae, is present in high concentrations in the placentae of cattle, sheep, goats and pigs. This growth factor is also found in other organs such as the mammary gland and epididymis, which are targets for brucellae. In chronic brucellosis, organisms may localize in joints or inter vertebral discs.

Diagnostic Procedures

The diagnosis of brucellosis depends on serological testing and on the isolation and identification of the infecting Brucella species. Care should be taken during collection and transportation of specimens, which should be processed in a biohazard cabinet. Specimens for laboratory examination should relate to the specific clinical condition encountered. MZN-stained smears from specimens, particularly cotyledons, foetal abomasal contents and uterine discharges, often reveal characteristic MZN-positive coccobacilli. In specimens containing cells, the organisms appear in clusters.

A nutritious medium such as Columbia agar, supplemented with 5% serum and appropriate antimicrobial agents, is used for isolation. Plates are incubated at 37°C in 5 to 10% CO_2 for up to 5 days. Although CO_2 is a specific requirement for individual species, the majority of brucellae are capnophilic.

Serological testing is used for international trade and for identifying infected herds or flocks and individual animals in national eradication schemes. Brucellae share antigens with some Gram-negative bacteria such as *Yersinia enterocolitica* serotype 0:9, and consequently cross-reactions can occur in agglutination tests. The polymerase chain reaction can be used to detect brucellae in tissues.

Clinical Infections

Although each *Brucella* species has its own natural host, *B.abortus, B.melitensis* and biotypes of *B.suis* can infect animals other than their preferred hosts.

Bovine Brucellosis

Bovine brucellosis, caused by B.*abortus* and formerly worldwide in distribution, has been eradicated or reduced to a low prevalence in many countries through national eradication programmes. Although acquired most often by ingestion, infection can occasionally follow venereal contact, penetration through skin abrasions, inhalation or transplacental transmission. Abortion storms may be encountered in herds with a high percentage of susceptible pregnant cows.

Abortion usually occurs after the fifth month of gestation and subsequent pregnancies are usually carried to term. Large numbers ofbrucellae are excreted in foetal fluids for about 2 to 4 weeks following an abortion and at subsequent parturitions, although infected calves appear normal. Infection in calves is of limited duration in contrast to cows in which infection of the mammary glands and associated lymph nodes persists for many years. Brucellae may be excreted intermittently in milk for a number of years. In bulls, the structures targeted include seminal vesicles, ampullae, testicles and epididymides. In tropical countries, hygromas involvingthe limb joints are often observed when the disease is endemic in a herd.

In affected herds, brucellosis can result in decreased fertility, reduced milk production, abortions in susceptible replacement animals and testicular degeneration in bulls. Abortion is a consequence of placentitis involving both cotyledons and intercotyledonary tissues. In the bull, necrotizing orchitis occasionally results in localized fibrotic lesions.

Diagnosis

Clinical signs are not specific although abortions in first-calf heifers and replacement animals may suggest the presence of the disease.

Clusters of MZN-positive coccobacilli may be evident in smears of cotyledons and MZN-positive organisms may also be detected in foetal abomasal contents and uterine discharges.

Isolation and identification of B. *abortus* is confirmatory.

Identification criteria for isolates:

- Colonial appearance
- MZN-positive organisms
- Bacterial cell agglutination with a high-titered antiserum
- Rapid urease activity
- Biotyping
- A range of serological tests, varying in sensitivity and specificity, is available for the identification of infected animals.
- Brucellin, an extract of *B.abortus,* has been used for intradermal testing.

- Molecular methods, such as PCR-based techniques, for the detection of brucellae in tissues and fluids have been described.

Treatment and Control

- Treatment of cattle with brucellosis is not practical. Three types of vaccines are used in cattle, attenuated strain 19 (S19) vaccine, adjuvanted 45120 vaccine and the more recent RB51 vaccine:
- S19 vaccine is administered to female calves up to *5* months of age. Vaccination of mature animals leadsto persistent antibody titres.
- 45120 bacterin although less effective, has been used in some national eradication schemes. Even when administered to adult animals, this vaccine does not induce persistent antibody titres.
- RB51 strain is a stable, rough mutant which induces good protection against abortion and does not result in serological responses detectable in tests used in conventional brucellosis surveillance programmes.

Diseases and Principal Hosts of the *Brucella* Species

Species	Host(s)	Diseases
B. *abortus*	CATTLE Sheep, goats and pigs Horses Humans	Abortion and orchitis Sporadic abortion Associated with bursitis (poll evil and fistulous withers) Undulant fever
B. *melitensis*	GOATS, Sheep Cattle Humans	Abortion Occasional abortion and excretion in milk Malta fever
B.suis	PIGS Humans	Abortion, orchitis, arthritis, spondylitis and herd infertility Undulant fever
B. ovis	SHEEP	Epididymitis in rams and sporadic abortion in ewes
B.canis	DOGS Humans	Abortion, epididymitis, discospondylitis and permanent infertility in males Undulant fever
B. *neotomae*	Desert wood rat *(Neotoma lepida)*	Non-pathogenic for the wood rat and has not been recovered from any other animal species

36

Vibrio and Related Organism

Introduction

Vibrios are Gram-negative, rigid, curved rods that are actively motile by means of a polar flagellum. The name 'Vibrio' is derived from the characteristic vibratory motility (from vibrare, meaning to vibrate). They are asporogenous and noncapsulated. Vibrios are present in marine environments and surface waters worldwide. The most important member of the genus is *Vibrio cholerae*, the causative agent of cholera. It was first isolated by Koch (1883) from cholera patients in Egypt, though it has been observed earlier by Pacini (1884) and others.

Natural Habitat

Vibrio spp. can be present in both freshwater and seawater as well as in the alimentary tracts of animals and man.

At least five *Vibrio* spp. are human pathogens including V. *cholerae,* the cholera bacillus, and V.*parahaemolyticus* which causes food poisoning. Only V. *metschnikovii* is associated with disease in domestic animals. It causes a cholera-like disease in chickens and other birds but its geographical distribution is very limited.

V. *anguillarum* causes infections in many species of fish especially in salt or brackish water. It causes high mortality in salt water eels.

Vibrio Cholerae

Morphology

The cholera vibrio is a short, curved, cylindrical rod, about 1.5 × 0.2-0.4 mm in size, with rounded or slightly pointed ends. The cell is typically comma shaped (hence the old name V comma) but the curvature is often lost on subculture. S-shaped or spiral form may be seen due to two or more cells lying end to end. Pleomorphism is frequent in old cultures. In stained films of mucus flakes from acute cholera cases, the vibrios are seen arranged in parallel rows, described by Koch as the 'fish in stream' appearance. It is actively motile, with a single sheathed polar flagellum. The motility is of the darting type, and when acute cholera stool or a young culture is examined under the microscope, the actively motile vibrios suggest a 'swarm of gnats'. The vibrios stain readily with aniline dyes and are Gram negative and non-acid fast.

Culture Characteristics

The cholera vibrio is strongly aerobic, growth being scanty and slow anaerobically. It grows within a temperature range of 16-40°C (optimum 37°C), Growth is better in an alkaline medium, the range of pH being 6.4-9.6 (optimum 8.2). NaCl (0.5-1%) is required for optimal growth though high concentration (6% and above) are inhibitory. It grows well on ordinary media. On MacConkey's agar, the colonies are colourless at first but become reddish on prolonged incubation due to the late fermentation of lactose. On blood agar, colonies are initially surrounded by a zone of greening, which later become clear due to haemodigestion. In peptone water, growth occurs in about six hours as a fine surface pellicle, which on shaking breaks up into membranous pieces.

V.*metschnikovii:* smooth, transparent colonies that are 2-4 mm in diameter at 48 hours. It may be haemolytic on blood agarand grows poorly on MacConkey agar.

V.anguillarum: small, smooth colonies within 48 hours. It is haemolytic on blood agar.

V.parahaemolyticus: moderate-sized (about 2 mm diameter) colonies in 24 hours, non-haemolytic on sheep blood agar. It forms greenish colonies on TCBS agar (BBL) and ferments mannitol causing an acid reaction (yellow) on mannitol salt agar.

A number of special media have been employed for the cultivation of cholera vibrios. They may be classified as follows:

Holding or Transport Media

1. Venkatraman – Ramakrishnan (VR) medium: In this medium vibrios do not multiply but remain viable for several weeks.
2. Cary-Blair medium: It is a suitable transport medium for Salmonella and Shigella as well as for vibrios.
3. V. *anguillarum:* blood agar with nutrient agar base and 2.0 per cent NaCl at 20°C for 48 hours.

4. V. *merschnikovii* and V. *parahaemoLyticus:* nutrient agar or blood agar at 37°Cfor 24 48 hours.

Enrichment Media

1. Alkaline peptone water at a pH of 8.6
2. Monsur's taurocholate tellurite peptone water at pH 9.2

   Both these are good transport as well as enrichment media.

Plating Media

1. Alkaline bile salt agar (BSA) at pH 8.2: The colonies after overnight growth, colonies are moist, translucent, round discs, about 1-2mm in diameter; with a bluish tinge in transmitted light.
2. TCBS medium: This medium, containing thiosulfate, citrate, bile salts and sucrose, is available commercially and is very widely used at present. Cholera vibrios produce large yellow convex colonies which may become green on continued incubation.
3. *V. parahaemolyticus* will grow on TCBS agar (BBL) formulated for V. *cholerae* and other enteric vibrios. Being halophilic, it also grows on mannitol salt agar (BBL) containing 7 .5 per cent NaCl intended for the pathogenic staphylococci.

Biochemical Reactions

Carbohydrate metabolism is fermentative, producing acid, but no gas. Cholera vibrios ferment glucose, mannitol, maltose, mannose and sucrose but not inositol, arabinose or lactose, though lactose may be split very slowly. Indole is formed and nitrates are reduced to nitrites. These two properties contribute to the 'cholera red reaction' which is tested by adding a few drops of concentrated sulphuric acid to a 24-hour peptone water culture.

With cholera vibrios, a reddish pink colour develops due to the formation of nitroso-indole, Catalase and oxidase testes are positive. Methyl red and urease tests are negative. Vibrios decarboxylate lysine and ornithine but not utilize arginine.

V. cholerae and *V.mimicus* have only a slight requirement for Na+ (NaCI), but most of the halophilic *Vibrio* spp. requires the supplementation of biochemical tests with 1 per cent NaCl. *V.parahaemolyticus* belongs to the lysine-decarboxylase-positive, arginine-dihydrolase- negative group of *Vibrio* spp. It is distinguished from other members of the group by negative reactions for sucrose, salicin and cellobiose fennentation but a positive reaction for arabinose. *V.metschnikovii* is the only clinically significant *Vibrio* sp. that is oxidase negative and nitrate-reductase-negative.

Resistance

Cholera vibrios are susceptible to heat, drying and acids, but resist high alkalinity. They are destroyed at 55°C in 15 minutes. Dried on linen or thread, they survive for 1-3 days but die in about three hours on cover slips. Survival in water is influenced by its pH temperature, salinity, presence of organic pollution and other facts.

Classification

1. Serological Classification:

VIBRIOS are classified according to biochemical properties & H antigen into Group A and Group B.

- Group A: This group includes vibrio cholarae and have similar biochemical properties and Common H (flagellar) Antigen. Group A is again classified based on somatic O antigen into O1 & Non-O1
- O1 (agglutinable vibrio): These vibrios agglutinate with O1 anitserum.
- (ii) Non-O1 (non-agglutinable vibrio): This are not agglutinated by O1 antiserum. They are currently upto O-139.
- O1 are classified into Biotypes-CLASSICAL and ELTOR and these are further classified into serotypes- OGAWA, INABA, HIKOJIMA
- Subgroup O1 can further be classified into two groups as follows;

Characteristics	Classical	Eltor
Hemolysis	-	+
Voges Proskauer test	-	+
Chick erythrocyte agglutination	-	+
Polymyxin B Sensitivity	+	-
Group IV phage susceptibility	+	-
Group V phage susceptibility	-	+

2. According to growth requirement

- Genus Vibrio includes 33 species the important are:
- Non halophillic: It includes following & they are able to grow in media without salts. *Eg. V. cholera*

- Halophillic: They are unable to grow in media without salts – have a high requirement of NaCl.

Eg.

V. Parahemolyticus is responsible for food poisoning

V. alginoltyticus is responsible for diarrheal disease

V. vulnificus is responsible for diarrheal disease

Pathogenecity of Vibrio Cholerae

Cholera is an acute diarrheal disease caused by *V.cholerae*. In its most severe form, cholera is a dramatic and terrifying illness in which profuse painless watery diarrhea and copious effortless vomiting may lead to hypovolemic shock and death in less than 24 hours. In treated cases, the disease may last 4-6 days, during which period the patient may pass a total volume a liquid stool equal to twice his body weight.

Clinical Feature of Cholera

- All the clinical features of severe cholera result from this massive loss of fluid and electrolytes.
- The cholera stool is typically a colourless watery fluid with flecks of mucus, said to resemble water in which rice has been washed (hence called 'rice water stools).
- The clinical severity of cholera varies widely, from the rapidly fatal disease to a transient asymptomatic colonization of the intestine by the vibrios.
- The incidence of mild and asymptomatic infections is more with ELTOR vibrios than with the classical cholera vibrios.
- The incubation period varies from less than 24 hours to about five days. The clinical illness may begin slowly with mild diarrhea and vomiting in 1-3 days or abruptly with sudden massive diarrhea.

Pathogenesis

- Vibrio Cholerae enters through orally contaminated food & water.
- If gastric acidity is less then they will be destroyed otherwise they passes to intestine attaches the mucus membrane breach the epithelial cell & attach to produce toxin.

Cholera Toxin

- It is very similarly to heat labile toxin of E.coli in structural, biochemical, biological & antigenic properties but is more potent.
- It's production is determined by filamentous phage integrated with bacterial chromosome.

Mode of Action

- Cholera Toxin has 2 sub units Sub-Unit A and Sub-Unit B
- Sub-Unit B binds to GM1 ganglioside on gut epithelial cells.
- This binding activates A subunit & it divides into A1 & A2.
- A1 activates adenly cyclase wich converts ATP to CAMP.
- CAMP increases the outflow of electrolytes & water to gut lumen & causes diarrhea.
- Sub-Unit A2 act as binding agent for A1 to B

Laboratory Diagnosis

Specimen

1. Stool

- Collected in acute stage of the disease before administration of Antibiotics.
- Collected by introducing lubricated catheter into rectum or directly in screw capped bottle.

2. Rectal Swab

- Very useful in convalescent carrier in which there is no watery diarrhea.
- Transport Media for Cholera are as follows If chances of delay occur then use V-R medium, Bile peptone transport medium, Monsur's medium, Carry Blair medium, Autoclaved sea water

Methods

The Methods for demonstration of the disease are as follows:

1. Direct Demonstration of bacilli
2. Microscopy: For rapid diagnosis, the characteristic darting motility of vibrios and its inhibition by antiserum can be demonstrated hanging

drop method using cholera stool from acute cases or more reliably after enrichment for 6 hours.

3. Culture Media

- Enrichment media – Alkaline peptone water
- For the plating media are – Bile Salt Agar, & TCBS

Method

If the laboratory facilities are available then the specimen which is collected is directly inoculated on plating media otherwise it is transported by holding media then inoculated on enrichment media for 6-8 hrs. Then subculture on plating media after colonies appear the slide agglutination test with O1 antiserum is done after the species is biochemically identified.

4. Biochemical Reactions

V. Cholera is identified from other by Biochemical reaction as follows it shows Oxidase positive, Catalese Positive, Indole Positive, V.P. Positive test & Citrate Positive but it shows Urease & Methyl red test negative.

5. For isolation of carrier

- More than one cycle of enrichment is required.
- As the excretion of bacilli is intermittent repeated stool examination is required.

Prophylaxis

- Provision of protected water supply and improvement of environmental sanitation
- An ideal cholera vaccine is yet to be found.

Diseases and Hosts of *Aeromonas, Plesiomonas* and *Vibrio* Species

Species	Host(s)	Disease
Vibrio parahaemolyticus	Humans	Food poisoning, associated with seafoods
V. metschnikovii	Chickens and other Birds	Cholera-like enteric disease
V. anguillarum	Salt·water eels and other fish	Skin necrosis, red areas on skin and generalisation with high mortality

37

Campylobacter

Campylobacter species are slender, curved motile Gram negative rods (0.2 to 0.5 pm wide) with polar flagella. Daughter cells which remain joined have a characteristic gull-winged appearance and long spirals formed by joined cells also occur. These microaerophilic organisms grow best on enriched media in an atmosphere of increased CO_2 and decreased oxygen tension. Many Campylobacter species grow on MacConkey agar. They are non-fermentative and oxidase-positive and have variable catalase reactions.

Usual Habitat

Many Campylobacter species are commensals in the intestinal tracts of warm-blooded animals. Campylobacter jejuni and C. lari (formerly C. laridis) colonize the intestines of birds, which can result in faecal contamination of water courses and stored food. A number of Campylobacter species are excreted in the faeces of pigs. Campylobacter fetus subspecies venerealis appears to be adapted principally to bovine preputial mucosa.

Differentiation of Campylobacter Species

Campylobacter species are strictly microaerophilic, requiring an atmosphere of 5 to 10% oxygen and 1 to 10% CO_2 for growth. A selective enriched medium such as Skirrow agar is usually used for primary isolation. Differentiation of isolates is based on colonial morphology and certain cultural, biochemical and antibiotic-susceptibility characteristics.

Colonial Morphology

- Campylobacter fetus subspecies venerealis and C.fetus subspecies fetus have small, round, smooth, translucent colonies with a dewdrop appearance.
- Campylobacter jejuni produces small, flat, grey colonies with a spreading, watery appearance.
- Other Campylobacler species vary somewhat in colonial appearance. C. coli produces a pink-tan pigment and the growth of C. mucosalis may

appear as a dirty-yellow colour on the inoculating loop.

- Colonies of some Campylobacter species, which may contaminate clinical specimens, can be slightly pigmented.
- Because Campylobacter species do not ferment carbohydrates, other metabolic activities of these organisms must be used for identification.

Biochemical and other Tests

Susceptibility or resistance to nalidixic acid or cephalothin, hydrogen sulphide production, nitrate reduction, growth at 25°C or 45°C and the catalase reaction are some of the criteria on which a definitive identification of the Campylobaeter species is based.

C. jejuni is the only species that hydrolyses sodium hippurate. For this test a large loopful of a 24 48 hour culture of C. jejuni is emulsified in 0.4 ml of a 1 per cent aqueous solution of sodium hippurate and incubated at 37°C for 2 hours. Then 0.2 ml of a ninhydrin solution is added to the tubes at 37°C. A positive reactionis given by a deep purple colour developing after10 minutes. The ninhydrin solution is prepared by adding 3.5 g of ninhydrin to 100 ml of a 1:1 mixture of acetone and butanol.

Pathogenesis and Pathogenicity

Campylobacter fetus subspecies venerealis and C.fetus subspecies fetus are structurally unusual in that they possess a microcapsule or S layer, which consists of high molecular weight proteins arranged in a lattice formation. This S layer confers resistance to serum-mediated destruction and phagocytosis and enhances survival in the genital tract. Virulence factors attributed to C.jejuni include soluble components with enterotoxin-like activity and poorly defined mechanisms for attaching to and invading host enterocytes. The contribution of heat stable endotoxin to the pathogenesis of campylobacteriosis is uncertain.

C. jejuni produces an adhesion, a cytotoxin and a heat-labile toxin similar to that of Escherichia coli. Transmission of many of the Campyloabacter species, including C. fetus subsp. fetus, is by the faecal-oral route. C. fetus subsp. vellerealis is transmitted by coitus and infection of the female genital tract may lead to metritis with resulting death and resorption of the embryo (infertility) or occasionally to abortion.

Diagnostic Procedures

Details of the diagnostic methods for individual clinical conditions are presented in relevant sections. Irrespective of the source of specimens for bacterial isolation, certain general principles relating to culture techniques apply. Campylobacter species require microaerophilic conditions for growth, usually supplied by commercially-available generator envelopes which deliver 6% oxygen, 10% carbon dioxide and 84% nitrogen. Although most pathogenic species grow optimally at 37'C, C.jejuni requires up to 5 days at 42°C for optimum growth.

Smears from cultures and from clinical specimens should be stained with dilute carbol fuchsin (DCF) for 4 minutes. This method stains the organisms more intensely than the Gram method.

Identification criteria for isolates:

- Growth only under rnicroaerophilic conditions
- Colonial morphology
- Cell morphology in smears stained with DCF or by immunofluorescence
- Metabolic characteristics and antibiotic susceptibility pattern.

Clinical Infections

The most important consequences of infections with organisms in this group are infertility in cattle due to C.fetus subspecies venerealis and abortion in ewes caused either by C.fetus subspecies fetus or by C.Jejuni.

Bovine Genital Campylobacteriosis

Campylobacter fetus subspecies venerealis, the principal cause of bovine genital campylobacteriosis, is transmitted during coitus to susceptible cows by asymptomatic carrier bulls. The bacteria survive in the glandular crypts of the prepuce and bulls may remain infected indefinitely. The disease is characterized by temporary infertility associated with early embryonic death; return lo oestrus at irregular periods and, occasionally, by sporadic abortion. About one-third of infected cows become carriers.

Campylobacter fetus subspecies venerealis persists in the vagina of carrier cows, a feature attributed to antigenic shifts in the immunodominant antigens of the S layer proteins. Extension of infection to the uterus with the development of endometritis and salpingitis can occur during the progestational phase of the oestrus cycle. Campylobacter fetus subspecies fetus, an enteric organism acquired by ingestion, can cause sporadic abortions in cows.

Diagnosis

Investigation of the breeding records and vaccination history of an affected herd may suggest campylobacteriosis. Campylobacter species can be detected by the fluorescent antibody technique in sheath washings from bulls or cervicovaginal mucus from cows. Isolation and identification of C.fetus subspecies venerealis from preputial or vaginal mucus is confirmatory.

Vaginal mucus agglutination test detects about 50% of infected, infertile cows on a herd basis. An ELISA can be used to demonstrate IgA antibodies in vaginal mucus after an abortion. A polymerase chain reaction has been developed as a rapid screening test for the detection of C. fetus subspecies venerealis in bull's semen.

Treatment and Control

Dihydrostreptomycin, administered either systemically or topically into the prepuce, is used for treating bulls. Intrauterine administration of dihydrostreptomycin can be used therapeutically. Vaccination with bacterins in an oil emulsion adjuvant is used therapeutically and prophylactically in problem herds.

Pathogenic and Nonpathogenic Campylobacter Species

Species	Principal host(s)	Disease
C. fetus subsp. *venerealis*	Cattle	Bovine genital campylobacteriasis (epizootic bovine infertility): infertility, early embryonic death and occasional abortion. Prepuce of asymptomatic bulls
C. fetus subsp. fetus	Sheep Cattle Man Cattle, sheep	Ovine genital campylobacteriosis: outbreaks of abortion Occasional abortions Occasional infections Commensal in the intestinal tract
C. jejuni	Sheep Dogs, cats, other animals and man Poultry Many domestic and wild animals and birds	Outbreaks of abortion Enteritis with diarrhoea Avian vibrionic hepatitis Commensal in the intestinal tract

C. mucosalis	Pigs	Associated with the porcine intestinal adenomatosis complex. Present in intestinal tract of normal pigs
C. hyointestinafis	Pigs	Associated with the porcine intestinal adenomatosis complex. Commensal in the intestine of normal pigs
C. coli	Pigs, man	May cause mild diarrhoea in pigs and enteritis in man. Commensal in the intestine of pigs. Tends toNincrease in numbers in pigs with swine dysentery caused by Serpulina hyodysenteriae
C. cryaerophila	Cattle, pigs, sheep, horses	Isolated from faeces of normal animals and infrequently from aborted foetuses. Significance Unknown
C. laridis	Dogs, horses, birds	Isolated from faeces. Disease status uncertain
C. sputorum biovar sputorum	Man	Commensal in oral cavity. Considered nonpathogenic
C. spulorum biovar bubulus	Cattle	Commensal in preputial cavity of bulls and genital tract of cows. Considered to be non-pathogenic
C. sputorum biovar fecalis	Sheep, cattle	Present in the intestinal tract and has been isolated from semen and vagina of cattle. Considered to be Non-pathogenic
C. upsaliensis	Dogs, man	Isolated from diarrhoeic and normal individuals. Disease status uncertain

38

Bordetella

The bordetellae are small (0.2- 0.5 x 0.5- 1.0 μm), Gram-negative rods that tend to be coccobacillary.They are strict aerobes and do not attack carbohydratesbut derive energy by the oxidation of amino acids. B. avium and B. bronchiseptica are motile by peritricholls flagella but B. pertussis and B. parapertussis are non-motile. All are catalase-positive and oxidase-positive. B. bronchiseptica and B. aium willgrow on MacConkey agar.

The genus Bordetella contains four species, B.pertussis, B.parapertussis, B. bronchiseptica and B. avium. Bordetella pertussis, the type species, and B.parapertussis are human pathogens associated with whooping cough in children. Bordetella bronchiseptica infects a wide range of animal species including man, while B.avium is a pathogen of avian species. The bordetellae are occasional pathogens which have an affinity for ciliated respiratory epithelium.

Bordetella bronchiseptica and B.avium are small (0.2 to 0.5 x 0.5 to 1.5 pm), Gramnegative rods with a coccobacillary appearance. They are catalase-positive, oxidase-positive aerobes and are motile peritrichous bacteria. Because they cannot utilize carbohydrates, they derive their energy mainly from the oxidation of amino acids and have no special growth requirements. They grow on MacConkey agar.

Natural Habitat

The bordetellae are inhabitants primarily of the upper respiratory tract of healthy and diseased humans, animals and birds. B. pertussis and B. parapertussis are human pathogens causing whooping cough and a mild form of whooping cough, respectively. B. bronchiseptica can be present in the upper respiratory tract of pigs, dogs, cats, rabbits, guinea-pigs, rats, horses and possibly other animals. B. avium inhabits the respiratory tract of infected poultry, principally turkeys. Mammalian infections are mainly transmitted by aerosols but in turkeys indirect spread can occur via water and litter.

Culture

The routine med ia used are sheep blood and MacConkey agars. B. avium and B. hronchiseptica grow well on both media. The plates are incubated aerobically at 37°C for 24-48 hours.If isolations are to be attempted from specimens containing a large number of bacterial contaminants, such as nasal swabs, a selective medium is required. The reason is two-fold - to prevent overgrowth and to maintain the alkaline to neutral conditions for the bordetellae. Even a few fermentative bacteria on a medium containing carbohydrates can produce sufficient acid to inhibit the Borderella species.

Several selective media have been described, such as MacConkey agar with 1 per cent glucose and 20 µg/ml furaltadone or blood agar with 20 µg/ml of clindamycin and 4 µg/ml neomycin.

B. avium will grow well on Smith-Baskerville (SB) medium medium,with or without the antibiotic supplement. The inoculatedSB medium is incubated aerobically at 37°C for 48 hours.

Colonial Morphology

On sheep or horse blood agar B. bronchiseptica forms very small, convex, smooth colonies with an entire edge after 24 hours. Some strains may be haemolytic. The colonies of B.avium are similar but are non-haemolytic. Phase modulation occurs in both species and this is thought to be due to loss of a capsule-like structure on subculture. The virulent, encapsulated phase I colonies are convex and shiny, those of phase II are larger, circular and convex with a smooth surface and the avirulent phase III colonies are large, flat and granular with an irregular edge.

The colonies on MacConkey agar are small, pale with a pinkish hue and amber discolouration of the underlying medium. B.avium and B. bronchiseptica have colonies of similar appearance on MacConkey agar.

Smith- Baskerville (SB) medium contains the pH indicator bromothymol blue and the agar is green at pH 6.8. The colonies of B. avium and B. bronchiseptica,after 24 hours incubation, are small (0.5 mm in diameteror less), blue colonies with a lighter blue (alkaline)reaction in the medium around them. After 48hours' incubation, the colonies are 1.0-2.0 mm in diameter,blue or blue with a green centre and the surroundingmedium is blue.

Biochemical Reactions

B. bronchiseptica is positive to the oxidase, catalase, citrate, urease and nitrate tests. It is motile and carbohydrates are not utilized. B. avium and Alcaligenes faecalis have similar reactions to those of B. bronchiseptica but are urease-negative and nitrate-negative.

Pathogenesis and Pathogenicity

The bordetellae exhibit phase changes, which correlate with virulence and are identifiable by colonial appearance. Virulence is mediated by several factors including a filamentous haemagglutinin, pertactin and fimbriae which allow attachment to the cilia of the upper respiratory tract. These factors are only expressed in the virulent phase (phase 1) and are controlled by a virulence gene regulatory system. After repeated subculture, isolates change to an avirulent form (phase 4) and colonies exhibit a different morphology which reflects alterations in bacterial structure. Phases 2 and 3 are poorly defined.

The tracheal cytotoxin inhibits ciliary motility and tracheobronchial clearance. In addition, B.bronchiseptica produces an adenylate cyclase-haemolysin which primarily targets phagocytic cells. This toxin is unique as it has the features of a repeats-in-structural-toxin but with an extra domain for an adenylate cyclase enzyme. Although B.avium lacks the filamentous haemagglutinin, the organism does produce a haemagglutinin, which specifically agglutinates guinea-pig red cells and correlates with pathogenicity for turkey poults. Two other toxins, dermonecrotic toxin and osteotoxin, may be of significance in atrophic rhinitis by contributing to nasal turbinate atrophy.

Diagnostic Procedures

Specimens for laboratory examination include nasal swabs, tracheal aspirates and exudates. Bordetellae are cultured on blood agar and MacConkey agar or on selective media. Plates are incubated aerobically at 37°C for 24 to 48 hours.

Identification criteria for isolates:

- Colonial appearance on blood agar or selective media.
- Growth on MacConkey agar.
- Biochemical profile.
- Slide baemagglutination tests correlating with the virulence of isolates.
- Serological tests which have been developed are of limited diagnostic value.

Clinical Infections

Clinical signs associated with bordetellae usually relate to upper respiratory tract infection. Young animals are most susceptible and infections in adults are usually mild or subclinical. Predisposing factors such as stress or concurrent infections contribute to field outbreaks of disease. Although morbidity rates may be high, mortality rates are usually low. Bordetella parapertussis, a recognised human pathogen, has been isolated from lambs with chronic nonprogressive pneumonia.

Bordetella bronchiseptica is implicated in a mild form of atrophic rhinitis in pigs and in canine infectious tracheobronchitis (kennel cough). Oropharyngeal swabs from healthy cats may yield B.bronchiseptica and severe bronchopneumonia associated with the organism has been reported in kittens. Bordetella bronchiseptica may occasionally cause outbreaks of respiratory disease in rabbits and in laboratory rodents. Bordetella avium causes turkey coryza and respiratory disease in quails.

Canine Infectious Tracheobronchitis

Canine infectious tracheobronchitis, also known as kennel cough, is one of the most prevalent respiratory complexes of dogs. Although Bordetella bronchiseptica, canine parainfluenzavirus 2 (PI-2) and canine adenovirus 2 (CAV 2) are considered to be the most important participating pathogens, other microbial pathogens may also be involved.

Clinical signs

Clinical signs of infection with B.bronchiseptica develop within 3 to 4 days of exposure and, without complications, persist for up to 14 days. They include coughing, gagging or retching and mild serous oculonasal discharge. Affected dogs usually remain active, alert and non-febrile. The disease is self-limiting unless complicated by bronchopneumonia which may develop in unvaccinated pups or in older immunosuppressed animals.

Diagnosis

Diagnosis is based on a history of recent exposure to carrier dogs and characteristic clinical signs. The appropriate specimen for laboratory examination is transtracheal aspiration fluid. Virulent isolates of B.bronchiseptica haemagglutinate ovine and bovine red cells. Serology, in association with vaccination history, may be of value for determining the involvement of respiratory viruses.

Treatment

Dogs with mild clinical signs do not require specific therapy. If coughing persists for more than 2 weeks or if bronchopneumonia is present, antibiotic therapy may be required. Amoxicillin has proved effective in field trials. Tetracyclines and fluoroquinolones may also be effective.

Control

Affected dogs should be isolated immediately. If predisposing factors are identified, they should be corrected. lntranasal vaccines containing B. bronchiseptica and

PI-2 antigens induce local protective immunity and are not affected by maternal antibodies. Modified live B. bronchiseptica vaccines decrease the severity of clinical signs but may not prevent infection. Modified live vaccines are available for many of the viruses associated with respiratory disease in dogs.

Turkey Coryza

Turkey coryza, caused by B.avium, is a highly contagious upper respiratory tract disease of poults with high morbidity and low mortality. Infection is spread through direct contact, by aerosols and from environmental sources. Mucus accumulates in the nares with swelling in the submaxillary sinuses. Beak-breathing, excessive lacrimation and sneezing may be evident. Infection with B. avium predisposes to secondary infections with bacteria such as Escherichia coli. Once E. coli becomes established, a more serious disease with high mortality can develop.

Diagnosis

Clinical signs and gross pathological features may be indicative of the disease. Isolation and identification of B. avium from sinus and tracheal exudates is confirmatory.

Virulent isolates agglutinate guinea-pig red blood cells. Microagglutination and ELlSA techniques may be of diagnostic value.

Treatment and Control

Broad spectrum antibiotics early in the course of disease may be beneficial. Commercially available bacterins and modified live vaccines may be used in susceptible flocks. Thorough cleaning and disinfection of turkey houses after an outbreak of disease are essential for the elimination of B. avium.

Diseases and Hosts of the Bordetellae

Species	Host(s)	Disease
Bordetella pertussis	Humans Chimpanzees	Classical form of whooping cough Rare case of whooping cough·like disease in captive Animals
B. parapertussis	Humans Lambs	Mild form of whooping cough Isolated from healthy and pneumonic lambs. Significance not known
B. bronchiseptica	Pigs Dogs Rabbits Guinea pigs and rats Horses, cats Humans	Atrophic rhinitis (with or without AR+ strains of Pasteurella multocida) Bronchopneumonia seen in young pigs Canine infectious tracheobronchitis (kennel cough) with or without concurrent respiratory viruses. Secondary invader in canine distemper 'Snuffles' like syndrome with upper respiratory tract infection, bronchopneumonia or septicaemia Similar to the disease in rabbits Respiratory infections (not common) Occasionally isolated from wounds and body flu ids (presumed to be zoonotic infections)
B. avium	Turkeys and less commonly other birds	Turkey coryza: rhinotracheitis and sinusitis in young poults. Morbidity high but mortality low

39

Moraxella

Moraxella bovis occurs as short (1.0 to 1.5 x 1.5 to 2.5 μm), plump Gram-negative rods or, occasionally, cocci which typically occur in pairs. This organism is non-motile, aerobic and usually catalase-positive and oxidase-positive. Although proteolytic, it is unable to utilize sugars. Growth, which is enhanced by the addition of blood or serum to media, does not occur on MacConkey agar. The optimal temperature for growth is 33-35°C. Most *M.phenylpyruvica* strains will grow on MacConkey agar. But *M.bovis* and *M.lacunata* are unable to do so.

Virulent strains, when isolated from cases of infectious bovine keratoconjunctivitis, are fimbriate, haemolytic and grow into the agar. Species, other than M.bovis, which are periodically isolated from clinical specimens, are generally regarded as non-pathogenic.

Usual Habitat

Moraxella bovis is found on mucous membranes of carrier cattle. The organism is susceptible to desiccation and is short-lived in the environment. It can survive for up to 72 hours in the salivary organs and on the body surface of flies, which can act as vectors.

Colonial Appearance

On blood agar after 48 hours' incubation the colonies of *M.bovis* are flat, round, small (1 mm diameter), greyish-white and friable, surrounded by a narrow zone of complete haemolysis. The appearance is not unlike that of a beta-haemolytic streptococcus. New isolates are often piliated and erode the agar, sinking into it. Colonial growths will autoagglutinate when suspended in saline. On subculture, colonial variation is common, pili are no longer formed and the colonies are butyrous and less likely to autoagglutinate. Some colonies can become non-haemolytic. The strains of *M. hovis (M.equi)* isolated from horses are non-haemolytic even on primary isolation. *M.lacunata* and *M. phenylpyruvica* are non-haemolytic on blood agar and some strains of *M. phenylpyruvica* will grow on MacConkey agar.

Biochemical Reactions

None of the moraxellae attack carbohydrates with the formation of acid. They are non-motile, indole-negative and all are sensitive to penicillin. *M.bovis* will slowly pit a Loeffler serum slope, is relatively salt tolerant and grows on media containing 5 per cent Nacl. Litmus milk medium inoculated with *M.bovis* becomes alkaline (blue) and develops three zones: a deep blue upper layer, a soft blue curd in the centrewith the bottom white (reduced) and coagulated. The *Moraxella* species havealso been differentiated by the analysis of the cellular fatty acid.

Clinical Infections

Moraxella bovis causes infectious bovine keratoconjunctivitis, an important ocular disease of cattle which occurs worldwide. Variants of M.bovis have been isolated from horses with conjunctivitis.

Infectious Bovine Keratoconjunctivitis

Infectious bovine keratoconjunctivitis (IBK), sometimes referred to as 'pink-eye' or New Forest disease, is a highly contagious condition affecting the superficial structures of the eyes, usually in animals under 2 years of age. The disease causes economic losses arising from decreased weight gain in beef breeds, loss of milk production, short term disruption of breeding programmes and treatment costs.

There appears to be an age-related immunity, probably as a result of previous exposure. Asymptomatic carrier animals harbour M.bovis in the nasolacrimal ducts, nasopharynx and vagina. Transmission can occur by direct contact, by aerosols and through flies acting as vectors.

Pathogenesis and Pathogenicity

The virulence of M.bovis is attributed to fimbriae, which allow adherence of the organisms to the cornea, circumventing the protective effects of lacrimal secretions and blinking. Two types of fimbriae are recognized, namely, Q fimbriae (pili) which are specific for colonization and 'I' fimbriae which allow local persistence of infection. Fimbrial antigens stimulate type-specific protective immunity.

During bacterial replication, haemolysin and other lytic enzymes such as fibrolysin, phosphatase, hyaluronidase and aminopeptidase are produced. Lipopolysaccharides, associated with O antigens, also appear to play a role in virulence. The haemolysin is a calcium-dependent, pore-forming cytolysin which damages the cell membranes of neutrophils. Release of hydrolytic enzymes from neutrophils on the corneal surface contributes to breakdown of its collagen matrix.

Strains which lack either haemolysin or fimbriae are avirulent. Isolates from carrier animals are often non-haemolytic and non-fimbriate but reversion to virulence can occur. It has been suggested that the deficiency of lysozyme in the lacrimal secretions of cattle may account for their susceptibility to M.bovis.

Clinical Signs

Infectious bovine keratoconjunctivitis initially manifests as blepharospasm, conjunctivitis and lacrimation. Progression of the condition through keratitis to corneal ulceration, opacity and abscessation may occasionally lead to panophthalmitis and permanent blindness. Following ulceration, vascularisation extends from the limbus and stromal oedema develops. There may be weakening of the cornea with the development of coning. In most mild cases, the cornea heals within a few weeks although there may be permanent scarring of the structure. Some carrier animals may exhibit persistent lacrimation. Following infection with a virulent strain of M.bovis, neutralizing antibodies develop which are active against haemolysin produced by other strains. In contrast, antibodies which block fimbrial-mediated adherence are type-specific and exposure to M. bovis possessing a different fimbrial type may result in disease.

Diagnostic Procedures

- The disease characteristically affects a number of animals in a herd. Lacrimal secretion is the most suitable specimen for laboratory examination. Because M.bovis is extremely susceptible to desiccation, specimens must be processed promptly. For transportation, swabs of lacrimal secretions should be placed in 1 to 2 ml of sterile water. Ideally, specimens should be cultured within 2 hours of collection.
- A fluorescent antibody technique for demonstrating M.bovis in smears from lacrimal secretions is available.
- Specimens should be cultured on blood agar and MacConkey agar and incubated aerobically at 37°C for 48 to 72 hours.
- Identification criteria for isolates:
    - Round, small, shiny, friable, colonies appear after 48 hours. Colonies of virulent strains are surrounded by a zone of complete haemolysis and are embedded in the agar.
    - No growth occurs on MacConkey agar.

- Cultures of virulent strains autoagglutinate in saline.
- Smears from colonies reveal short Gram-negative rods in pairs.
- Reactions in the catalase and oxidase tests are positive.
- A Loeffler's serum slope may be pitted after 10 days.

Isolation

Lacrimal secretions should be inoculated as soon as possible after collection on blood agar and incubated at 35 °C for 48-72 hours. The inoculation of a MacConkey plate is useful to gauge the degree of contamination by other Gram- negative bacteria. *M. bovis* is unable to grow on MacConkey agar.

Treatment

Antimicrobial therapy should be administered subconjunctivally or topically early in the disease.

Control

Fimbriae-derived bacterins which are available commercially in some countries are of uncertain efficacy. Management-related methods are important in the control of IBK. These include isolation of affected animals, reduction of exposure to mechanical irritants, the use of insecticidal ear tags and the control of concurrent diseases, such as infectious bovine rhinotracheitis or Thelazia infestation. The prophylactic use of intramuscular oxytetracycline can be considered for animals at risk. Animals which are blind should be housed. Vitamin A supplementation may be beneficial.

Summary of the Diseases and Sites of Isolation of the *Moraxella* species

Species	Host	Disease and/or site of isolation
Moraxella bovis	Cattle Horses	Infectious bovine keratoconjunctivitis ['pink eye' or 'New Forest disease· (UK)] Rare isolates from horses with conjunctivitis
M.lacunata	Many animal species Humans	Isolated from guinea·pigs, aborted equine foetuses, goats with viral pneumonia and encephalitis, a goat with septicaemia and from various pathological specimens from dogs and pigs. Its role in disease processes in animals is not known May cause conjunctivitis
M. phenylpyruvica	Sheep and cattle Pigs Goats	Recovered from the urinogenital tract and brain Isolated from the urinogenital tract Obtained from the intestinal tract Pathogenicity for animals is unknown

40

Bacteroides spp.

Bacteroides spp are non-spore forming gram-negative bacilli that are part of the human resident flora. Microbiologically, they are distinguished from other genera by growth in 20% bile. At present, the *Bacteroides fragilis* group consists of ten species: *B. fragilis* (the most frequent isolate), *B. distasonis*, *B. thetaiotaomicron*, *B. vulgatus*, *B. ovatus*, *B. eggerrthii*, *B. merdae*, *B. stercoris*, *B. uniformis*, and *B. caccae*. Since 1990, many organisms previously designated as *Bacteroides* have been reclassified (see chapter on Anaerobes other than *Bacteroides*). *Bacteroides*, the predominant genus in the human intestine, are important in numerous metabolic activities and may provide some level of protection from invasive pathogens. All 10 species are usually isolated from the colon, although infections caused by or associated with them can include virtually any organ.

Characteristics

The gram-negative Bacteroides spp. or closely related genera are capsulated obligatory anaerobic bacilli that are non-spore forming, pale-staining, and some are motile by peritrichous flagella, while other taxa are non-motile. Bacteroides, Parabacteroides, Odoribacter are generally bile resistant, distinguished from genera which are bile sensitive. They are normally commensal, found in the intestinal tract of humans (mouth, colon, urogenital tract) and other animals. Many cultures of Bacteroides strains display brown to black pigmentation on blood agar media caused by esculin hydrolysis.

Morphology

In general, all strains were stained poorly with safranin, but well with dilute carbolfuchsin. Motility was not observed in any strains.

The morphology of Bacteroides fragilis in fresh cultures and in pus is not especially characteristic. In this form, the bacillus measures 0.5 by 2.0 to 2.5 micra. Older cultures tend to develop larger forms (0.6 to 0.8 by 2.0 to 9.0 micra) but even in this state the organism cannot be distinguished from aerobic Gram-negative bacilli, by its morphology.

Bacteroides funduliformis, on the other hand, is characteristic morphologically because of its extreme pleomorphism. In the necrotic centers of lesions, growth occurs in such dense clumps of small bacillary forms that it is difficult to pick out the individual bacilli. Bipolar staining is a prominent feature and gives the bacilli a diplococcus-like appearance. Filaments 1.0 by 10.0 micra have been observed in microscopic sections of the granulating periphery of abscesses. On solid mediums, long unbranched filaments develop after six or seven days of cultivation. In solid mediums, there are fusiform swellings 4.0 to 5.0 micra wide.

In brain broth the earliest forms are coccobacillary, similar to those seen in pus, or rarely, short curved "comma" shaped rods 0.3 to 0.5 by 2.0 to 4.0 micra. Within two days the characteristic "funduliform,""ball,""leukocyte," or "thetoid" forms appear, together with straight bipolar staining rods which vary in size from coccobacillary up to filaments 20.0 to 40.0 micra long. The ball-shaped organisms (1.5 to 3.0 micra in diameter) seem to be formed by bacilli, swollen so that the bipolar dark staining beads appear as crescentic caps opposite each other on the ball.

Bacteroides A is slightly shorter and more plump (1.0 by 2.0 micra) than Bacteroides fragilis. Bipolar staining in day-old broth cultures is prominent. A characteristic feature of this bacillus is its tendency to develop tremendously swollen forms even in forty-eight hour cultures. These swollen bacilli are less definite in outline than the "funduliforms" of Bacteroidesfunduliformis, and look like bacteria which are undergoing lysis.

Grampositive granules are present in the bacilli only in very fresh cultures. The strain of Fusiformis dentium was Gram-positive only in the pus from which it was obtained. The long filaments of the Actinomycete develop in cultures seven or eight days old. Fresher cultures contain only bacillary forms. The anaerobic diphtheroid loses its diphtheroid appearance when it becomes Gram-negative after three to four days of incubation. Its morphology is not characteristic.

Biologic Characteristics

In general, all strains have been able to survive in glucose brain broth for several weeks at room temperature. None of the organisms survive long, however, if exposed to air. They are able to grow in an atmosphere of nitrogen. Although all strains seem to grow better in brain broth which contains fermentable carbohydrate, growth is not improved on solid mediums by the addition of higher concentrations of glucose.

Bacteroides fragilis grows on the surface of plain blood agar as very fine, transparent, moist, grayish white, fimbriated or spreading colonies, without production of hemolysis. Isolated surface colonies do not exceed 2 mm. in diameter. On this medium, growth is seldom perceptible before five days of incubation.

Growth on hormone blood agar is more profuse and more rapid, and the colonies are yellow and opaque. Deep colonies are yellow, opaque, and lens-shaped. They do not exceed 1 mm.

Growth of Bacteroides fragilis in glucose-brain broth takes place in twenty-four hours and produces a diffuse clouding of the medium. A characteristic feature of the growth in fermentable carbohydrate mediums is the slight but definite production of gas.

In an anaerobic jar, growth is obtained in peptone-water sugar mediums, but not in plain broth. The biologic reactions of the first three strains are uniformly similar, and those of the fourth strain, which is a subspecies, vary only slightly.

The production of indol by Bacteroides fragilis is listed as either negative or positive. For the most part, tests for indol were negative but we have encountered an occasional faintly positive reaction in fourteen-day old cultures.

Bacteroides funduliformis grows on plain blood agar with as much difficulty as does Bacteroides fragilis, but the colonies are larger (3 mm.) and more opaque. The centers of the colonies are grayish-white and opaque, the borders are fimbriated and may spread to coalesce with other colonies on a sufficiently moist medium. On slants, when the moisture has drained from the surface, colonies are discoid, discrete, and have slightly raised centers and edges.

The colonies when they first appear are fairly moist, but after incubation for ten days they are hard and dry, and can be picked entire from the medium. Deep colonies are lens-shaped or triangular and measure 2 mm. in their greatest diameters. They occasionally produce gas in blood agar which contains glucose, particularly if the medium is not too stiff. The hemolytic action of this species is an important characteristic. The radius of the zone of hemolysis measures 4 to 5 mm.

In brain broth, Bacteroides funduliformis grows in clumps at the bottom of the tube. In fresh cultures, a few bacilli are diffused throughout the broth, but they are scarcely sufficient in number to produce appreciable clouding. Often, the only indication of growth is the presence of bubbles of gas caught in the particles of brain at the bottom of the tube.

General Characteristics

- Gram negative anaerobic rod
- Shape: Pleuromorphic
- Size: (0.5-1.5)µm wide and (2-6)µm long
- Non motile except B. polypragmatus, B. xylanolyticus
- Non capsulated except fragilis
- Non spore forming

Classification

On the basis of medical importance bacteriodes is classified into two group;

1. Bacteroides fragilis group:

Examples:

- B.fragilis
- B. distasonis
- B. ovatus
- B. thetaiotaomicron
- B. vulgatus
- B. idgatus
- B. uniformis
- B. tiariabilisleggerthii
- B. splanchnicus

These are commensals of GI tract

2. Bacteroides melaninogenicus group

Examples:

- *B. melaninogenicus* sub spp *intermedius*
- *B. ruminicola*
- Now the name has changed to *Prevotella melaningenica*
- These are Norma flora of URT, GIT, vagina

Susceptibility to Disinfectants

More specific information on Bacteroides spp. is not available, but most bacteria have been shown to be susceptible to low concentration of chlorine, 1% sodium hypochlorite, 70% ethanol, phenolics such as orthophenylphenol and ortho-benzyl-paua-chlorophenol, 2% aqueous glutaraldehyde, iodine, formaldehyde, and peracetic acid (0.001% to 0.2%).

Physical Inactivation

Information specific to Bacteroides and like genera is not available, but most bacteria can be inactivated by moist heat (121°C for 15 min - 30 min) and dry heat (160-170°C for 1-2 hours).

Survival Outside Host

Bacteroides and like genera have been detected in faeces infected water by PCR for at least 2 weeks at 4°C; 4 to 5 days at 14°C; 1 to 2 days at 24°C; and 1 day at 30°C.

Epidemiology

Bacteroides fragilis are endogenous organisms of the GI tract. Spread of strains among patients is not known, although this topic has not been well studied. Thus, infections due to this organism are most likely caused by endogenous strains.

Mode of Transmission

Infection results from displacement of Bacteroides spp. or closely related genera from normal mucosal location as a result of trauma such as animal/ human bites, burns, cuts, or penetration of foreign objects, including those involved in surgery. There is no evidence that organisms are invasive on their own.

Pathogenesis

Bacteroides fragilis is the most common opportunistic pathogen of Bacteroides spp. Spread to bloodstream (bacteremia) is more common for B. fragilis than any other anaerobe. Deep pain and tenderness below the diaphragm are typical of B. fragilis infection. Widespread intra-abdominal abscesses may be associated with fever and abdominal pain.

Multiple virulence factors have been implicated in the pathogenesis of this organism. They include the capsular polysaccharide (which inhibits opsonophagocytosis and promotes abscess formation), pili and fimbriae (promotes adherence), and production of a number of different enzymes (hyaluronidase, hemolysin, peroxidase, collagenase, protease, heparinase, and neuraminidase). In addition, superoxide dismutase and catalase also considered virulence factors. These enzymes defend *B. fragilis* against oxygen radicals and increase aerotolerance.

Bacteroides spp. represents an important anaerobic bacterial genus associated with human infections. In combination with other facultative/strict anaerobes, they are responsible for the majority of localized abscesses within the cranium, thorax, peritoneum, liver, and female genital tract. They can cause pulmonary abscesses when naturally-occurring oropharangeal Bacteroides and closely related genera are aspirated into the lung. These taxa can lead to many types of diseases, some of which can be fatal, including noma (cancrum oris), human apical periodontitis, endocarditis, pelvic inflammatory disease, suppurative thrombophelebitis, and wound infections. Organisms from oral flora also have a role in dental abscesses and infectivity of human bites.

Clinical Manifestations

The hallmarks of nearly any infection involving *Bacteroides* spp include abscess formation, and they are frequent isolates in polymicrobial infections. Typical sites of polymicrobial infections involving *Bacteroides* include the abdomen and pelvis, perirectal, skin and soft tissue, and solid organs. Although isolation of *Bacteroides* spp as the sole pathogen can occur, it is unusual. Infections where single organism isolation is most commonly associated with *Bacteroides* include endocarditis, meningitis, septic arthritis and osteomyelitis.

Laboratory Diagnosis

Bacteroides fragilis may be isolated as a single agent, such as in blood cultures, or more typically from mixed infections. The organism is aerotolerant, but requires an anaerobic environment to propagate. Simple identification from blood cultures includes Gram stain and growth on blood agar and *Bacteroides*-bile-esculin (BBE) agar for isolation and presumptive identification of*Bacteroides fragilis* group (as well as *Bilophila wadsworthia*). *B. fragilis* will appear as dark colonies with brown-black halos on BBE agar due to the hydrolysis of esculin. *B. fragilis* can be further presumptively identified by resistance to kanamycin, vancomycin and colistin, using a disk test, and will grow in 20% bile, produce catalase (most strains), and is variably indole positive.

Laboratory diagnosis:

Samples: pus, exudates, biopsy

1. Microscopy

Gram negative, non motile, non sporing pleuromorphic rod

2. Culture

i) Blood Agar (BA)

Kanamycin or neomycin Blood agar selective for Anaerobes

Inoculates anaerobically at 37C for 48 hours

Fragilis Form non haemolytic grey colony of 1-3 mm diameter

Melanogenicus form black brown haemolytic colony in 3-5 days

ii) Bactrroides Bile Esculin Agar (BBE)

Hydrolyses esculin,

Colony surrounded by dark zone

3. Biochemical tests for ***fragilis***

i) Esculin hydrolyses: positive (+)

ii) Indole: negative (-)

iii) Urease: Negative (-)

iv) Catalase: Positive (+)

v) Oxidase: variable (+/-)

vi) Fermentative:

- Glucose +ve
- Lactose + ve
- Maltose + ve
- Sucrose + ve
- Rhamnose - ve
- Arabinose - ve
- Salicin - ve
- Trehalose - ve

vii) Bile resistant: grow in 20% bile

viii) Thioglycolate with bile: Positive (+)

ix) Beta lactamase production: *fragilis* is resistant to penicillin

4. Serology

5. PCR

Drug Resistance

Resistance against antibiotics is increasingly common, and frequently observed with penicillin, ampicillin, cephalothin, tetracycline, piperacillin, chloramphenicol, kanamycin, colistin, rifampicin, vancomycin, and the aminoglycosides. The B. fragilis group is commonly resistant to expanded and broad spectrum cephalosporins, including β-lactamase-resistant drugs such as cefoxitin, and clindamycin.

Strain resistance to imipenem and metronizadole, although it has been detected worldwide, are rarely encountered. Resistance to quinolones is increasing. Some members of B. fragilis group have shown resistance to ampicillin-sulbactam and amoxicillin-clavulanante combination therapies. Isolates often produce β-lactamase, often making penicillin based antibiotics ineffective.

41

Dichelobacter

Many non-spore-forming, anaerobic, Gram-negative bacteria cause opportunistic mixed infections, often in association with facultative anaerobes. Synergistic interactions between the organisms in these mixed infections are common. *Fusobacterium* species and bacteria formerly referred to as *Bacteroides* species account for more than 50%of the anaerobic organisms isolated from these infections.

Usual Habitat

Nan-spore-forming, Gram-negative anaerobes are often found on muwus membranes, particularly in the digestive tract, of animals and man. They are excreted in the faeces and they can survive for short periods in the environment. *Dichelobacter nodosus,* a primary pathogen of the epidermal tissues of the hoof region of ruminants, survives for less than 4 days in mud.

Diagnostic Procedures

In order to ensure that isolates of anaerobes are aetiologically significant, specimens for isolation procedures should be obtained by direct sampling from discharges or lesions and by supra pubic puncture in urinary infections.

Specimens should be processed promptly after collection. Commercial kits and transport media are available for specimens from suspected anaerobic infections. In the core of a tissue specimen over 2-3 cm, an anaerobic microenvironment is usually maintained. Samples of fluid in a syringe remain suitable for anaerobic culture if air is expelled from the syringe and the needle is plugged.

Anaerobic jars with an atmosphere of hydrogen and 10%CO2 are used for incubating cultures at 37°C for up to 7 days.

Enriched blood agars for the isolation of anaerobes are supplemented with 5 to 10% ruminant red cells, yeast extract, vitamin K and haemin. Selective media can be prepared by adding appropriate antimicrobial agents. Media must be pre-reduced by storing them in an anaerobic atmosphere for at least 6 hours before inoculation. Liquid media, such as cooked meat broth or thioglycollate medium supplemented with vitamin K and haemin, are useful for subculturing but are unsuitable for primary isolation.

Special selective media are required for the isolation of *Dichelobacler nodosus* from ruminant footrot. In some media formulations, powdered ovine hoof is added to promote enhanced growth.

Pathogenesis and Pathogenicity

Non-spore-forming anaerobes usually exert pathogenic effects when anatomical barriers are breached allowing invasion of underlying tissues. They replicate only at low or negative reduction potentials (Eh). Most of those involved in opportunistic infections produce superoxide dismutase which allows them to survive in oxygenated tissues until the *Eh* reaches levels favouring their growth.

Tissue trauma and necrosis followed by multiplication of facultatively anaerobic bacteria can lower Eh levels to a range suitable for the proliferation of non-spore-forming anaerobes. Most infections involving these organisms are mixed. Two or more bacterial species, interacting synergistically, may produce lesions which the individual organisms cannot. A relevant example of this type of synergism is the production by *Arcanobacterium pyogenes* of a heatlabile factor which stimulates *F. Necrophorum* replication. In turn, F. *necrophorwn* produces a leukotoxin which correlates with the strain virulence and aids survival of *A. pyogenes*.

Synergism between *F. necrophorum* and *Dichelobacter nodosus* is important in the pathogenesis of ruminant pedallesions. In this instance, *F. Necrophorum* facilitates tissue invasion by *D. nodosus* and is itselfstimulated by a growth factor elaborated by *D.nodosus.*

Three biotypes of *F. necrophorum* are recognised. Biotype A, designated *F. necrophorum* subspecies *necrophorum* has greater haemolytic activity and is more virulent than biotype B, F. *necrophorum* subspecies *funduliforme.* Biotype C, reclassified as *F. Pseudonecrophorum* appears to be non-pathogenic. Characteristics of *D. nodosus* which correlate with its ability to damage tissues include the production of thermostable proteases and elastase and the presence of agarolytic activity on agar-based media containing powdered hoof.

Clinical Infections

Calf diphtheria

This condition usually presents as necrotic pharyngitis or laryngitis in calves less than 3 months of age. The aetiological agent, *F. necrophorum,* can enter through abrasions in the mucosa of the pharynx or larynx often caused by ingestion of coarse feed. Clinical signs include fever, depression, anorexia, excessive salivation, respiratory distress and a foul smell from the mouth. Untreated calves may develop a fatal necrotizing pneumonia. Treatment with potentiated sulphonamides or tetracyclines early in the course of the disease is usually effective.

Bovine Liver Abscess

Hepatic abscessation in cattle, secondary to rumenitis, is encountered most commonly in feedlot animals. The feeding of rations high in carbohydrates and the resulting rapid intraruminal fermentation can lead to the development of ulcers. *Fusobacterium necrophorum* together with other anaerobes and *Arcanobacterium pyogenes* invade the tissues, and occasional emboli which reach the liver via the portal vein initiate abscess formation.

Affected cattle rarely show clinical signs and lesions are usually detected at slaughter. Management techniques in feedlots should be aimed at reducing the incidence of rumenitis. Chlortetracycline in feed during the finishing period can reduce the prevalence of liver abscess.

Necrotic Rhinitis of Pigs

This sporadic condition, primarily affecting young pigs, is characterised by suppuration and necrosis of the snout as a result of infection with *F. necrophorum,* often in association with other anaerobes.

These organisms enter through abrasions in the nasal mucosa. Signs include swelling of the face, sneezing and a foul-smelling nasal discharge. In chronic infections, involvement of the nasal and facial bones can result in permanent facial deformity ('bull nose'). Potentiated sulphonamides administered early in the course of the infection may be beneficial.

Thrush of the Hoof

This necrotic condition of the horse's hoof is associated with poor hygiene, wet conditions and lack of regular cleaning of the hooves. Infection with F. necrophorum, secondary to hoof damage, results in a localized inflammatory response. Thrush, which commonly affects the hind feet, is characterized by a foul-smelling discharge in the sulci close to the frog. The aim of therapy is to encourage regeneration of the frog by providing dry, clean stabling, regular attention to the hooves and exercise.

Black Spot of Bovine Teats

Black spot or black pox of the teat orifice and sphincter of dairy cows presents as a localized area of necrosis with black scab formation due to invasion by Fusobacterium necrophorum. The condition can contribute to stenosis of the sphincter and may predispose to mastitis.

42

Spirochaetes

The order *Spirochaetales* contains two families, *Leptospiraceae* and *Spirochaetaceae*. It comprises spiral or helical bacteria (spirochaetes) which share some unique morphological and functional features. Members of the order are motile by means of **endoflagella** which are located within the periplasm.

The spirochactes are slender, motile, flexuous, unicellular, helically coiled bacteria ranging rrom 0.1 -3.0 μm in width. The outer sheath, the outermost layer of a spirochaete cell, is a multi layered membrane that completely surrounds the peri plasmic flagella (axial filament) and the helical protoplasmic cylinder. The cylinder consists or the nuclear material, cytoplasm, cytoplasmic membraneand the peptidoglycan portion of the cell wall.

The periplasmic flagella are wrapped around the cylinder and are in the peri plasmic space of these Gram-negative bacteria. One end of each flagellum is inserted near a pole of the protoplasmic cylinder and attached by plate-like structures called insertion discs. The distal end of each flagellum is not inserted and extends to the centre of the cell and may overlap the flagellum from the opposite end. The periplasmic flagella facilitate the motility of the bacteria in viscid environments.

Leptospira species

Members of this species (leptospires) are motile helical bacteria (0.1 x 6 to 12 μm) with hook-shaped ends. Although cytochemically Gram-negative, they do not stain well with conventional bacteriological dyes and are usually visualized using dark-field microscopy. Silver impregnation and immunological staining techniques are used to demonstrate leptospires in tissues. Leptospirosis, which can affect all domestic animals and humans, ranges in severity from mild infections of the urinary or genital systems to serious systemic disease.

Culture

The media used for the culture of leptospires include Korthof and Stuart broths, Fletcher semisolid medium, EMJH and Tween 80-albumin medium (OAC) that areliquid but can be made semi-solid by the addition of 1.5g agar/litre of medium. Contaminated samples (or cultures) can be filtered through a 0.45 µm bacteriological filter and inoculated into a medium contain ing 5-fluorouracil at 200 µg/ml. The media can be inoculated with 1-3 drops of carefully taken urine within a few minutes of collection, or urine dilUled 1: 10 with 1 per cent BSA, as soon as possible after collection.

A 10 per cem tissue suspension in 1 per cent BSA (1-2 drops) or a few drops of oxalated or heparinised blood can also be inoculated into media. The cultures are incubated at 30°C for up to 8 weeks. A drop of the culture is examined by darkfield microscopy once weekly. *Leptospira hardjo* is one of the slowest growing serovars and *bratislava* is difficult to culture in laboratory media.

Usual Habitat

Leptospires can survive in ponds, rivers, surface waters, moist soil and mud when environmental temperatures are warm. Pathogenic leptospires can persist in the renal tubules or in the genital tract of carrier animals. Although indirect transmission can occur when environmental conditions are favourable, these fragile organisms are transmitted most effectively by direct contact.

Pathogenesis and Pathogenicity

Leptospires gain entry through mucous membranes or damaged skin from direct or indirect contact. After epithelial penetration there is haematogenous spread with localisat ion and proliferation in parenchymatous organs, particularly the liver, kidneys, spleen and sometimes meninges. In the kidneys the organisms reach and localise in the lumen of proximal convoluted tubules. Penetration and multiplication in the foetus can occur in pregnant animals leading to foetal death and resorption, abortion or weak offspring. The foetus, if infection occurs in the third trimester, can produce specific antibodies and may overcome the infection.

Antibody production in infected animals begins a few days after the onset of leptospiraemia. The leptospires tend to persist in sites such as renal tubules, eyes and uterus where antibody activity is minimal.

Leptospires damage vascular endothelium resulting in haemorrhages. Serovars in the serogroups Autumnalis, Grippotyphosa, Icterohaemorrhagiae and Pomona produce a haemolysin that is probably responsible for the haemoglobinuria (redwater) in young calves infected with these serovars. Cytotoxic protein is produced by virulent strains but the role of the toxin is unknown. Virulence varies between the serovars and between two genotypes of *L. interrogans* serovar *hardjo* known as *hardjobovis* and *hardjoprojitno.*

Diagnostic Procedures

Diagnosis of leptospirosis in maintenance hosts usually requires screening of a defined population. Clinical signs, together with a history suggestive of exposure to contaminated urine, may suggest acute leptospirosis.

Organisms may be detected in fresh urine by dark-field microscopy, but this technique is relatively insensitive. Slow-growing serovars such as hardjo may require incubation for six months in liquid media at 30°C. Commonly, EMJH (Ellinghausen, McCullough, Johnson and Harris) medium based on 1% bovine serum albumin and Tween 80 is used for isolation. Isolates should be identified using DNA profiles and serology.

Fluorescent antibody procedures are often used for the demonstration of leptospires in tissues. Suitable tissues include kidney, liver and lung. Silver impregnation techniques can also be used for demonstration of leptospires.

DNA hybridization, PCR, magnetic immunocapture PCR and immunomagnetic antigen capture systems have also been developed for the demonstration of leptospiral infection in tissues and urine. The standard serological reference test, the microscopic agglutination test, is potentially hazardous because it involves mixing live culture growing in liquid medium with equal volumes of doubling dilutions of test serum.

Clinical Infections

Leptospirosis in Cattle and Sheep

Cattle are maintenance hosts for L. borgpetersenii serovar hardjo and there is increasing evidence that this serovar is also host-adapted for sheep. Leptospira interrogans serovar hardjo is also host-adapted for cattle. Although L. interrogans serovar hardjo appears to cause only sporadic cases of disease in cattle, it may be more virulent than L. borgpetersenii serovar hardjo. Susceptible replacement heifers, reared separately and introduced into an infected dairy herd for the first time at calving, may develop acute disease with pyrexia and agalactia affecting all quarters. Infection may also result in abortions and stillbirths.

If management practices allow exposure to infection and the subsequent development of immunity before breeding age, reproductive problems may not develop. Agalactia caused by leptospiral infection can be confirmed by demonstrating a rising antibody titre in paired serum samples. Infection with serovar hardjo in sheep, particularly in intensively managed lowland flocks, can cause abortions and agalactia. Dihydrostreptomycin or amoxycillin can be used for reducing or eliminating urinary excretion of the organisms.

Both monovalent and multivalent inactivated vaccines, which are commercially available may not always be effective. Serovars incorporated into vaccines should be those which are associated with disease in a particular region. Infection with serovarspomona, grippotyphosa, and icterohaemorrhagiae can cause serious disease, particularly in calves and lambs. Infection is usually accompanied by pyrexia, haemoglobinuria, jaundice and anorexia. Extensive renal damage with resultant uraemia often precedes death. Vaccination is used for control of serovar pomona which is an important cause of bovine abortion in some countries.

Leptospirosis in Domestic Animals

Host	Disease syndrome
Cattle	• Milk·drop syndrome, with or without any other clinical signs (often *hardjo)* • Abortion and neonatal mortality: abortion 'storms' *(pomona)* and sporadic abortions *(hardjo)* • Infertility (often *hardjo)* • Haemoglobinuria, jaundice and fever in calves and, less commonly, in young adults. • Serovars commonly involved are *pomona, grippotyphosa* and *icterohaemorrhagiae.* • Occasionally, some animals show signs of meningitis
Pigs	• Subclinical with or without leptospiruria • Subclinical, often with leptospiruria: especially with *pomona.* Pigs are considered to be the maintenance host for this serovar • Fever and focal non-suppurative mastitis and leptospiruria • Infertility, abortions and stillbirths: often *canicola, pomona* or *icterohaemorrhagiae* • Fever, anorexia, jaundice, haemoglobinuria and high mortality in young pigs: often *icterohaemorrhagiae*
Dogs	• Subclinical with leptospiru ria: often *canicola* • Acute haemorrhagic disease: high fever, vomiting, prostration and often early death; usually *icferohaemorrhagiae* • Less acute icteric type: intense icterus, depression, fever, haemorrhages with blood in faeces and urine; *canicola* or *icterohaemorrhagiae* • Uraemic type: uraemia associated with extensive kidney damage, ulcerative stomatitis and uraemic breath. Death occurs in a high percentage of cases. These severe signs can occur 1-3 years after the initial infection: often *canicola* • Rarely a chronic, active hepatitis: seen in a *grippotyphosa* infection
Horses	• Recurrent iridocyclitis ('periodic ophthalmia' or 'moon blindness') which can result in blindness. Aetiology has not been conclusively determined • Occasionally abortion with foetuses of 6 months to term • Rarely fever, anorexia, depression and icterus
Sheep	• Mainly subclinical infections with leptospiruria: serovars such as *hardjo* • Occasionally, acute leptospirosis with depression, dyspnoea, haemoglobinuria, anaemia and high mortality in lambs: often p*omona*

43

Serpulina (Treponema) Species

Members of the genus *Serpulina (Treponema)* are host-associated spirochaetes found in the oral cavity, intestinal tract and genital region of animals and humans. The cells are wider, are not as tightly coiled as th e leptospires and can be stained by aniline dyes.

The species of veterinary significance are S. *Hyodysenleriae* (swine dysentery) and *T. paraluiscunicllii* (ventdisease of rabbits). S. *innocens,* present in the faecesof pigs and dogs, is thought to be non-pathogenic,although some workers regard S. *innocens* as a strainof S. *hyodysenteriae* of lower virulence. S. *hyodysenteriae* and S. *innocens* are s imilarmorphologically, culturally as well as biochemicallyand can be grown on laboratory media. *T. Paraluiscuniculi* has not been cultured *in vitro.*

Normal Habitat

*T. paraluisclinicul*i produces a benign venereal disease of rabbits and is present in lesions in the genito-perineal area of rabbits. It causes latentin fection in mice, guinea- pigs and hamsters. The treponemes can be found in the lymph nodes of these animals. The reservoir of S. *hyodysenteriae* is the intestinal tract of pigs, wild rats and mice. Recovered, asymptomatic pigs can excrete these organisms in faeces for 3 months or more. Survival of S. *hyodysenteriae* in soil or voided pig faeces is short, about 24 48 hours. Infection is by the faecal-oral route.

Pathogenesis

After S. *hyodysenteriae* enters a susceptible pig the spirochaete invades goblet cells of the colonic mucosa, multiplies in the crypts of Lieberkuhn and causes necrosis and erosion of the mucosal cells. The faecal material is watery and contains mucus, blood and necrotic debris. Only the large intestine is involved and colonic malabsorption occurs. The Lipopolysaccharides of the bacterium are thought to play a part in the pathogenicity. S. *hyodysenteriae* is unable to produce dysentery in gnotobioticpigs and it appears that the pathogen requires the interaction of other bacteria of the normal flora such as *Fusobacterium necrophorum, Bacteroides vulgatus, B.fragilis* and *Campylobacter coli.* The disease was once thought to be due to C. *coli* as these curved rods can often be seen in significant numbers in faecal smears from pigs with swine dysentery.

Laboratory Diagnosis

Specimens

Deep mucosal scrapings should be taken from a portion of the affected large intestine from a dead pig, or rectal swabs and faeces from several live affected pigs. The numbers of S. *hyodysenteriae* can sometimes be low in faeces and are difficult to see. Sections of affected colon in 10 per cent formalin should be taken for histopathology.

Direct microscopy

Various methods can be used to visualise S. *hyodysenteriae:*

- A portion of deep scrapings from the mucosa or faecal material is examined in a drop of water under darkfield microscopy.
- Fixed smears of mucosal scrapings or faeces are stained by dilute carbolfuchsin (D CF) for 4-6 minutes, by Victoria blue 4-R or by a silver impregnation technique.
- Histological sections of colon can be stained by a silver impregnation stain or Victoria blue 4-R.
- Fluorescent anti body technique can be carried out on mucosal scrapings or faeces.
- Three to 5 serpulinas per high power field is considered significant. S. *hyodysenteriae* is 0.3-0.4 µm in width, 6-8 µm in length, is loosely coiled with 2-4 coils and tapered ends. It is motile by flexing movements under the darkfield. Pigs normally have other non -pathogenic treponemes such as S. *innocens* and smaller, tightly coiled spirochaetes in the intestinal tract that can complicate the interpretation of darkfield examination or stained smears.

Isolation

S. *hyodysenteriae* is anaerobic but oxygen-tolerant and growth is enhanced by CO2. It grows well onfreshly poured, or pre-reduced, trypticase soy agar with

5 per cent blood and 400 μg/ml spectinomycin at 42°C for 2-3 days under an atmosphere delivered by an H_2+ CO_2 envelope. Increased growth has been obtained using Fastidious Anaerobe Agar(Lab M) with 5 percent sheep blood and also by theincorporation of 1 per cent sodium RNA (BHD Chemicals) into media. Sodium RNA can be sterilized by filtration or by autoclaving. It is a growth enhancer and also increases the difference in haemolysis between S. *hyodysenteriae* and S.*innocens*. Colonic mucosal scrapings or faeces can be prepared as a 1:10 suspension in saline and clarified by centrifugation at slow speed. The supernatant can be passed, seri ally, through 0.8, 0.65 and 0.45 μm filters and this material is used to streak the culture medium.

Identification

Colonial characteristics

After about 48 hours' incubation *S. Hyodysenteriae* appears as small, translucent colonies with a zone ofclear haemolysis. *S. innocens* is weakly beta haemolytic.

Microscopic appearance

DCF-stained smears or dark field microscopy from the colony reveals the typical helical serpulinas.

Biochemical characteristics

S. *hyodysenteriae* must be distinguished from S. *innocens.* Belanger and Jacques (1991) suggest that a rapidand simple differentiation can be based on the combinedresults of the haemolysis, the haemolysis intensificationtest and an indole spot test.

- Haemotysis: a distinctive characteristic of S. *Hyodysenteriae* is the production of a strong betahaemolysis that is best demonstrated on sheepblood agar. S. *innocens* gives a weak reaction.
- Haemolysis intensification: this test can be carriedout by cutting an agar block (0.5 cm2) from a 4-dayculture and by incubating the plate for another 4days. The intensification of haemolysis is recognisedby a lighter zone, 1- 2mm wide, along the cutline. S. *hyodysemeriae* gives this reaction (ringphenomenon) but S. *innocens* does not.
- Indole spot test for anaerobes: the reagent is 1 percent p-dimethyl-aminocinamaldehyde in 10 percent concentrated HCl. This is used to saturate a strip of filter paper in a Petri dish and growth from a culture of the test bacterium is smeared on the filter paper. S. *hyodysemeriae*

gives a positive reaction indicated by a blue colour usually within 1- 3 minutes. *S. innocens* is negative and the filter paper remains pink.

Neither *S. hyodysenteriae* nor S. *innocens* are very reactive biochemically and, with field strains particularly they have many reactions in common.

Tests for S. *hyodysenteriae* antigens

- The fluorescent antibody test is useful as a screening test foriden tification of *S. hyodysenteriae*
- A slide agglutination testand a microscopic agglutination lest using absorbed polyclonal serum have been described.

Tests for S. *hyodysenteriae* antibodies

Several serological procedures have been developed for the serodiagnos is of swine dysentery and for the detection of carrier animals. The ELISA appears to be the most sensitive but should be used as a herd test, rather than for individual animals, as false-positive and false-negative results can occur.

44

Borrelia

Borreliae, which are longer and wider than other spirochaetes, have a similar helical shape. In addition to a linear chromosome, which is unique among bacteria, borreliae possess linear and circular plasmids. Although these spirochaetes can cause disease in animals and humans, subclinical infections are also common. Borreliae are transmitted by arthropod vectors.

Usual Habitat

Borreliae are obligate parasites in a variety of vertebrate hosts. Although these organisms persist in the environment for short periods, they depend on vertebrate reservoir hosts and arthropod vectors for long-term survival.

Differentiation of *Borrelia* Species

Borreliae can be differentiated from other spirochaetes by their morphology, by the low guanine and cytosine content of their genomic DNA and by ecological, cultural and biochemical characteristics. Identification of Borrelia species depends mainly on genetic analysis. At least nine genospecies or genomic groups of B.burgdorferi sensulato, have been identified using DNA-DNA hybridization, 16s rRNA sequencing and other molecular techniques.

Clinical Infections

The species of particular veterinary importance are B. burgdorferi sensu lato, the cause of Lyme disease in animals and humans, and B. anserina which causes avian borreliosis. The significance of two other species, B.theileri and B.coriaceae, as animal pathogens, is uncertain.

Lyme Disease

This condition, also known as Lyme borreliosis, was first identified in 1975 following investigation of a cluster of arthritis cases in children near the town of Old Lyme, Connecticut. The causative agent, a spirochaete, was named Borrelia burgdorferi.

Epidemiology

Lyme disease has been reported in humans, dogs, horses and cattle, and infection has been documented in sheep. Ticks are the only competent vectors of *B. burgdorferi* sensu lato. Infection is usually acquired by larval stages of ticks feeding on small rodents.

The most common tick vector for B. *burgdorferi* sensu lato in Europe is *Ixodes ricinus;* in central and eastern USA it is *I. scapularis;* on the west coast of the USA, it is *I. pacificus,* and in Eurasia it is *I. persulcatus.* Transovarial transmission of the spirochaete in the tick, which may occur infrequently, is not epidemiologically important. Occasional transmission of borreliae from infected incidental hosts to uninfected ticks may occur. Although B. *burgdorferi* sensulato has been demonstrated in the urine of dogs and horses, infected urine is an unlikely source of infection.

Pathogenesis

Transmission of *B. burgdorferi* sensulato occurs when an infected tick feeds on a susceptible animal. Prior to feeding, the spirochaetes are restricted to the midgut of the ticks and, following ingestion of blood, they are found in the salivary glands. Following ingestion of blood by the tick, a change occurs in the expression of the outer surface protein (Osp) of the borreliae. This change in Osp expression from OspA to OspC appears to be essential for virulence.

After entering the bloodstream of a susceptible host, borreliae multiply and are disseminated throughout the body. Organisms may be demonstrated in joints, brain, nerves, eyes and heart. Whether disease is caused by active infection or by host immune responses to the organism is unclear. Recent studies suggest that persistent infection, leading to the induction of cytokines, contributes to the development of lesions. There may be an association between different genotypes of *B.burgdorferi* and particular clinical syndromes in humans.

Clinical Signs

Most infections are subclinical. Serological surveys demonstrate that exposure is common in both animal and human populations in endemic areas. The clinical manifestations of Lyme disease relate mainly to the sites of localization of the organisms. Clinical disease is reported frequently in dogs. Signs include fever, lethargy, arthritis and evidence of cardiac, renal or neurological disturbance. The clinical signs in horses are similar to those in dogs and include lameness, uveitis, nephritis, hepatitis and encephalitis. Lameness in cattle and sheep associated with *B.burgdorferi* sensu lato infection has been reported.

Diagnosis

Laboratory confirmation of Lyme disease may prove difficult because the spirochaetes may be present in low numbers in specimens from clinically affected animals. In addition, the organism is fastidious in its cultural requirements.

A history of exposure to tick infestation in an endemic area in association with characteristic clinical signs may suggest Lyme disease. The ELISA is extensively used for antibody detection; Western immunoblotting is sometimes used for confirmation of ELISA results. Immunofluorescence assays may also be used but, in common with ELISA, the results of these methods may be difficult to interpret.

Culture of borreliae from clinically affected animals is confirmatory. Cultures in Barbour-Stoenner-Kelly medium should be incubated for six weeks under microaerophilic conditions and should be carried out in specialized laboratories. Low numbers of borreliae can be detected in samples by PCR techniques. These techniques can also be used for identifying genospecies and for epidemiological investigations.

Treatment and Control

Acute Lyme disease responds to treatment with amoxycillin and oxytetracycline. In chronic disease prolonged or repeated courses of treatment may be required. Acaricidal sprays, baths or dips should be used to control tick infestation. Where feasible, tick habitats such as rough brush and scrub should be cleared. Prompt removal of ticks from companion animals may prevent infection. However, because some tick species can transmit spirochaetes shortly after attachment, it cannot be assumed that daily removal of ticks will prevent infection. A number of vaccines, including whole cell bacterins and a recombinant subunit vaccine, are commercially available for use in dogs.

Avian Spirochaetosis

This acute disease of birds, caused by *Borrelia anserina,* can result in significant economic loss in flocks in tropicaland subtropical regions where the disease is endemic.Chickens, turkeys, pheasants, ducks and geese are susceptible to infection. Soft ticks of the genus *Argas* frequently transmit the disease. However, when there is contact between susceptible birds and infected material such as blood, tissues or excreta, transmission may occur. Because *B.anserina* survives poorly in the environment and for a limited time in infected birds, *Argas* ticks are important reservoirs of the organisms.

The borreliae survive trans-stadial moulting in ticks and can be transmitted transovarially between tick generations. Outbreaks of avian spirochaetosis coincide with periods of peak tick activity during warm, humid seasons. Morbidity and mortality are low in flocks continually exposed to infection. The disease is characterized by fever, marked anaemia and weight loss. Paralysis may develop as the disease progresses. Immunity, which follows recovery, is serotype specific. Several serotypes may be present in a particular region.

Diagnosis can be confirmed by demonstration of the spirochaetes in buffy coat smears using dark-field microscopy. Blood or tissue smears can also be examined using immunofluorescence. Giemsa-stained smears or silver impregnation techniques can be used to demonstrate the borreliae in tissues. The organisms are usually isolated by inoculating embryonated eggs or young chicks with infected blood or homogenized tissues.

Treatment with antibiotics is effective. Inactivated vaccines and tick eradication are the main control measures.

45

Brachyspira

Five genospecies of intestinal spirochaetes have been isolated from pigs namely *Brachyspira hyodysenteriae,B. pilosicoli, B. innocens, Serpulina intermedia* and *S. murdochii.* The genera *Serpulina* and *Brachyspira* were recently combined. These anaerobic spirochaetes have six to fourteen spirals and are 0.1 to 0.5 μm in width.

Usual Habitat

Pathogenic *Brachyspira* species are found in the intestinal tract of both clinically affected and normal pigs. Carrier pigs can shed *B.hyodysenteriae* for up to three months and are the principal source of infection for healthy pigs.

Differentiation of *Brachyspira* Species

The differentiation of *B.hyodysenteriae* from other intestinal spirochaetes is based on its pattern of haemolysis on blood agar. Tests for detecting indole production or the hydrolysis of hippurate are also useful diagnostically. Restriction endonnclease analysis, restriction fragment length polymorphism, ribotyping using 16s rRNA analysis, PCR-based assays and multilocus enzyme electrophoresis have been developed both for differentiating species and for distinguishing strains of organisms within species. *Brachyspira hyodysenteriae* isolates can also he allocated to several serogroups and serotypes.

Pathogenesis

Most information on the pathogenesis of *Brachyspira* species derives from studies of *B.hyodysenteriae.* Motility in mucus is an essential virulence factor of thisorganism; mutant strains with altered motility are lesscapable of colonizing the pig intestine. Colonization may be enhanced by factors in mucuswith chemotactic activity for the organisms. Factors with such chemotactic activities have been demonstrated *in vitro*. Haemolytic activity,demonstrated *in vitro,* correlates with pathogenicity andthree genes encoding haemolytic and cytotoxic activityhave been cloned and sequenced.

The pathogenesis of infection with *B.pilosicoli* differs from that of *B. hyodysenteriae* in that attachment of the spirochaetes to the intestinal mucosa appears to be important. Attachment of B.*pilosicoli* to the epithelial cells of the colonic mucosa leads to disruption of function with resultant cell shedding and oedema.

Clinical Infections

Infections with *Brachyspira* species are of importance in pigs. *Brachyspira hyodysenteriae,* the cause of swine dysentery, and *B. pilosicoli,* the cause of porcine intestinal spirochaetosis, are recognized pathogens. There is evidence that *Serpulina intermedia* may be associated with porcine spirochaetal colitis, but this has not been confirmed experimentally. Pigs acquire infection through exposure to contaminated faeces. The disease usually spreads slowly through a herd, affecting only one or two pens at a time. Dogs, rats, mice and flies may act as transport hosts for the spirochaetes. Mice populations can maintain *B. hyodysenteriae.* Although strains of *B. pilosicoli* have been found in many species including humans, dogs, chickens and pheasants, cross-infection between species has not been clearly demonstrated. *Brachyspira* species can survive in the environment for limited periods only if protected from desiccation. *Brachyspira hyodysenteriae* can persist for several weeks in moist faeces and for at least three days in slurry.

Clinical Signs

Infection with *B. hyodysenteriae* causes dysentery which is most often encountered in weaned pigs from six to twelve weeks of age. Affected pigs lose condition and become emaciated. Appetite is decreased and thirst may be evident. During recovery, there may be large amounts of mucus in the faeces. Although mortality is low, reduced weight gains due to poor food conversion cause major economic loss.

Brachyspira pilosicoli was identified in 1996 as the cause of porcine intestinal spirochaetosis. Previously, enteric disease had been produced experimentally by infecting pigs with a weakly-haemolytic spirochaete. The clinical signs in porcine intestinal spirocbaetosis are similar to those of swine dysentery but are less severe. Diarrhoea contains mucus rather than blood. Reduced feed conversion efficiency with poor weight gain has a major effect on production.

Diagnosis

History, clinical signs and gross lesions may indicate swine dysentery. Blood agar with added antibiotics is used for the culture of *Brachyspira* species.

Cultures are incubated anaerobically at 42°C for at least three days. Complete haemolysis is present around colonies of B. *hyodysenteriae;* other enteric spirochaetes are weakly haemolytic. Definitive identification can be made using immnnofluorescence, DNA probes or biochemical tests. Serological tests such as ELISA can be used to investigate infection in herds. PCR-based techniques have been developed and may be useful for laboratory confirmation.

Treatment and Control

Medication of drinking water is a useful method of treatment. Drugs commonly used include tiamulin, lincomycin and the nitroimidazoles. Improved hygiene, medication of feed and alteration of the diet may assist in controlling infection. Depopulation, thorough cleaning and disinfection of premises and strict rodent control are required for eradication of the disease.

46

Mycoplasma

The mycoplasmas are microorganisms in the class Mollicutes. Of the nine genera in this class, five contain species of veterinary interest. The genus Mycoplasma, in which there are about 100 species, contains most of the animal pathogens. The first mycoplasma identified in 1890 was Mycoplasma mycoides subspecies mycoides, the cause of contagious bovine pleuropneumonia. Similar types of mycoplasmas which were subsequently identified were called pleuropneumonia- like organisms (PPLO).

Mycoplasmas, the smallest prokaryotic cells capable of self-replication, are pleomorphic organisms ranging from spherical (0.3 to 0.9 pm in diameter) to filamentous (up to 1.0 µm long). Because they cannot synthesize peptidoglycan or its precursors, they do not possess rigid cell walls but have flexible, triple-layged outer membranes.

Their flexibility allows them to pass through bacterial membrane filters of pore size 0.22 to 0.45 µm. Mycoplasmas are susceptible to desiccation, heat, detergents and disinfectants. However, they are resistant to antibiotics such as penicillin which interfere with the synthesis of bacterial cell walls. Based on 5s rRNA sequence analyses, the mycoplasmas have been shown to be linked phylogenetically to Gram-positive bacteria such as Clostridium species which have low guanine-cytosine content in their DNA. They require enriched media for growth, characteristically forming umbonate micro-colonies when illuminated obliquely and microcolonies with a 'fried egg' dppearance in transmitted light. The dense central zone is due to extension of the microcolony into the agar. Mycoplasmas, which have relatively small genomes (approximately 800 genes) are fastidious in their growth requirements.

Most mycoplasmas are facultative anaerobes and some grow optimally in an atmosphere of 5 to 10% CO_2. Nonpathogenic anaerobic mycoplasmas are found in the rumens of sheep and cattle. The genera Mycoplasma and Ureaplasma contain animal pathogens.

Usual Habitat

Mycoplasmas are found on mucosal surfaces of the conjunctiva, nasal cavity, oropharynx and intestinal and genital tracts of animals and humans. Some species have tropisms for particular anatomical sites while others are found in many locations. In general, they are host-specific and suwive for short periods in the environment.

Colonial Morphology

When examined microscopically at low magnification, unstained microcolonies of Mycoplasma species are 0.1 to 0.6 mm in diameter and have a 'fried-egg' appearance. Some species produce colonies up to 1.5 mm in diameter which can be seen without magnification.

Colonies of Ureaplasma species are usually 0.02 to 0.06 mm in diameter and often lack a typical peripheral zone. Because their colonies are tiny, these organisms were formerly referred to as T-mycoplasmas.

Dienes stain facilitates recognition of microcolonies by staining the central zone dark blue and the peripheral zone a lighter blue.

Microcolonies of Mycoplasma species require differentiation from colonies of bacterial L-forms. However, L-forms often revert to normal and produce cell walls and typical bacterial colonies when subcultured on non-inhibitory media.

Mycoplasma species and Ureaplasma species require sterols for growth, and this is reflected in their sensitivity to inhibition by digitonin. As Acholeplasma species are sterol-independent, they are resistant to inhibition by digitonin. In the digitonin sensitivity test, a filter paper disc impregnated with digitonin is placed on medium inoculated with the isolate. A zone of growth inhibition exceeding 5 mm around the disc indicates sensitivity to digitonin.

Pathogenesis and Pathogenicity

Mycoplasmas adhere to host cells, an attribute essential for pathogenicity. This close contact facilitates toxic damage to the host cells by soluble factors produced by the pathogen. Some pathogenic species possess structures composed of unique adhesion proteins which promote attachment to mammalian cells. Mycoplasmas can adhere to neutrophils and macrophages and can also impair phagocytic functions. In addition, individual species damage cells by active penetration.

Modulation or activation of host immune responses is critical in the pathogenesis of mycoplasmal diseases. Some pathogenic mycoplasmas, including those involved in pulmonary diseases, are mitogenic for B and T lymphocytes. Activation of macrophages and monocytes leads to the release of cytokines including tumour necrosis factor and interleukins, resulting in the initiation of inflammation.

Pneumonia-producing mycoplasmas, which adhere to ciliated respiratory epithelium, can induce ciliostasis, loss of cilia and cytopathic change. Inflammation can also be induced in the bovine mammary gland by a membrane associated toxin of M. *bovis*.

Diagnostic Procedures

The presence of Mycoplasma species or mycoplasmal antigens in samples can be demonstrated immunologically or by nucleic acid procedures:

- Fluorescent antibody techniques
- Peroxidase-antiperoxidase procedures on paraffin embedded tissues
- Polymerase chain reaction techniques
- Inoculated mycoplasmal medium is incubated, aerobically or in 10% CO_2, in a humid atmosphere at 37°C for up to 14 days.
- Fluid samples can be inoculated directly onto agar or into broth media. Tissue specimens such as lung should he freshly sampled and a cut surface moved across the surface of a solid medium. Alternatively, the tissue can be homogenized in broth and samples of the suspension used for inoculation of liquid or solid media.

Identification criteria for isolates:

- 'Fried-egg' microcolonies
- Microcolony size
- Cholesterol requirement for growth (digitonin sensitivity test)
- Biochemical profile including urease production
- Fluorescent antibody technique on microcolonies
- Growth inhibition test with specific antisera

Serological tests

- Complement fixation tests for the major mycoplasmal diseases of ruminants are used for certification when animals are traded internationally.
- Tests based on ELISA are being developed for the diagnosis of economically important mycoplasmal diseases.
- Rapid plate agglutination tests are used for screening poultry flocks and for the field diagnosis of contagious bovine pleuropneumonia.
- Haemagglutination-inhibition tests can be used to determine the antibody levels in avian mycoplasmal diseases.

Clinical Infections

Mycoplasmas are often involved in disease Erocesses affecting mucosal surfaces. Factors such as extremes of age, stress and intercurrent infection may predispose to tissue invasion. In addition, mycoplasmas may exacerbate disease initiated by other pathogens, particularly in the respiratory tract.

Mycoplasmal infections cause respiratory diseases of major economic importance in farm animals especially in ruminants, pigs and poultry. Infections associated with mastitis or conjunctivitis in cattle and with disease conditions in domestic carnivores are usually of lesser importance. Several mycoplasmas have been isolated from dogs and cats but their precise role in disease has not been clearly defined. They have been implicated in respiratory and urinary tract disease in dogs. In cats, M. felis can occasionally cause conjunctivitis and M.gateae is associated with arthritis.

Contagious Bovine Pleuropneumonia

Contagious bovine pleuropneumonia (CBPP) is a severe contagious disease of cattle which has been recognized for more than 200 years and formerly had a worldwide distribution. It is caused by *M.* mycoides subspecies mycoides (small colony type), a member of the 'mycoides cluster'. This cluster is composed of six closely related members including the *M.* mycoides and *M.* capricolwn subspecies of sheep and goats and bovine mycoplasma group 7. Members of the cluster share biochemical, immunological and genetic characteristics which render individual species and subspecies difficult to differentiate.

The main method of transmission is by aerosols. Transmission of the disease requires close contact with clinically affected animals or asymptomatic carriers. Clinical signs become apparent three weeks after infection. The severity of the disease relates to strain virulence and the immune status of the host. Spread of infection can be relatively slow with peak morbidity (about 50%) at 7 to 8 months after introduction of infection into a herd. In severe outbreaks the mortality rate may be high.

Clinical Signs and Pathology

Clinical signs in the acute form of CBPP include sudden onset of high fever, anorexia, depression, drop in milk yield, accelerated respiration and coughing. Animals adopt a characteristic stance with the head and neck extended and elbows abducted. Expiratory grunting and mucopurulent nasal discharge may be present. Death can occur 1 to 3 weeks after the onset of clinical signs.

Arthritis, synovitis and endocarditis may be present in affected calves. At postmortem, the pneumonic lungs have a marbled appearance. Grey and red consolidated lobules alternate irregularly with pink emphysematous lobules and the interlobular septa are distended and oedematous. There may be abundant serofibrinous exudate in the pleural cavity. In chronic cases, fibrous encapsulation of necrotic foci is commonly found. These necrotic foci contain viable mycoplasmas and breakdown of the capsules in chronically affected animals is a major factor in the persistence and spread of CBPP in endemic areas.

Diagnosis

In endemic regions, clinical signs and characteristic postmortem findings allow a presumptive diagnosis. Techniques, such as the polymerase chain reaction, based on the detection of specific DNA in tissue samples can be used to differentiate M. Mycoides subspecies mycoides (small colony type) from other members of the 'mycoides cluster'. The fluorescent antibody test can be used on pleural fluid to confirm the presence of the pathogen. Isolation and definitive identification of the pathogen from broncho-alveolar lavage, pleural fluid, lung tissue or the broncho-pulmonary lymph nodes is confirmatory. Polymerase chain reaction-based tests may be useful confirmatory tests.

Serological Tests

- Rapid field serum agglutination test
- Passive haemagglutination screening test
- Complement fixation test for determining disease status of animals crossing national boundaries

- Dot-blot technique for confirmation
- A competitive ELlSA is presently under development.

Treatment and Control

Although treatment with antimicrobial drugs may be attempted in countries where the disease is endemic, it is generally unsatisfactory especially for chronically affected animals. In countries where CBPP is exotic, slaughter of affected and in-contact cattle is mandatory. In endemic regions, control strategies are based on prohibiting movement of suspect animals, mandatory quarantine and the elimination of carrier animals by serological testing and slaughter.

Annual vaccination with attenuated vaccines is carried out to stimulate effective immunity in cattle in endemic areas. The virulence of attenuated vaccines varies with the strain of mycoplasma employed. Annual vaccination may be discontinued as eradication of the disease progresses.

Infections with *Mycoplasma Bovis*

Strains of *M.bovis,* which is worldwide in distribution, can cause severe pneumonia in calves in the absence of other respiratory pathogens and can exacerbate respiratory disease caused by *Pasteurella* and *Mannheimia* species. *Mycoplasmabovis* has also been associated with mastitis and polyarthritis.

Diagnostic techniques are similar to those used for other mycoplasmas. A number of other *Mycoplasma* species cause sporadic cases of mastitis in cattle. Although the mastitis may be severe, systemic involvement is uncommon. There is often a dramatic loss of milk production and the serous or purulent mastitic exudate has a high leukocyte count. Mycoplasmal mastitis should be considered when other common bacterial causal agents have been excluded.

Contagious Agalactia of Sheep and Goats

This severe febrile disease of sheep and goats, caused by *M.agalactiae,* is prevalent in parts of Europe, northern Africa and parts of Asia. It usually becomes evident immediately after parturition and is characterized by mastitis, arthritis and conjunctivitis. Pregnant animals may abort and the disease can be fatal in young animals due to pneumonic complications. The organism is shed in milk and may remain localized in the supramammary lymph nodes between lactations. Disease due to *M. agalactiae* must be distinguished from mastitis and arthritis associated with *M. capricolum* snbspecies *capricolum,M. mycoides* subspecies *mycoides* (large colony type) and *M. mycoides* subspecies *capri*. Inactivated and attenuated vaccines for *M.agalactiae* are commercially available.

Contagious Caprine Pleuropneumonia

Contagious caprine pleuropneumonia (CCPP), caused by *M. capricolum* subspecies *capripneumoniae* (formerly *Mycoplasma* strain F38), is present in northern and eastern Africa and in Turkey. The disease is characterized by pneumonia, fibrinous pleurisy, profuse pleural exudate and a marbled appearance on the cut surface of affected lungs.

Although similar in many respects to CBPP, well developed necrotic areas in the lungs in chronic CCPP are rare. The disease is highly contagious and is transmitted by aerosols. Nomadic herds often carry infection to regions free of the disease. Pleuropneumonia in goats can occasionally be caused by *M.mycoides* subspecies *capri* or M.*mycoides* subspecies *mycoides* (large colony type). However, monoclonal antibody to *M.capricolum* subspecies *capripneumoniae* is specific for this organism in a growth inhibition disc test. Inactivated vaccines give satisfactory protection.

Mycoplasmal Diseases of Poultry

Mycoplasma gallisepticum causes chronic respiratory disease in chickens and infectious sinusitis in turkeys. The organism is transmitted through infection of the embryo in the egg or by aerosols. Clinical signs are consistent with upper respiratory tract involvement in chickens. In turkeys, there is swelling of the paranasal sinuses.

Reduced egg production may be evident. Diagnosis is based on isolation and identification of the pathogen and on flock testing using the serum plate agglutination test. Haemagglutination inhibition and ELISA tests are also used in flocks to confirm infection. Although antimicrobial medication of feed is used during outbreaks, the establishment of specific-pathogen-free flocks is the preferred method for controlling the disease. Eggs used for hatching should be dipped in a tylosin solution to eliminate the pathogen. Modified live vaccines and bacterins are available.

Mycoplasma meleagridis may be egg-transmitted and may be present in turkey semen. Aerosol transmission is less important with this pathogen than with *M. gallisepticum.*

The clinical features of the infection include reduced egg hatchability, airsacculitis in young poults and joint and bone deformities in growers. Confirmation requires isolation and identification of the pathogen. The serum plate agglutination test is used for flock testing. Tylosin, administered in the water for the first 10 days of life is of therapeutic value. Eggs used for hatching should be dipped in tylosin solution. Semen should be obtained from *M. meleagridis-free* toms.

Mycoplasma synoviae, the cause of infectious synovitis in chickens and turkeys, is transmitted mainly by aerosols. Egg transmission is much less important than in *M. gallisepticum* and *M.meleagridis* infections. Synovitis, arthritis and respiratory signs are the main clinical features.

Confirmation requires isolation and identification of the pathogen or positive serological tests. Tetracycline medication of the feed is used for treatment and control. Eradication is possible through the development of specific-pathogen-free flocks.

47

Coxiella Burnet II

Introduction

Q fever is a zoonotic bacterial infection caused by Coxiella burnetii, an obligate intracellular parasite, classified within the family Rickettsiaccae. Q fever (Q for "query") was first used in 1937 to describe a mysterious febrile illness of packing house workers in Brisbane, Australia. The causative agent was isolated from infected workers and later was identified as Rickettsiae. Almost simultaneously the same organism was identified wood tick collected in Montana. Cox, an American, and Burnet, an Australian, where honored for their early work with this organism, hence the name "Coxiella burnetii".

Host and Susceptibility

Coxiella burnetii is a well-established infectious agent that has reached a state of balanced pathogenicity in a plethora of host. Human are unnatural and usually dead-end-host. The microorganism generally maintained in a less pathogenic form in separate cycle existing independently among wild mammals and their haematophagus arthropods, principally tick and in domestic animals.

Morphology

It is an intracellular organism which is a polymorphic bacillus (0.2-0.4mm width, 0.4-1.0 mm length), which has a cell membrane, like Gram negative bacteria. However, it stains poorly with pigmented Gram stain, but gimenez staining is traditionally used to stain the *Coxiella burnetii* pathogen from pathological materials and crop.

Coxiella burnetii has several distinctive characteristics, including a sporulation-like process that protects the organism against the external environment, where it can survive for long periods. In mammals, the usual host cell of *C. burnetii* is the macrophage, which is unable to kill the bacterium. The other an important characteristic of *C. burnetii* is its antigenic variations, those called phase variation. This antigenic shift can be measured and is valuable for differentiating acute from chronic Q fever. Thus, it displays two antigenic phases those are phase-I and phase- II that are liable to the Lipopolysaccharide

(LPS) of the membrane.

Phase I *Coxiella burnetii* antigens are corresponds to the smooth phase of Gram-negative bacteria and are more highly infectious and phase-II antigen is corresponding to the granular (Rough) phase which has a lower virulence. The strain of *Coixella burnetii* pathogens are grouped into six (I VI) genomic group based on the Restriction Fragment Length Polymorphism (RFLP). The pathogenicity and the virulence of the *C. burnetii* are associated with genetic characteristics, type of strains and plasmid groups and with host factors such as pregnancy.

Characteristics

- Small, Gram-negative, pleomorphic coccobacillus; obligate intracellular bacterium that replicates in macrophages and monocytes.
- Order: Legionellales; Family: Coxiellaceae
- Size: 0.4-1.2 mm in length and 0.2-0.4 mm in width
- Nucleic acid: Coxiella genome is approximately 2000 kb.

Physicochemical Properties

Resistant to heat, low or high pH, 0.5% sodium hypochlorite, UV irradiation, and environmental conditions such as desiccation, extreme temperatures and sunlight because of the presence of a spore-like stage. Reported to survive for 7-10 months on wool at 15-20°C, for more than 1 month on fresh meat in cold storage, and for 40 months in skim milk at room temperature.

The microorganism has two antigenic forms: phase I and phase II. Phase I is the highly infectious form found in nature and has intact lipopolysaccharide (LPS) on the cell membrane, whereas phase II is laboratory-grown, attenuated, avirulent in animals, and has truncated LPS.

Infected cells contain 2 structural forms of the bacteria: large cell variant (LCV) or vegetative forms, and small cell variant (SCV) or condensed forms. SCV released during lysis of infected cells result in the spore-like form found in the environment.

Coxiella burnetii does not grow in artificial media, but can be cultured in embryonated eggs, primary cells, several cell lines including Vero cells and a number of macrophage-like cell lines.

Stain red with modified Ziehl Nelson and Machiavello procedures, and purple with Giemsa stain. They are gram negative, non-motile and varies in shape from pleomorphic to ovoid even to rod like.

On adaptation to laboratory culture, Coxiella burnetii undergoes modification of antigenicity and virulence from phase I (analogous to smooth gram negative bacteria) found in animals to phase II (rough from). However, this variation of antigenicity has relevance for vaccine formation and serological diagnosis.

Phase variation

Coxiella burnetii has two antigenic surface phases; phase I and phase II, which vary in their pathogenic and immunogenic properties and undergo antigenic phase variation due no Lipopolysaccharide truncation when serially passaged in embryonated eggs or cell culture. The phase I form is extremely infectious and exists in human and other animals. Passaging can also result in a shift to form spore which allows the organism to survive in harsh environment. Indeed, it can survive for greater than 40 months in skim milk at room temperature and is readily recovered from soil up to one month after contamination. It has high level of resistance to chemical and physical agents. High level of resistance to heat has been attributed to the formation of endospore.

Disinfection

C. burnetii is relatively resistant to disinfectants. The infective dose is also reported to be low. Agents reported to be effective with a contact time of 30 minutes include 70% ethanol and some quaternary ammonium-based disinfectants (e.g., MicroChem-Plus®, 5% Enviro-Chem®). This organism can also be inactivated with 5% hydrogen peroxide, or by gaseous formaldehyde, 5% chloroform or ethylene gas in a sealed, humidified chamber. Variable susceptibility has been reported for hypochlorite, phenolic disinfectants and formalin.

Sodium hypochlorite (1:100 dilution of household bleach) or 1% Virkon S® result in greater than 90% reduction in infectivity. Although 2% formaldehyde is reported to destroy C. burnetii, it has been isolated from tissues stored in formaldehyde for several months. Sources also differ on the effectiveness of Lysol®, which has changed its formulation a few times. Physical inactivation can be accomplished by gamma irradiation or high heat, including high temperature pasteurization of milk (e.g., 161°F/72°C for 15 seconds).

Source of Infection and Transmission

Infection of non-pregnant animals is clinically silent and is followed by latent infection until pregnancy when there is recrudescence with infection in the intestine, uterus, placenta and udder, and excretion from this site at parturition.

The organism is present in high concentration in the placenta and fetal fluids, and subsequent vaginal fluids. It is also excreted in urine, and through milk. Cattle and small ruminants can shade the organism through faeces, sperm and reproductive discharge. Parturient cats and dogs have also been implicated as source of human infection.

Transmission is by direct contact and through inhalation. There is significant contamination of the environment of infected animals at the time of parturition and this is probably a critical period for transmission of the disease within herds and flocks.

Coxiella burnetii is transmitted from various reservoir hosts to human through direct contact as well as air born, vector born, and through contaminated vehicle. In ticks, there is transovarian (transmission of Coxiella burnetii from adult tick to egg) and transstadial (transmission of Coxiella burnetii from larvae to nymph, and adult). The organism is present in the semen of seropositive bulls and venereal transmission is suspected.

Many arthropods including cockroaches, beetles, flies, fleas, bugs, lice, mite and ticks are naturally or experimentally infected; human rarely acquired the disease by bites of arthropods. Most human cases of Q fever can be traced to exposure to infected sheep, cattle and goats either directly or through unpasteurized milk.

Pathogenesis and Pathogenicity

The organism probably follows the oropharyngeal route as its port of entry into the lungs and intestines of both humans and animals. It is highly infectious, and a very low dose is sufficient to initiate infection. Primary multiplication takes place in the regional lymph nodes after the initial entry, and a transient bacteraemia develops which persists for five to seven days.

Coxiellaburnetii has two morphologically distinct cell variants; an intracellular and metabolically active large cell variant (LCV) and a spore like small cell variant (SCV). These two forms are morphologically and functionally distinct. The LCV is larger, elongated less electron-dense bacteria and metabolically active and replicating large bacteria. While, the SCV presents a compact rod-shaped with a very dense central region and it is considered the metabolically dormant and less replicating. The SCV are shed by infected animals.

After infection the organism attaches to the cell membrane of phagocytic cell. After phagocytosis, the phagosome containing the SCV fuses with the lysosome. The SCV are metabolically activated in the acidic phagolysosomes and can undergo vegetative growth to form LCV.

The LCV and the activated SCV can both divide by binary fission and the LCV can also undergo sporogenic differentiation. The spores that are produced can undergo further development to become metabolically inactive SCV. And both spores and SCV can then be released from the infected host cell by either cell lysis or exocytosis.

The acidic environment also protects Coxiella burnetii from the effects of antibiotics, as the efficacy of antibioticsis decreased in the acidic PH. The SCV and sporeforms are more difficult to denature than LCV.

Coxiella burnetii also has two distinct antigenic phases, Phase I and Phase II, based on changes that occur in the organism during in vitro culture. The primary significance of these two phases is that antibodies to phase II antigens are made during the early stages of the infection, but antibodies to phaseI antigens predominate if the organism persists longer. This switch is used to distinguish acute from chronic infections in people, although it is not currently employed in animals. Phase I bacteria (wild virulent type) with a smooth full length LPS were isolated from infected humans, animals and arthropods. Phase I bacterium converts to an avirulent phase II with rough LPS after several passages in embryonated egg or cell cultures. The virulence and the pathogenicity of the Coxiella burnetiiare associated with genetic characteristics, plasmid groups and type of strains and also with host factors such as pregnancy.

After Coxiella burnetii enter to the host and gain access to vascular endothelium and respiratory and renal epithelium, it multiplies in phagosomes; due to an enzyme system adapted low pH (5.0). It causes necrosis and hemorrhage of many other organs including the liver, central nervous system and mononuclear phagocytic system. In animals the pattern may be similar, although mildly affected animals may have latent infection. Latent infection can persist particularly in the locating mammary gland and the pregnant uterus but become reactivated during parturition. Abortion (sporadically) may result from placentitis or the delivery may be normal and produce viable young. Immune complex pneumonia can develop in many organs.

Clinical Signs

In ruminants, significant clinical signs seem to be limited to pregnant animals, and are characterized by abortions, stillbirths, and the birth of small or weak offspring. Reproductive losses may occur as outbreaks in sheep and goats, but they seem to be sporadic in cattle. Most abortions are reported to occur near term. Anorexia, depression, agalactia and retained fetal membranes are possible, but they seem to be uncommon, and most abortions have no significant premonitory signs. Subsequent pregnancies might sometimes be affected. Links between infection with C. burnetii and endometritis/ metritis or infertility have been suggested in cattle and sheep, and a possible link with subclinical mastitis has been proposed in cattle.

Post Mortem Lesions

C. burnetii abortions in ruminants are characterized by placentitis primarily affecting the intercotyledonary areas. The placenta is typically leathery and thickened, and it may contain large amounts of mucopurulent or purulent exudates, especially at the edges of the cotyledons and in the intercotyledonary areas. Severe vasculitis is uncommon, but thrombi and some degree of vascular inflammation may be noted. Aborted fetuses tend to be fresh, though they are occasionally autolyzed. Fetal lesions are usually non-specific, although pneumonia and microscopic evidence of hepatic necrosis or granulomatous inflammation have been reported.

Diagnosis

Collection of Samples

In the acute-phase of the disease, immediately serum sample should be collected after the onset of symptoms (within the first 2 weeks) with a convalescent-phase specimen collected 3–6 weeks later. Before antibiotic administration the whole blood should be collected in EDTA-treated anticoagulant tubes and shipped refrigerated on the frozen gel packs by overnight. The Buffy coat can be saved for DNA amplification and stored frozen in a non–frost-free freezer if samples are to be prepared for other laboratory tests. The most commonly evaluated sample used for confirmation for chronic Q fever is heart valve tissue.

Staining and Related Measures

The diagnosis of coxiellosis in aborted small ruminants by an ordinary method is to detect the pathogen using staining techniques. The smears prepared from the samples are usually stained by Gimenez, Stamp, Giemsa stain or Machiavello.

Isolation and Cultivation

For routine diagnosis of Q fever cultivation of *C. burnetii* is not recommended, since the process is very difficult, time consuming, and dangerous. In addition to this culture of *C. burnetii* requires a bio-safety level 3 (BSL- 3) laboratory because bacteria are highly infective and can be hazardous for laboratory workers. Patients with chronic Q fever have already received antibiotics which can further complicate isolation attempts; a negative culture does not rule out a *C. burnetii* infection. Specimens can be referred to CDC through state public health laboratories for culture.

Isolation must be made in cell culture of macrophage or fibroblast lines, embryonated eggs or laboratory rodents, but cell dissociated to phase II because phase I is highly infectious.

Phase I antigens are isolates from organisms taken directly from animals or their patients. This natural phase is highly infectious, contain large amount of lipopolysaccharides (LPs), and form smooth colonies in culture. Phase II antigens are round in organisms that have been passed serially in embryonated eggs, have a truncated, and lack some cell surface antigens.

Because of laboratory acquired infection caused by C. burnetii, cultivation of the organism has been discouraged. However, the use of a shell vial assay with human lung fibroblast to isolate the organism from buffy coat and biopsy specimens has not resulted in any laboratory acquired infection.

Isolation of C. burnetii is dangerous to laboratory personnel, requiring BSL 3 conditions, and culture is rarely used for diagnosis. Embryonated chicken eggs do not work as well as cells, and are no longer recommended for the initial isolation. Laboratory animals, such as mice and guinea pigs, have occasionally been employed to isolate C. burnetii, mostly in the past. Various genotyping methods, such as multiple-locus variable-number tandem repeat analysis (MLVA), multispacer sequence typing (MST) and single-nucleotide-polymorphisms can be useful for linking outbreaks to their source.

Serology

Serological tests including compliment fixation test (CFT), Indirect Fluorescence Antibody (IFA) and enzyme linked immunosorbent assays (ELISA) and Microagglutination can be used to help diagnose Q fever.

Molecular Method

Multispacer Sequence Typing, depending on DNA sequence variations in10 short intergenic regions, can be performed on isolated *C. burnetii* strains or directly on extracted DNA from clinical samples. Polymerase Chain Reaction (PCR) or immuno histochemical methods can be used for detection of Coxiella burnetii in tissue culture or tissue specimens derived from patients.

Diagnostic laboratories usually use PCR to detect C. burnetii in secretions, excretions and tissues. Loop-mediated isothermal amplification (LAMP) assays have also been published. Nucleic acids of C. burnetii can occur in the placenta after a normal delivery, or concurrently with other pathogens; thus, caution should be used when attributing a clinical case to this organism. Histopathology and quantitative PCR may be helpful in establishing a causative role. Recently vaccinated animals can excrete vaccine strains during the first month.

Differential Diagnosis

In animals the differential diagnosis includes other causes of abortion and infertility like leptospirosis, brucellosis, Listeriosis, and salmonellosis.

Treatments

Treatment is indicated for all infections, even for those that are subclinical. For domestic small ruminants' oral therapeutic dose may be given for 24 weeks.

Doxycycline is the most effective drug against C. burnetii, in most reports. In early studies, the fluoroquinolones were found to be one of the most effective agents in eliminating C. burnetii from L929 cells. However, C. burnetii remain susceptible to Levofloxacin, Moxifloxacin, and to a lesser extent Ciprofloxacin. Erythromycin was proposed as an empirical treatment for C. burnetii pneumonia. Although C. burnetii is susceptible to a number of antibiotics, as discussed earlier, Oxytetracycline is the preferred choice.

Control and Prevention

Phase I formalin inactivated Q fever vaccine which provide effective in the immunization of dairy cattle. Administration of formalin inactivated phase I Coxiella burnetii vaccine in naturally infected ewes and cows eliminate shedding of the organism in milk. Aborted animals should be isolated for 3 weeks and aborted and placental contaminated material should be burned. Ideally, manure should be composted for 6 months before application to fields. Feed areas should be raised to keep them free from contamination with faeces and urine. Milk and milk products should be pasteurized.

48

Ehrlichia (Heartwater – Cowdriosis)

Aetiology

Classification of the causative agent

- Ehrlichia ruminantium (formerly Cowdria ruminantium) Order Rickettsiales, Family Anaplasmataceae
- Small, Gram negative, pleomorphic coccus, and obligate intracellular parasite.
- Strains of E. ruminantium are very diverse and vary in virulence: while some strains are highly virulent, others appear to be less-pathogenic.
- E. ruminantium has a high level of genomic plasticity. Several different genotypes can co-exist in a geographical area, and may recombine to form new strains.
- E. ruminantium multiplies in vascular endothelial cells throughout the body to cause severe vascular compromise.
- It usually occurs in clumps of from less than five to several thousand organisms within the cytoplasm of infected capillary endothelial cells, and can be detected in brain smears by light microscopy.

Resistance to Physical and Chemical Action

Temperature

Heat labile and loses its viability within 12–38 hours at room temperature. Infective stabilities can be cryopreserved in DMSO (dimethyl sulphide) or better yet in sucrosepotassium phosphate-glutamate medium (SPG). Infective half-life of thawed stabilate kept on ice is only 20–30 minutes.

pH: Not applicable.

Disinfectants: Not applicable.

Survival:

The heartwater organism is extremely fragile and cannot persist outside of a host for more than a few hours. Because of its fragility, the organism must be stored in dry ice or liquid nitrogen to preserve its infectivity.

Epidemiology

- Heartwater occurs only where its Amblyomma tick vectors are present.
- Epidemiology depends on interaction of tick vector, causative agent, and vertebrate hosts.
    - Tick vector: tick infection rates, seasonal changes influencing abundance and activity, and intensity of tick control.
    - Causative Agent: differing genotypes affecting virulence or stimulation of crossprotection.
    - Vertebrate Host: availability of wild animal reservoirs, and age and genetic resistance.
- Because of its extreme fragility, the principal mode of bringing the disease into an area is by introduction of infected ticks or carrier animals. It is not known for how long wild or domestic ruminants can be a source of infection for ticks in nature, but it may be many months. Ticks are a robust reservoir of E. ruminantium, and infection can persist in them for at least 15 months. Careful dipping and hand-dressing followed by inspection to ensure the absence of ticks is recommended for animals in transit to heartwater free areas.

Hosts

- All domestic and wild ruminants can be infected, but the former appear to be the most susceptible. Indigenous domestic ruminants are usually more resistant to the disease. Wild animals could play a role as reservoir.
- Heartwater causes severe disease in cattle, sheep, and goats, with milder disease in some indigenous African breeds of sheep and goats, and inapparent disease in several species of antelope indigenous to Africa. Bos indicus (Zebu) cattle breeds are in general more resistant than Bos taurus (European) breeds.

Transmission

- Heartwater is transmitted transstadially by ticks of the genus Amblyomma, which are biological vectors of heartwater. Ticks become infected by feeding on acutely ill or subclinically infected animals. Of the 13 species capable of transmitting the disease, A. variegatum (tropical bont tick) is by far the most important because it is the most widespread. Other major vector species are the bont tick A. hebraeum (in southern Africa), A. gemma and A. lepidum (in Somalia, East Africa and the Sudan).

- A. astrion (mainly feed on buffalo) and A. pomposum (distributed in Angola, Congo and Central African Republic) are also natural vectors of the disease. Four other African ticks, A. sparsum (feed on reptiles and buffalo mainly), A. cohaerans (feed on African buffalo), A. marmoreum (adults occur on tortoises and immature stages on goats) and A. tholloni (adults feed on elephants) experimentally transmit heartwater.
- Amblyomma ticks are three-host ticks whose life cycles may take from 5 months to 4 years to complete. Because the ticks may pick up the infection as larvae or nymphs and transmit it as nymphs or as adults, the infection can persist in the tick for at least 15 months. Infection does not pass transovarially. While transmission of heartwater can be by adult and nymphal ticks in the field, in general adults prefer to feed on cattle and nymphs on sheep and goats.
- Cattle egrets have been implicated in the dispersal of Amblyomma ticks in the Caribbean.
- Heartwater can be transmitted vertically and through colostrum of carrier dams.
- Transmission can also occur by intravenous inoculation of blood, tick homogenates or cell culture material containing E. ruminatium.

Sources of the Agent

- Amblyomma ticks fed on an infected vertebrate host.
- Whole blood or plasma of vertebrate host during the febrile reaction, but highest levels of agent occur during the second or third day of fever.
- Colostrum containing infected cells (reticulo-endothelial cells and macrophages) has been speculated.

Diagnosis

The average incubation period in natural infections is 2–3 weeks, but can vary from 10 days to 1 month. The incubation period after intravenous blood inoculation is seven to 10 days in sheep and goats, and 10 to16 days in cattle. However the incubation is strongly dependent on the dose of **elementary bodies** inoculated as shown experimentally using in vitro cultivated E. ruminantium.

Lesions

The gross lesions in cattle, sheep, and goats are very similar. Heartwater derives its name from one of the prominent lesions observed in the disease, namely pronounced hydropericardium. The most common macroscopic lesions are hydropericardium, hydrothorax, pulmonary oedema, intestinal congestion, oedema of the mediastinal and bronchial lymph nodes, petechiae on the epicardium and endocardium, congestion of the brain, and moderate splenomegaly.

- Accumulation of straw-coloured to reddish fluid in the pericardium is more consistently observed in sheep and goats than in cattle.
- Hydropericardium, hydrothorax, ascites (mild), mediastinal oedema, and pulmonary oedema are common and result from increased vascular permeability.
- Oedema of the mediastinal and bronchial lymph nodes may occur.
- Froth in the trachea is often seen, reflecting terminal dyspnoea due to pulmonary oedema.
- Subendocardial petechial haemorrhages are usually present.
- Submucosal and subserosal haemorrhages may occur elsewhere in the body.
- Gross brain lesions are usually absent except for subtle swelling of the brain, which may result in conal herniation.
- Nephritis of varying degree, especially in Angora goats.
- Congestion and/or oedema of the abomasal folds are a regular finding in cattle, but less so in sheep and goats.

Diagnosis

Identification of the Agent

- Heartwater is often diagnosed by observing E. ruminantium colonies in the brain or intima of blood vessels. Brain smears are air dried, fixed with methanol and stained with Giemsa. E. ruminantium can also be detected in formalin-fixed brain sections using immunoperoxidase techniques.
- Polymerase chain reaction (PCR) techniques are available to reveal the presence of E. ruminantium in the blood of animals with clinical signs, in the tick vector, and to a lesser extent in the blood or bone marrow of carrier animals. In cultures, the organism is identified by microscopic

examination, or by immunofluorescence/immunoperoxidase staining. In some cases, heartwater may be diagnosed by inoculating fresh blood into a susceptible sheep or goat.

Serological tests

- Serological tests available include indirect fluorescent antibody tests, enzyme linked immunosorbent assays (ELISAs) and Western blot. However, when the whole E. ruminantium is used as antigen, cross-reactions with Ehrlichia spp. occur in all of these tests. Serology has limited diagnostic applications.

Prevention and Control

- As E. ruminantium cannot survive outside a living host for more than a few hours at room temperature, heartwater is usually introduced into free areas by infected animals, including subclinical carriers, or by ticks.
- In heartwater-free countries, susceptible ruminants from endemic regions are tested before importation. All animals that may carry Amblyomma, including non-ruminant species, must be inspected for ticks before entry.
- Ticks may be carried into a country on illegally imported animals or migrating birds.
- Outbreaks are usually controlled with quarantines, euthanasia of infected animals and tick control.
- During an outbreak, ticks should not be allowed to feed on infected animals.
- Iatrogenic transfer of blood between animals must also be avoided.
- In endemic regions, heartwater can be prevented by tick control and vaccination.
- Animals moved into endemic areas may be protected by prophylactic treatment with tetracycline.
- Vaccination currently consists of infection with a live E. ruminantium strain, then treatment with antibiotics when a fever develops. Alternatively, the vaccine may be given to young kids or lambs during their first week of life, or to calves less than 5 to 8 weeks of age.
- Intensive tick control may increase the susceptibility of animals to heartwater, because it eliminates the immune boosting effect of persistent exposure to small doses of organisms.

- In endemic areas, animals with heartwater can be treated with antibiotics.
- Tetracycline (oxytetracycline at 10 mg/kg or doxycycline at 2 mg/kg) is effective during the early, febrile stages of this disease, but animals often die before treatment can be administered. Antibiotic treatment alone is not always successful in later stages.
- Vector control measures aimed at eradication of Amblyomma ticks by acaricide treatment of cattle and small ruminants

Inactivated vaccines

- Inactivated vaccines using the Gardel strain and subsequently other strains have been produced in bioreactors for an industrial production following the development of the whole production process. The Mbizi strain inactivated vaccine is being developed commercially by Onderstepoort Biological Products in South Africa.

Live attenuated vaccines

- Isolates of attenuated virulence that do not necessitate treatment of animals would be ideal but a limited number of such attenuated isolates are available.
- An attenuated Senegal isolate has been obtained and shown to confer 100% protection against an homologous lethal challenge, but very poor protection against a heterologous challenge.

Recombinant vaccines

- Several reports show partial protection of mice using map1 DNA vaccination and an improvement of protection by vaccination following a prime (plasmid) – boost (recombinant MAP1) protocol. However, protection of ruminants has never been demonstrated using this strategy.
- In opposition, significant protection of sheep was reported against homologous and heterologous experimental challenge following plasmid vaccination using a cocktail of four ORFs (open reading frames) from the 1H12 locus in the E. ruminantium genome. No further results have been described since then.
- Recombinant vaccines will probably not be available in the near future.

Ehrlichiosis in Dogs

Signs, Diagnosis and Treatment of Ehrlichiosis in Dogs

Ehrlichia is a type of bacteria that infect dogs and other species worldwide, causing a disease called ehrlichiosis. Ehrlichiosis has also been called tropical canine pancytopenia (and several other names). Ehrlichia is commonly transmitted by ticks.

Cause

Ehrlichia bacteria infect white blood cells. There are many species of Ehrlichia, which infect a wide variety of animals, but there are only a few species that affect dogs. A closely related infection affecting platelets is caused by a bacteria called Anaplasma platys and is sometimes referred to as ehrlichiosis as well (Anaplasma platys used to be called Ehrlichia platys until recently). Most Ehrlichia infections are acquired through tick bites. Infection is also possible via blood transfusions.

Risk Factors

Ehrlichiosis occurs worldwide in areas where the ticks that carry the disease are common. While any dog can be infected, some breeds, most notably German shepherds, are prone to more serious chronic infections. Retired racing greyhounds from areas where ehrlichiosis is common may suffer from chronic, undetected infections and should be checked for ehrlichiosis and other tick-borne diseases when adopted.

Signs and Symptoms of Ehrlichiosis

The symptoms and severity of illness seen with ehrlichiosis depends on the species of Ehrlichia involved and the immune response of the dog. Generally, Erlichia canis appears to produce the most severe illness, and infections tend to progress through various stages.

The acute phase occurs within the first few weeks of being infected and is rarely fatal. Recovery can occur, or the dog can enter a "subclinical phase" which can last for years, where there are no symptoms. Some dogs, but not all, eventually progress to the chronic phase, where very severe illness can develop. However, in practice is is difficult to distinguish these phases.

Signs and symptoms of ehrlichiosis may include:

- Fever
- lethargy

- loss of appetite
- weight loss
- abnormal bleeding (e.g., nosebleeds, bleeding under skin - looks like little spots or patches of bruising) enlarged lymph nodes
- enlarged spleen
- pain and stiffness (due to arthritis and muscle pain)
- coughing discharge from the eyes and/or nose
- vomiting and diarrhea
- inflammation of the eye
- Neurological symptoms (e.g., incoordination, depression, paralysis, etc.)
- Signs of other organ involvement can appear in the chronic form, especially kidney disease.

Note: Anaplasma platys causes recurrent low platelet counts but tends to produce only mild symptoms, if any.

Diagnosis of Ehrlichiosis

It can be difficult to confirm a diagnosis of ehrlichiosis. Blood tests typically show a decreased number of platelets ("thrombocytopenia") and sometimes decreased numbers of red blood cells (anemia) and/or white blood cells.

Changes in the protein levels in the blood may also occur. Blood smears can be examined for the presence of the Ehrlichia organisms. If they are present, the diagnosis can be confirmed, but they may not always show up on a smear.

Blood can also be tested for antibodies to Ehrlichia, though this can sometimes produce incorrect results. Specialized testing can check for genetic material from Ehrlichia, and while this is the most sensitive test, it is not widely available and has some limitations as well. Generally, a combination of lab tests along with clinical signs and history are used to make a diagnosis.

The diagnosis is further complicated by the fact that dogs infected with Ehrlichia may also be infected with other diseases carried by ticks, such as Babesia, Lyme disease, or Rocky Mountain Spotted Fever. Infection with a bacteria called Bartonella has also been found in conjunction with Erlichiosis and other tick borne diseases.The presence of these other diseases can make symptoms more severe and and the diagnosis more complicated.

Treatment of Ehrlichiosis

Ehrlichiosis responds well to treatment with the anitbiotic doxycycline. Improvement in symptoms is usually very quick, but several weeks of treatment is usually needed to ensure a full recovery. In severe cases where blood cell counts are very low, blood transfusions may be needed. Reinfection is possible as immunity to Ehrlichia bacteria is not long lasting.

Prevention of Ehrlichiosis

Preventing exposure to the ticks that carry Ehrlichia is the best means of preventing ehrlichiosis. Check your dog daily for ticks and remove them as soon as possible (it is believed that ticks must feed for at least 24-48 hours to spread Ehrlichia).

This is especially important in peak tick season or if your dog spends time in the woods or tall grass (consider avoiding these areas in tick season). Products that prevent ticks such as monthly parasite preventatives (e.g., Frontline®, Revolution®) or tick collars (e.g., Preventic®) can be used; be sure to follow your veterinarian's advice when using these products. Keep grass and brush trimmed in your yard, and in areas where ticks are a serious problem, you may also consider treating the yard and kennel area for ticks.

49

Anaplasma

Bacteria from the genus *Anaplasma* belong to the Procaryota kingdom, the family Anaplasmataceae. *Anaplasma* belong to obligate intracellular microorganisms, Gram-negative bacteria, living in the blood cells of mammals. Vertebrates can be their reservoir, i.e. an environment where the pathogen can live and proliferate for many years. However, in many cases bacteria from the genus *Anaplasma* cause diseases in domestic animals and people.

Taxonomical Position of Bacteria from the Genus *Anaplasma*

The genera *Anaplasma, Ehrlichia, Neorickettsia* and *Wolbachia* include obligate intracellular bacteria, parasitizing in the vacuoles of cells in eukarytic hosts. Animal pathogens were attributed to the genus *Anaplasma*, such as *A. centrale*, *A. marginale*, *A. platys*, *A. ovis* and *A. bovis*, and also the aetiological factor of human anaplasmosis, *A. phagocytophilum.*

Anaplasma marginale

A. marginale are obligate intracellular bacteria parasitizing in erythrocytes of higher vertebrates, mostly ruminants.

Anaplasma Centrale

A. centrale is a parasite in the erythrocytes of ruminants, mainly cattle. As opposed to *A. marginale*, it creates concentrations in the central part of the cell. *A. centrale* is less pathogenic to cattle than *A. marginale* but, most importantly, occasionally gives resistance against the latter. Hence it is used for the preparation of live vaccine strains, assuring immunological protection against bovine anaplasmosis.

Anaplasma Bovis

A. bovis is a bacterium detected mainly in cattle, but also observed in small mammals which are probably a reservoir of this bacterium. It occurs in monocytes, and the disease it causes is called monocytic anaplasmosis. The symptoms of the disease are most visible in calves, but also in adult animals;

symptoms include weakening of the body, marked reduction in weight, elevated temperature, enlargement of prescapular lymph nodes, paling of the mucous membranes, and in many cases an elevated amount of secreted mucus.

Anaplasma Ovis

A. ovis mainly parasitizes small ruminants, i.e. sheep and goats, and its presence has been confirmed in most regions of the world both in farm and wild animals. Similarly to *A. marginale* and *A. centrale*, these bacteria live in erythrocytes, hence the anaemia of the infected animals.

In the case of *A. ovis*, bacterial inclusions are found 35–40% of the time in the central or submarginal part of the erythrocyte of the host and the remaining 60–65% of the time in the marginal part.

Anaplasma Platys

A. platys is mainly a pathogen of canines, usually dogs. It lives in blood cells, causing Canine Cyclic Thrombocytopenia (CCT). Clinical symptoms of this disease vary depending on the species of the dog, but fundamentally CCT causes weakening of the animal, lethargy, anorexia, respiratory distress, fever, increased mucous secretion, purulent ocular discharge, splenomegaly and muzzle hyperkeratosis. These bacteria do not cause death, and treatment with suitable antibiotics is usually successful after four weeks. Doxycycline is one of such effective antibiotic.

Anaplasma Phagocytophilum

A. phagocytophilum is a frequently described bacterium from the genus *Anaplasma*. It has been detected in domestic animals, mostly small ruminants, but also in people. These bacteria are an aetiological factor of granulocytic anaplasmosis which in the case of human infections is defined as Human Granulocytic Anaplasmosis – HGA (Human Anaplasmosis – HA).

In the case of farm animals anaplasmosis caused by *A. phagocytophilum* is a disease with subclinical or heavy symptoms. It has unspecific symptoms, such as fever, drowsiness, anorexia, abortions, drop in body weight and reduction in milk production. Left unattended, especially in weaker individuals, it can lead to death.

Anaplasmosis in Beef Cattle

Anaplasmosis is an infectious disease of cattle caused by several species of the blood parasite *Anaplasma*. *A. marginale* is the most common pathogen of cattle. Sheep and goats are much less commonly affected. Anaplasmosis is also called "yellow bag" or "yellow fever" as affected animals can develop a jaundiced appearance.

Transmission

A. marginale can be transmitted two different ways. First, it can be transmitted mechanically when red blood cells infected with *A. marginale* are inoculated into susceptible cattle. This can occur through needles, dehorners, ear taggers, castrating knives or other surgical instruments, and tattoo instruments.

Mechanical transmission can also occur through the mouthparts of biting insects, such as biting flies. Face flies, houseflies, and other non-biting insects do not transmit the disease. Horn flies, although they bite, typically do not go from animal to animal so they are not thought to spread *Anaplasma*. Mechanical transmission of infected red blood cells must occur within a few minutes of the blood leaving the infected animal, as the blood parasite does not survive more than a few minutes outside the animal.

Once susceptible cattle are infected with *Anaplasma*, the organism multiplies in the bloodstream and attaches to the animal's red blood cells. The animal's immune system destroys the infected red blood cells in an attempt to fight off the infection. Unfortunately, uninfected blood cells are also destroyed. When the number of blood cells being destroyed exceeds the number of blood cells that the body can produce, the animal becomes anaemic. It takes 3 to 6 weeks for clinical signs to appear after the animal is infected.

Outbreaks

Although many outbreaks of anaplasmosis occur in the spring and summer, they can occur at any time of the year. The many ways it can be transmitted and the potential for carrier animals makes the source of an outbreak confusing.

Pathogenesis

Stages of the Disease

Anaplasmosis can be divided into four stages: incubation, developmental, convalescent, and carrier. These stages and the symptoms associated with them are described below.

Incubation Stage

The incubation stage begins with the original infection with *A. marginale* and lasts until 1 percent of the animal's red blood cells (RBCs) are infected. The average incubation stage ranges from 3 to 8 weeks, but wide variations have been documented.

Most variation is directly related to the number of organisms introduced into the animal. After gaining entry into a susceptible animal, the anaplasma parasite slowly reproduces in the animal's blood during the incubation phase. During this period the animal remains healthy and shows no signs of being infected. Finally, after the parasite has reproduced many times and established itself in the RBCs of the animal, the body attempts to destroy the parasite.

Developmental Stage

During the developmental stage, which normally lasts from 4 to 9 days, most of the characteristic signs of anaplasmosis appear. Clinical signs begin to be expressed about halfway through this phase. As the infected animal's body destroys the parasite, RBCs are destroyed as well. When a substantial loss of RBCs has occurred, the animal will show signs of clinical anemia. The body temperature will commonly rise to 104° to 107° F (40° to 41°C), and a rapid decrease in milk production will occur in lactating cows.

Cattle producers first notice the anemic, anaplasmosis-infected animal when it becomes weak and lags behind the herd. It refuses to eat or drink water. The skin becomes pale around the eyes and on the muzzle, lips, and teats. Later, the animal may show constipation, excitement, rapid weight loss, and yellow tinged skin. The animal may fall or lie down and be unable to rise. Affected cattle either die or begin a recovery 1 to 4 days after the first signs of the disease.

As a general rule, unless infected cattle can be detected during the early developmental stage, they should not be treated. There are two primary reasons for this practice. First, if the animal is forced to move or becomes excited, it may die of anoxia (lack of oxygen in the animal's system). Second, antibiotic treatments do little or nothing to affect the outcome of the disease when given during the late developmental or convalescent stage.

Convalescent Stage

Cattle that survive the clinical disease lose weight, abort calves, and recover slowly over a 2- or 3-month period. This is known as the convalescent stage, which lasts until normal blood values return. This stage is differentiated from the developmental stage by an increase in the production of RBCs (erythropoiesis) in the peripheral blood, shown in an increase in hemoglobin levels and high total white blood cell counts, among other characteristics. Death losses normally occur during the late developmental stage or early convalescent stage.

Cattle of all ages may become infected with anaplasmosis, but the severity of illness increases with age. Calves under 6 months of age seldom show enough signs to indicate that they are infected. Cattle 6 months to 3 years of age become increasingly ill, and more deaths occur with advancing age. After 3 years of age, 30 to 50 percent of cattle with clinical anaplasmosis die if untreated.

Carrier Stage

Unless adequately medicated, cattle that recover from anaplasmosis remain reservoirs (carriers) of the disease for the rest of their lives. During the carrier stage, an animal will not exhibit any clinical signs associated with the persistent low-level *A. marginale* infection.

Nevertheless, the blood from these recovered animals will cause anaplasmosis if introduced into susceptible cattle. Carriers very rarely become ill with anaplasmosis a second time. Unidentified carriers in a herd are the most likely source of infection for future outbreaks of the disease.

Clinical Signs

Anaplasmosis is unusual because the clinical signs are most severe in adult animals. Calves less than a year old that are infected with *A. marginale* usually do not show clinical signs of the disease, but will become carriers. Carrier animals have immunity against anaplasmosis, so even if they are infected later in life, they will generally not get sick. Cattle 1 to 3 years old will show increasingly more severe clinical signs. Recovered animals will also become carriers. Newly infected adult cattle over 3 years will show the most severe clinical signs, and 30 percent to 50 percent will die if they are not treated early.

Unless cattle are being watched carefully, dead cows are frequently the first thing noticed with an anaplasmosis outbreak. If cattle are carefully observed, weakness may be the first clinical sign that is noticed with anaplasmosis. Infected cattle will fall behind the rest of the herd and will not eat or drink. Cows with light skin will initially look pale around the eyes and muzzle, but later this can change to a yellowish colour (jaundice). This jaundice is due to the destruction of the blood cells and their contents being released into the blood stream. Weight loss is rapid. Cattle can become extremely aggressive if they are oxygen deprived due to the severe anaemia. Oxygen deprivation can also result in abortions in pregnant cows. Constipation, high fever, and laboured breathing can also be seen. The most critical period is the first 4 to 9 days after clinical signs appear. Cows that survive this period have an increased chance of survival.

Diagnostic Tests

Three tests are routinely used to detect the presence of anaplasma antibodies in blood. They are the complement fixation (CF), fluorescent antibody (FA), and the rapid card agglutination (RCA) tests.

Treatment

Treatment of anaplasmosis is most effective if given in the early stage of the disease. A single dose of long-acting oxytetracycline (ex. LA-200®) is administered subcutaneously at 9 mg per pound of body weight. Blood transfusions are occasionally used. Animals in later stages of the disease may be so anemic that the stress of handling them will kill them. There is also evidence that antibiotics at this stage are not effective. Therefore, for very weakened or belligerent cattle, antibiotic treatment is not recommended.

All affected animals should be provided with easy access to food and water and a low-stress environment. It may take surviving animals up to 3 months to completely recover from the disease. Animals treated with a single dose of antibiotics and those not treated at all will both become carrier animals. Carrier animals can be cleared of anaplasmosis with repeated injections of long-acting oxytetracycline or prolonged feeding of chlortetracycline.

Possible antibiotic treatment protocols for eliminating the carrier state are described below:

- **Oxytetracycline (50 to 100 mg/ml)**: 5 or 10 day treatment. 10 mg per pound of body weight (BW) daily for 5 days or 5 mg per pound of BW for 10 days.
- **Intramuscular**: To ensure adequate absorption of the medication and prevent excessive muscle inflammation, do not inject more than 10 ml per injection site.
- **Intravenous**: Oxytetracycline should be diluted with physiological saline or administered by a veterinarian.

Oxytetracycline (LA-200): 4 treatments at 3-day intervals.

Each animal receives 4 treatments of LA- 200 at 3-day intervals at a dosage of 9 mg per pound of BW. The total dose should be divided between two injection sites and given by deep intramuscular injection.

Chlortetracycline: 60-day treatment.

Chlortetracycline fed at a level of 5 mg per pound of BW daily for 60 days will eliminate the carrier state of anaplasmosis. Oral administration permits treatment on a herd basis and the use of economical antibiotic premixes.

Chlortetracycline: 120-day treatment.

Chlortetracycline fed at the rate of 0.5 mg per pound of BW per day for 120 days will eliminate the carrier state of anaplasmosis. Attempts to eliminate the carrier state of anaplasmosis by feeding chlortetracycline at the rate of 1 mg per pound of BW every other day for 60 feedings (120 days) did not consistently rid animals of the *A. marginale* infections.

General Control Programs

Vaccinating will not prevent susceptible cattle from becoming infected either, but will reduce the clinical signs of the disease. Vaccination requires a first injection and a booster 4 weeks later. Both injections must be completed 2 weeks before the vector season, and the manufacturer recommends an annual booster vaccine.

50

Rickettsiales

Organisms in the order *Rickettsiales* form a diverse group of small (0.3 to 0.5 x 0.8 to 2.0 pm), non-motile, pleomorphic Gram-negative bacteria which replicate only in host cells. They can be cultured in the yolk sac of embryonated eggs or in selected tissue culture cell lines. Because they stain poorly with aniline dyes, these organisms should be stained by Romanowsky methods such as Giemsa or Leishman. In addition to host-cell dependence and poor affinity for basic dyes, a requirement for an invertebrate vector distinguishes them from conventional bacteria and the *Chlamydiales.*

Epidemiology

Animal hosts and arthropod vectors are the reservoirs for most rickettsiae. A number of rickettsia1 organisms, including *Ehrlichia canis, Anaplasma marginale* and *Haemobartonella felis* produce latent infections. In arthropods, rickettsiae replicate in the epithelial cells of the gut before spreading to other organs, including the salivary glands and ovaries where further replication may occur.

Organisms are transmitted when the arthropod feeds on the animal host. Some organisms such as *Rickettsia rickettsii* are maintained in a tick population by transovarial transmission. Trans-stadial but not transovarial transmission of *E. canis* and *E. phagocytophila* occurs in ticks.

With the exception of *Coxiella burnetii,* which produces endospore-like forms and can remain viable in dust for up to 50 days, most rickettsiae are labile outside host cells. Aerosol transmission of *C.burnetii* commonly occurs in domestic animals and humans. In addition, a silent cycle involving ticks and small wild mammals may constitute a possible source of infection for some domestic species.

Pathogenesis and Pathogenicity

Many Rickettsia species including the causal agents of typhus *(R.* prowuzekii), murine typhus *(R.* typhi) and scrub typhus (R. tsutsugamushi) are primarily human pathogens. Rocky Mountain spotted fever, caused by Rickettsia rickettsii, which is a common rickettsia1 disease of humans, also affects dogs.

These highly pathogenic organisms have a predilection for the endothelial cells of small blood vessels. Rickettsia species produce phospholipase which damages the membranes of phagosomes allowing the organisms to escape into the cytoplasm. Replication in the cytoplasm induces cytotoxic effects.

Ehrlichia species, with the exception of the human pathogens E. chaffeensis and E. sennetsu, are pathogens of domestic and feral animals. They have a predilection for either leukocytes or platelets and they survive and replicate in phagosomes by inhibiting phagosomellysosome fusion.

Cowdria ruminantium, the cause of heartwater in ruminants, probably parasitizes macrophages and other cell types in lymphoid tissues during the initial phase of infection. Organisms finally localize in membrane-bound vacuoles in endothelial cells throughout the body.

Two species in the genus **Neorickettsia** cause acute febrile disease in dogs. These organisms, which localize predominantly in lymph nodes, produce a generalized lymphadenopathy. Coxiella burnetii grows preferentially in the acid environment of phagolysosomes and many of its metabolic activities are detectable only at pH 5 or lower. This pathogen localizes and replicates in cells of the female reproductive tract and mammary glands of ruminants.

Members of the Anaplasmataceae have a predilection for erythrocytes. Anaplasma species and Aegyptianella pullorum are found within vacuoles in red blood cells, whereas Haemobartonella and Eperythrozoon species are located on erythrocyte surfaces. Although H. felis does not usually penetrate red cells, it may erode surface membranes, increasing osmotic fragility of the erythrocytes and shortening their life span. Attachment of the organism to the surface of red cells appears to alter their surface antigens, stimulating autoantibody production and immune-mediated injury to the red cells. Anaemia, due to *H.* felis infection results from a combination of haemolysis and premature removal of red cells from the circulation.

Clinical Infections

Rickettsia1 organisms are relatively host-specific. Because definitive arthropod or fluke vectors are involved in the transmission of most rickettsiae, diseases associated with these organisms tend to occur in defined geographical regions. In many instances the clinical signs reflect the targeting of a particular cell type by the causal agent of rickettsia1 disease. Q fever and Rocky Mountain spotted fever are important zoonotic diseases.

Rocky Mountain Spotted Fever in Dogs

Rocky Mountain spotted fever, caused by Rickettsia ricketlsii, affects mainly humans and dogs. In North America, the main tick vectors are Dermacentor variabilis and D. andersoni. Rhipicephalus sanguineus and Amblyomma cajennense are the main vectors in Central and South America. Ticks acquire the pathogen while feeding on infected small wild mammals. Rickettsia rickettsii is maintained in the tick population by transovarial and transstadial transmission. An infected tick must remain attached for up to 20 hours before salivary transmission to the host occurs. The organisms, which replicate in endothelial cells of infected dogs, produce vasculitis, increased vascular permeability and haemorrhage.

Clinical Signs

The incubation period of the disease is 2 to 10 days and the course is usually less than 2 weeks. Clinical signs include fever, depression, conjunctivitis, retinal haemorrhages, muscle and joint pain, coughing, dyspnoea and oedema of the extremities. Neurological disturbance, which occurs in about 80% of affected dogs, presents as stupor, ataxia, neck rigidity, seizures and coma. Dogs with mild disease and those treated early in the infection usually recover. In severe disease, death may result from cardiovascular, neurological or renal damage. At postmortem there is widespread haemorrhage, splenomegaly and generalized lymphadenopathy.

Diagnosis

Rocky Mountain spotted fever should be considered in dogs with systemic disease which have been exposed to ticks in endemic areas. Indirect fluorescent antibody test or ELISA demonstrating a rising antibody titre to R. rickettsii is diagnostic. Antibodies are not demonstrable until at least 10 days after infection. A marked thrombocytopenia and leukopenia may be present during the acute phase of the disease. The disease must be differentiated from acute canine monocytic ehrlichiosis.

Treatment and Control

Tetracycline therapy, which usually produces clinical improvement within 24 hours, must be continued for 2 weeks. Supportive therapy is necessary for severely debilitated dogs. Frequent removal of ticks is recommended. Because the disease is zoonotic, gloves should be worn during this procedure.

Canine Monocytic Ehrlichiosis

Canine monocytic ehrlichiosis, a generalized disease of Canidae caused by Ehrlichia canis, is confined to tropical and subtropical regions. Rhipicephalus sanguineus, the brown tick, is one of the main vectors and trans-stadial transmission occurs. After detachment from an infected host, ticks can transmit the agent to susceptible dogs for up to 5 months. Dogs often remain carriers for more than 2 years after recovery from acute disease. Human ehrlichiosis is caused by E. chaffeensis which is closely related to E. canis.

Clinical Signs

Following an incubation period lasting up to 3 weeks, the disease can progress through acute, subclinical and chronic phases. The acute phase, in which signs range from mild to severe, is characterized by fever, thrombocytopenia, leukopenia and anaemia. Most affected dogs recover but some progress to a subclinical phase lasting months or years during which low blood cell values persist but clinical signs are minimal.

A minority of these dogs later develop a severe form of the disease known as tropical canine pancytopenia. Persistent bone marrow depression, along with haemorrhages, neurological distur-bance, peripheral oedema and emaciation are characteristic of this phase of the disease. Hypotensive shock may ultimately develop, leading to death. Progression to this chronic phase of the disease may be influenced by factors such as breed susceptibility, immunosuppression and the virulence of the infecting strain of E. canis.

Diagnosis

Typical clinical and haematological features in dogs exposed to ticks in an endemic area may suggest canine monocytic ehrlichiosis. Morulae of E. canis may be detected in mononuclear cells in Giemsa-stained smears of the buffy-mat layer prepared from peripheral blood.

Seroconversion can be demonstrated 3 weeks after infection using indirect immunofluorescence. Antibody titres of 1:10 or greater are considered to be indicative of infection. Ehrlichia canis can be cultured in a canine macrophage cell line.

Treatment and Control

Doxycycline therapy for 10 days is recommended. Tetracyclines and chlordmphenicol are also effective. Fluid replacement therapy or blood transfusions may be necessary. Tetracyclines can be administered to susceptible dogs entering an endemic area as a short-term prophylactic measure.

Canine Granulocytic Ehrlichiosis

This disease, recently described in the USA, is caused by Ehrlichia ewingii. Nentrophils are the primary target cells for the pathogen. Infected dogs, which exhibit mild clinical signs, recover uneventfully.

Canine Cyclic Thrombocytopenia

'Ehrlichia platys', the cause of this condition, parasitizes platelets. Infected dogs, with thrombocytopenia recurring at intervals of about 10 days, are usually asymptomatic. Seromnversion, detected by indirect immunofluorescence, can be demonstrated about 2 weeks after infection.

Potomac Horse Fever

Potomac horse fever, also known as equine monocytic ehrlichiosis and equine ehrlichial colitis, is caused by Ehrlichia risticii. Originally described in 1970 in horses near the Potomac River in Virginia and Maryland, the disease has now been reported throughout North America and in some European countries. Potomac horse fever occurs during summer months and a fluke vector has been suggested. Ehrlichia risticii infects epithelial cells of the crypts in the colon and also targets monocytes, tissue macrophages and mast cells.

Clinical Signs

Fever, anorexia, depression, diarrhoea, colic, leukopenia and laminitis may be evident. The case fatality rate can reach 30%. Transplacental transmission of E. risticii may occur and the agent may induce abortion. Patchy hyperaemia of the large intestine may be found at postmortem.

Diagnosis

- Clinical signs, although non-specific, may suggest the disease in endemic areas.
- A rising antibody titre detected by indirect immunofluorescence or ELISA tests is consistent with active infection.

Treatment and control

Oxytetracycline intravenously for 7 days is therapeutically effective. Inactivated vaccines are commercially available in Northern America.

Equine Granulocytic Ehrlichiosis

This disease, often known as equine ehrlichiosis, is caused by Ehrlichia equi. It has been reported from the USA, some European countries and Israel. Clinical signs include fever, depression, ataxia, limb oedema, icterus and petechial haemorrhages on mucous membranes. The disease is relatively mild, the mortality rate is low and cases tend to occur in late autumn and in winter. The mode of transmission is not known. Diagnosis is based on the demonstration of morulae of E. equi in neutrophils during the acute phase of the disease. Elevated antibody titres, demonstrated by indirect immunofluorescence, and marked leukopenia are additional indicators of infection. Tetracycline therapy is effective.

Bovine Petechial Fever

Bovine petechial fever, also called Ondiri disease, which occurs in both wild and domestic ruminants is caused by 'Ehrlichia ondiri'. Clinical disease is most common in cattle imported into endemic areas. The disease is limited to highland areas of Kenya and other East African countries, and the vector is considered to be a species of tick with restricted distribution. 'Ehrlichia ondiri' is thought to replicate initially in the spleen and to spread subsequently to other organs.

Clinical signs include high fluctuating fever, depressed milk yield and widespread petechiation of visible mucous membranes. Oedema and petechiation of the conjunctiva produces 'poached-egg' eye, a feature typical of severe cases. Death often results from pulmonary oedema. Recovered animals which become carriers are resistant to reinfection for at least 2 years. The organisms are often found in granulocytes in smears of peripheral blood stained by the Giemsa method. Tetracyclines are effective only when administered during the incubation period of the disease.

Tick-Borne Fever

Tick-borne fever is a rickettsial disease of domestic and wild ruminants caused by Ehrlichia phagocytophila. The disease tends to be endemic on certain tick-infected upland farms in some European countries. The main vector is the tick, Ixodes ricinus, in which trans-stadial transmission occurs. Transmission to ruminant hosts occurs through the bites of infected ticks and, less commonly, through contaminated instruments. Recovered animals remain infected for up to 2 years and act as reservoirs of infection for ticks. These carrier animals are immune to challenge with the homologous strain of E. phagocytophila. Since the maintenance of immunity appears to be related to repeated exposure to E. phagocytophila, removal of animals from tick-infected pastures results in a decline in protective immunity.

Clinical Signs

Clinical signs, which develop after an incubation period of up to 13 days, include fever, inappetence and a reduced growth rate in young animals. A drop in milk production and abortions or stillbirths may occur in naive, pregnant animals after transfer to farms where the disease is endemic. Most affected animals recover within hvo weeks. However, E. phagocytophila depresses both antibody-mediated and cell-mediated immune responses, increasing the susceptibility of young lambs to tick pyaemia and louping ill, diseases which are also tick-transmitted. Haematological changes in tickborne fever include leukopenia and transient thrombocytopenia.

Diagnosis

The disease should be considered in sick ruminants on tick-infected pastures in endemic regions. In Giemsa-stained blood smears, more than 70% of neutrophils contain intracytoplasmic blue morulae during the febrile period of the disease. Inciirect immunofluorescence is used to detect rising antibody titres.

Treatment and Control

Affected lactating cows should be treated with oxytetracycline. Tick control is an essential part of disease prevention. Long-acting tetracyclines, administered to lambs in the first 2 to 3 weeks of life, may protect against infection with E. phagocytophila.

Elokomin Fluke Fever

'Neorickettsia elokominica', the cause of Elokomin fluke fever, i s morphologically indistinguishable from N. *helminthoeca* and has the same fluke vector. The disease is milder than salmon poisoning disease and has a wider host range which includes *Canidae,* bears, racoons and ferrets. *'Neorickettsia elokominica'* infection may be concurrent with *N. helminthoeca* infection and there is no cross-protection between the two organisms.

Bovine Anaplasmosis

Bovine anaplasmosis, or gall sickness, caused by *Anaplasma marginale,* affects cattle in tropical and subendemic area may suggest the condition. Giemsa-stained blood smears may contain denselystained bodies (0.3 to 1.0 pm in diameter) located near the periphery of erythrocytes. The organisms are most numerous about 10 days after the onset of fever, when up to 50% of erythrocytes can be affected.

The organisms can be identified in blood smears by immunofluorescence. A radioactive RNA probe and a polymerase chain reaction-based method are sensitive techniques used for detecting the pathogen.

Serological tests are of particular value in detecting latent infections. These tests include complement fixation test, card agglutination test, ELISA and dot enzyme-linked immunosorbent assay.

Treatment and Control

Long-acting oxytetracycline or imidocarb dipropionate, administered early in the disease, are effective. Supportive therapy is essential in severe cases. In endemic areas, control measures are aimed at minimizing stress in indigenously-reared cattle. Prior to introduction into an endemic region, animals must be vaccinated. A live *A. centrale* vaccine, which provides partial protection against *A. marginale,* is used only in calves. Attenuated and inactivated *A. marginale* vaccines are also available.

51

Chlamydia and Chlamydophila Species

Chlamydiae are obligate intracellular bacteria with an unusual developmental cycle during which unique infectious forms are produced. They replicate within cytoplasmic vacuoles in host cells. On account of their apparent inability to generate ATP, with resultant dependence on host cell metabolism, they have been termed 'energy parasites'.

The family Chlamydiaceae belongs to the order Chlamydiales. Currently two genera, Chlamydia and Chlamydophila, and nine species are described. Formerly a single genus and four species, Chlamydia trachomatis, C. psittaci, C. pneumoniae and C. pecorum, were recognised. This classification was based on phenotypic characteristics such as host preference, inclusion morphology, iodine staining for the presence of glycogen, and sulphonamide susceptibility. However, recent nucleic acid sequencing studies of the 16s and 23s rRNA genes confirm two distinct lineages.

In the developmental cycle of chlamydiae, infectious and reproductive forms are morphologically distinct. Infectious extracellular forms, called elementary bodies (EBs) are small (200 to 300 nm), metabolically inert and osmotically stable. Each EB is surrounded by a conventional bacterial cytoplasmic membrane, a periplasmic space and an outer envelope containing lipopolysaccharide.

The periplasmic space does not contain a detectable peptidoglycan layer and the EB relies on disulphide cross-linked envelope proteins for osmotic stability. Elementary bodies enter host cells by receptor-mediated endocytosis. Acidification of the endosome and fusion with lysosomes are prevented by mechanisms which are not fully understood. A process of structural reorganization within the pathogen, of several hours duration, results in the conversion of an EB into a reticulate body (RB). The RB, about 1 μm in diameter, is metabolically active, osmotically fragile and replicates by binary fission within the endosome. The endosome and its contents, when stained, is called an inclusion. When a number of inclusions containing RBs of C.trachomatis are formed in an infected cell, fusion of these structures may occur.

Usual Habitat

The gastrointestinal tract appears to be the usual site of Chlamydophila species infection in animals. Intestinal infections are often subclinical and persistent. Faecal shedding of the organisms, which is typically prolonged, becomes intermittent with time. The EBs can survive in the environment for several days.

Pathogenesis and Pathogenicity

Chlamydiae infect over 130 species of birds and a large number of mammalian species including man. In recent years isolations have also been made from invertebrate species. Chlamydial species are usually associated with specific diseases in particular hosts. In sheep, C.abortus is an important cause of abortion whereas infections with C. pecorum are frequently inapparent. Interspecies transmission is uncommon. When it occurs, the outcome of infection in the secondary host may be either similar, as in transmission from sheep to cattle, or severe, as in transmission from sheep to pregnant women. Infection with C.pecorum is associated with conjunctivitis, arthritis and inapparent intestinal infection. The type of clinical presentation relates to the route of infection and the degree of exposure. Environmental factors and management practices can influence the prevalence of some chlamydial infections such as enzootic abortion in ewes, which tend to be more prevalent in intensively managed lowland flocks.

Many chlamydial infections are asymptomatic, particularly when they are localized in superficial epithelia. They may also persist for long periods without inducing protective immunity. However, chronic infections may repeatedly stimulate the host immune system. Chlamydiae possess a number of heat shock proteins which are partially homologous to heat shock proteins in other bacteria and to a number of human mitochondrial proteins.

Diagnostic Procedures

Specimens for isolation of the organism should be placed in suitable transport medium such as sucrosephosphate-glutamate medium supplemented with foetal calf serum, aminoglycoside antibiotics and an antifungal agent. As chlamydiae are thermolabile, samples should be kept at 4°C. For long term storage, samples should be frozen at -70°C. However, each cycle of freezing and thawing reduces the titre of the stored organisms.

Direct microscopy is suitable for the detection of the organisms in smears or tissue sections containing moderate numbers of organisms. Placental smears from cases of chlamydial abortion typically contain large numbers of organisms. Suitable chemical staining procedures include the modified Ziehl-Neelsen, Giemsa, modified Machiavello and Castaneda methods. Methylene blue-stained smears can be examined by darkfield microscopy.

Several commercial kit sets, employing ELISA methodology, have been developed for the detection of C. trachomatis. Many of these kit sets detect chlamydial lipopolysaccharide (LPS) which is common to all Chlamydia and Chlamydophila species. Consequently, they can be used to detect the LPS of species in both genera.

Chlamydiae can be isolated either in embryonated eggs, inoculated into the yolk sac, or in a number of continuous cell lines such as McCoy, L929, baby hamster kidney and Vero. Tissue culture cells are usually grown in flat-bottomed vials or in bottles containing coverslips for ease of fixing and subsequent staining.

Polymerase chain reaction techniques have been developed for the detection of chlamydial DNA in samples. Several serological procedures are available for the detection of antibodies to chlamydiae including complement fixation, ELISA, indirect immunofluorescence and micro-immunofluorescence.

Clinical Infections

A wide range of animal species are susceptible to infections with chlamydiae. Both the severity and the type of disease produced by chlamydiae are highly variable, ranging from clinically inapparent infections and local infections of epithelial surfaces to severe systemic infections. Diseases associated with chlamydial infections include conjunctivitis, arthritis, abortion, urethritis, enteritis, pneumonia and encephalomyelitis. Clinical signs and their severity are influenced by factors related to both host and pathogen, and one type of clinical presentation usually predominates in outbreaks of disease.

Enzootic Abortion of Ewes

Enzootic abortion of ewes (EAE) caused by C. abortus (formerly known as ovine strains of C. psittaci), is primarily a disease of intensively managed flocks. The disease is economically significant in most sheepproducing countries. Although abortion associated with C. abortus is best documented in sheep, it has also been reported in other domestic species including cattle, pigs and goats. Chlamydial infection in cattle and goats often originates from sheep. The source of infection in pigs is less clearly defined).

Epidemiology

Infection is usually introduced into clean flocks when infected replacement ewes abort. Large numbers of chlamydiae are shed in placentas and uterine discharges from affected ewes. Organisms can remain viable in the environment for several days at low temperatures.

Infection occurs by ingestion. The role of infected rams in venereal spread is uncertain. Ewes infected late in pregnancy do not usually abort but may do so in the next pregnancy. Infection early in pregnancy can result in abortion during that pregnancy. Ewe lambs may acquire infection during the neonatal period and abort during their first pregnancy. As a result, the most dramatic outbreaks of EAE often occur in the year following the introduction of infection into a flock.

Clinical Signs

Enzootic abortion of ewes is characterized by abortion during late pregnancy or by the birth of premature weak lambs. Aborted lambs are well developed and fresh. Necrosis of cotyledons and oedema of adjacent intercotyledonary tissue in affected placentas is often present along with a dirty pink uterine exudate. Aborting ewes rarely show evidence of clinical disease and their subsequent fertility is usually unimpaired. Although up to 30% of animals in a fully susceptible flock may abort, a rate of 5 to 10% is more usual in flocks in which the disease is endemic.

Diagnosis

Aborted lambs and evidence of necrotic placentitis are suggestive of EAE. Large numbers of EBs can be demonstrated in placental smears using suitable staining procedures. Commercial diagnostic kits are available for the detection of chlamydial antigen in samples. Isolation of chlamydiae in suitable cell lines or in the yolk sac of embryonated eggs is possible. Polymerase chain reaction techniques are available and can be carried out using species-specific primers to distinguish C. abortus and C. pecorum.

A number of different serological tests can be used for the detection of chlamydial antibodies including the complement fixation test, ELISA and indirect immunofluorescence. The use of recombinant antigens specific for C.abortus may improve the specificity of serological tests.

Treatment and Control

Chlamydiae are susceptible to a number of antibiotics which can be used during an outbreak. Administration of long-acting oxytetracycline to in-contact pregnant ewes has been shown to increase the number of live born lambs. However, antibiotic treatment does not eliminate the infection and treated ewes may shed chlamydiae at parturition.

Transmission of infection in an affected flock can be reduced by isolating all aborted ewes for 2 to 3 weeks, removing and destroying all placentas, thoroughly cleaning areas where abortions occurred and administering long-acting oxytetracycline to ewes which have not yet lambed.

A decision should be made either to vaccinate or attempt to eradicate the disease by culling. A live attenuated vaccine is available which should be administered to ewes prior to breeding. An inactivated vaccine is also available which can be used in pregnant animals. Chlamydophila abortus infection is serious and potentially life-threatening for pregnant women who should avoid contact with ewes during the lambing season.

Feline Chlamydiosis

Chlamydaphila felis (formerly known as feline strains of C.psittaci) is associated with conjunctivitis and less commonly rhinitis. Feline pneumonitis, the original name for feline chlamydiosis, is now considered a misnomer because of the rarity of lower respiratory tract infection caused by C.felis in cats.

Epidemiology

Serological surveys have revealed that up to 10% of cats become infected with C. felis. Infection is transmitted by direct or indirect contact with conjunctival or nasal secretions. Organisms may also be shed from the reproductive tract. Infection may be persistent with prolonged shedding of organisms and clinical relapses. The stress of parturition and lactation may trigger shedding of organisms by infected queens facilitating transmission to their offspring.

Clinical Signs

After an incubation period of about 5 days, unilateral or bilateral conjunctival congestion, clear ocular discharge and blepharospasm become evident. If secondary infection with organisms such as Mycoplasma felis and Staphylococcus species occurs, the ocular discharge may become mucopurulent. Conjunctivitis may be accompanied by sneezing and nasal discharge. The condition usually resolves without treatment in a few weeks. However, persistent infection with recurring clinical episodes also occurs.

Diagnosis

Stained conjunctival smears may reveal intracytoplasmic inclusions. The organism may be isolated in suitable cell lines or in embryonated eggs. Commercial diagnostic ELISA kits for detecting the genus-specific lipopolysaccharide antigen are available. Polymerase chain reaction protocols have been developed for samples. The complement fixation test or the indirect immunofluorescence test can be used to detect chlamydial antibody titres. However, the level of antibody does not necessarily correlate with active infection.

Treatment and Control

Chlamydiae are susceptible to several antibiotics. All in-contact cats should be treated at the same time. Modified live vaccines are available for parenteral inoculation. Vaccination reduces the clinical effects of natural infection but does not prevent infection or the shedding of organisms. Inadvertent intraocular administration of the vaccine can result in conjunctivitis. A small number of cases of conjunctivitis in humans involving C. felis have been reported.

Avian Chlamydiosis

Infections with C. psittaci in psittacine birds were origin-ally designated psittacosis and ornithosis was reserved for chlamydia1 infection in other avian species. Avian chlamydiosis is currently the preferred designation for the condition. The disease has been recorded worldwide.

Epidemiology

A wide range of both wild and domestic avian species are susceptible to infection. Isolates can be divided into several serovars on the basis of reactivity with monoclonal antibodies. The organism is present in respiratory discharges and faeces of infected birds. Infection is usually acquired by inhalation or by ingestion. Subclinical infection is common. Clinically affected and carrier birds may shed organisms intermittently for prolonged periods. Stress arising from captivity, transportation, egg-laying, overcrowding and intercurrent infection is important in precipitating disease outbreaks.

Clinical Signs

Avian chlamydiosis is a generalized infection, affecting particularly the digestive and respiratory tracts. The incubation period is up to 10 days. Clinical signs vary in nature and severity, depending on the strain of C.psittaci and the species and age of the affected birds. Signs include loss of condition, nasal and ocular discharges, diarrhoea and respiratory distress. The most frequent post-mortem findings are hepatosplenomegaly, airsacculitis, peritonitis.

Diagnosis

Organisms may be identified in stained impression smears of affected tissues. Chlamydial antigen may be detected using immuno histochemistry or ELISA kits. chlamydial DNA may be demonstrated by the polymerase chain reaction. Isolation of C.psittaci is carried out in cell culture or embryonated eggs.

Antibodies to C.psittaci may be detected using suitable serological tests including the complement fixation test and ELISA. However, interpretation of antibody titres can be difficult particularly when single samples are tested. Paired serum samples or samples from several birds in a flock are more reliable for diagnosis.

Treatment and Control

Tetracyclines are the antibiotics of choice. An extended course of treatment over several weeks is required. No commercial vaccines are available. Imported birds, particularly psittacine species, should be held in quarantine and receive tetracycline medicated feeds. Proper husbandry and suitable transportation minimize the occurrence of clinical disease. Avian chlamydial isolates are potentially zoonotic. Infection, which commonly follows aerosol inhalation, may be subclinical or result in systemic disease. Pulmonary involvement is common. Meningitis or meningoencephalitis may develop in severely affected individuals.

52

Emerging and Re-Emerging Diseases

Emerging Infectious Diseases

Emerging infectious diseases are those due to newly identified and previously unknown infections which cause public health problems either locally or internationally.

Re-emerging Infectious Diseases

Re-emerging infectious diseases are those due to the reappearance and increase of infections which are known, but had formerly fallen to levels so low that they were no longer considered a public health problem.

What Causes Emergence or Re-emergence of Infectious Diseases?

Several factors contribute to the emergence and re-emergence of infectious diseases, but most can be linked with the increasing number of people living and moving on earth: rapid and intense international travel; overcrowding in cities with poor sanitation; changes in handling and processing of large quantities of food; and increased exposure of humans to disease vectors and reservoirs in nature. Other factors include a deteriorating public health infrastructure which is unable to cope with population demands, and the emergence of resistance to antibiotics linked to their increased misuse.

Avian Influenza

Avian influenza is a zoonotic, globally important disease of birds that can be categorised as either low pathogenic (LPAI) or highly pathogenic (HPAI) according to the virulence of the virus in animals. Outbreaks of LPAI are common around the world but LPAI typically causes no clinical signs or only minor illness in infected birds. LPAI strains are generally less of a threat to human health; in addition, LPAI can have significant economic repercussions as a result of export restrictions and culling of birds, particularly in developing countries where the use of vaccines and other veterinary medicines is difficult due to weak veterinarian services and small farming units. Compared with LPAI, HPAI is more readily detected in poultry due to very high levels of mortality in infected birds, and can cause potentially catastrophic economic consequences, from significantly reduced livestock populations to lost export markets. Though relatively rare, sporadic human infections of HPAI have occurred in cases of close contact with infected birds and caused serious illness and even death.

Transmission and Spread

Influenza viruses are shed in the oral, respiratory and faecal secretions of infected birds and spread via direct contact between healthy and infected birds, or indirectly via contact with contaminated equipment and people. Transmission to humans leading to clinical disease is a rare event but can take place in cases of close contact with infected birds or in heavily contaminated environments. The ability of the virus to infect a diversity of wild birds and other vertebrate hosts (including pigs), its prolonged environmental survival in favourable conditions, its zoonotic potential and its capacity for mutation via antigenic drift or antigenic shift, as well as the lack of completely effective vaccines for use in poultry, make avian influenza a particularly dangerous disease for both animal and human populations.

Control Measures

Various controls are already in place for avian influenza across most of the world, including biosecurity and surveillance, stamping out, and regional vaccination programmes for poultry in high-risk areas. As there is no treatment for avian influenza once clinical signs appear, targeted stamping out without vaccination is the chosen control method in most developed countries, where early detection and compensation programmes are widespread. Vaccination is typically only implemented in endemic areas as a preventative or during an outbreak as an adjunct control measure when all other measures are insufficient. Moreover, vaccination has to be accompanied by surveillance and movement controls to be effective because although vaccines may reduce the susceptibility, morbidity and mortality of birds subsequently exposed to infection, they do not prevent the shedding of potentially infectious levels of the virus.

Bluetongue

Bluetongue is a non-zoonotic, non-contagious vector-borne disease, caused by a virus of the Reoviridae family with 25 known serotypes worldwide. It affects all domestic and wild ruminants, with sheep and cattle experiencing the highest rates of infection. The severity of the disease depends on the strain and morbidity can be very high in susceptible animals. Bluetongue outbreaks also cause direct economic losses through disease and mortality, loss of production, loss of milk yield, and declines in fertility. As bluetongue is not zoonotic, it poses no risks to human health and cannot be contracted or spread through food.

Transmission and Spread

Bluetongue is transmitted by biting midges that are infected with the virus after ingesting blood from infected animals. The spread of the disease depends mainly on those ecological and climatic factors that favour biting midge populations; outbreaks are often seasonal, occurring at or shortly after the season of peak midge activity. As climate change becomes more of a concern in the future, the risks of bluetongue causing an epidemic will continue to increase. According to a leading expert on vector-borne diseases, "through changes to climate, one of the most competent vector species of midge has spread around southern Europe and further north than it had before" while at the same time midge vectors have become "better at being able to transmit the virus."

Control Measures

Bluetongue control measures include the surveillance of susceptible animals, quarantine, zoning, insect control and vaccination. Since controlling midge populations is not possible, they are too numerous vaccination is the most effective practical measure to minimise losses related to the disease in endemic regions. Vaccines used against bluetongue are both live attenuated and inactivated. Live attenuated vaccines, until recently the only bluetongue vaccines commercially available, are relatively inexpensive and can provide long-lasting protection. However, they are not always sufficiently weakened and they can actually generate the disease they intend to prevent. Though more expensive, if properly produced and administered, inactivated vaccines can provide reliable and protective immunity from bluetongue. An efficient treatment method for infected animals is still lacking.

West Nile Fever

West Nile fever is the disease caused by West Nile virus (WNV), a mosquito borne virus that primarily affects birds but also horses and humans. Originally confined to tropical areas, the disease has recently spread to temperate zones largely as a result of climate-related vector expansion and is now endemic in most regions of the world. Because of its potential for permanent neurological problems or death in humans, WNV remains a significant public health risk in endemic areas.

Transmission and Spread

WNV is transmitted by mosquitoes, while birds are the most commonly infected animal and serve as the primary reservoir host. The disease is maintained between these two agents in endemic regions but it can spread to humans and horses under warm environmental conditions. The vast majority of human cases are caused by mosquito bites, and the virus cannot be transmitted directly from person to person. Recent climate change developments favour the emergence of WNV outbreaks by reducing the time that elapses between the mosquito bite and the transmission time, which is temperature dependent.

Control Measures

Mosquito control and surveillance remain the primary control measures for the prevention of WNV. There are effective, licensed vaccines for use in horses, including inactivated WNV formulations and 'chimeric' recombinant vaccines, which express WNV proteins from a different virus backbone. Currently no human vaccines are available, although several vaccine candidates are under development. Though vaccination of horses means that they are protected from the disease, it does not break the transmission cycle as mosquitoes (and vaccinated animals) continue to be infected and capable of spreading the disease.

Classical Swine Fever

Classical swine fever (CSF) is a highly contagious disease caused by a species specific virus that affects swine but is related to viruses affecting cattle and sheep. Highly virulent strains of the virus result in high levels of mortality among infected pigs. While the virus cannot be transmitted to humans and therefore poses no risk for human health, it can pose serious economic threats. In the developed world, the greatest costs result from trade restrictions imposed on both live pigs and pork products once an outbreak is reported.

Transmission and Spread

Transmission of CSF usually takes place through direct contact between infected animals. The virus can also be spread through the transportation of animals in contaminated vehicles and through feed since the virus can survive in pork products for months.

Control Measures

In areas where outbreaks have not occurred, control measures include early detection and reporting, movement controls including strict import policies and quarantining of pigs, and effective hygiene measures, including protecting domestic pigs from contact with wild boars. Stamping out remains the main control measure used to prevent the further spread of disease during outbreaks. While effective vaccines are available, vaccination is used only rarely to limit the spread of the disease after an outbreak has already occurred. In countries which are free of disease, or where eradication is in progress, vaccination is normally prohibited. There are no effective treatments for CSF in infected animals.

Equine Influenza

Equine influenza is a highly contagious respiratory disease of horses (and other members of the equidae family including donkeys and zebras) caused by two main subtypes of influenza type A viruses (H7N7 and H3N8). Equine influenza outbreaks are typically characterised by a rapid spread of the virus and very high infection rates in unvaccinated horses. However, clinical signs of illness usually resolve within a few days without complications and fatalities are rare. The virus does not cause disease in humans, and its economic impact is primarily due to the highly contagious nature of the virus and the disruption of equestrian activities.

Transmission and Spread

The transmission of the disease occurs through direct contact with infected animals or through the transmission of the virus on clothing or equipment carried by humans that work with horses. Equine influenza can spread quickly in susceptible populations and cause outbreaks where conditions are most favourable, for example when horses are kept in close quarters, during transportation, and where mixing of horses from different locations occurs. The virus is often spread across borders by the movement of infected horses.

Control Measures

Rapid diagnosis, movement restrictions and vaccination are the key control measures. Vaccines are widely available and routinely used for competition horses in Europe, the Americas, and Asia. In some countries vaccination is mandatory for horses that are competing in equestrian events and the International Federation for Equestrian Sports requires vaccination of horses every six months. Vaccines have been useful in limiting virus replication, reducing or eliminating clinical disease and reducing the risks of transmission. However, given strain variation, vaccines are not always successful in preventing infection. Moreover, vaccines are not always easy to obtain in developing countries.

53

Trans-Boundary Animal Disease and Their Impacts on International Trade

Trans-boundary Animal Diseases (TADs) are highly contagious diseases of livestock in the world and transmissible diseases which have the potential for very serious and rapid spread, irrespective of national borders, which has serious socio-economic or public health consequence and they are importance in the international trade of animals and animal products. With rapidly increasing globalization, an associated risk of movement of trans-boundary diseases is emerging. Trans-boundary animal diseases represent a serious threat.

They reduce production and productivity, disrupt local and national economies and also threaten human health (zoonotic). Trans-boundary animal disease is a concern globally; cumulative effort is needed at international level to minimize the spread of infectious diseases across the borders.

Considering that livestock rearing constitutes a significant share in the national economy of a developing country, it is imperative to take up disease control initiatives. Measures are required to safeguard the livestock industry from epidemics of infectious diseases and to uphold safe international trade of livestock and their products. In this regard, it is essential to develop scientific and risk-based standards that facilitate the international trade in animal commodities.

Introduction

Trans- boundary Animal Diseases (TADs) are highly contagious diseases of livestock in the world. Moreover, their economic importance is a major constraint in international trade. Their implication on human Health and National food security cannot be over emphasized. Zoonotic diseases among TAD's include diseases like West Nile Virus (WNV), Rift Valley Fever (RVF), Mad Cow disease (BSE), Bovine Tuberculosis and Highly Pathogenic Avian Influenza (HPAI). Other important TADs are Foot and Mouth Disease (FMD), Contagious Bovine Pleuropneumonia (CBPP), Lumpy Skin Disease, African Swine Fever (ASF) and Newcastle Disease (ND). They have the potential for very rapid spread, irrespective of national borders and these diseases can cause serious socio-economic and possibly public health consequences.

Trans-boundary animal disease have a multi casual origin; some factors associated with this process include: Trade and international travel (increased frequency and speed of local and international travel, fostered by the globalization process promotes the spread of microorganisms on a global scale), Changes of agricultural practices (animal domestication was one of the main promoters of microbial evolution by facilitating the availability of new susceptible hosts at high densities, due to the intensification of livestock systems), Climate change (which causes changes eco-geographical distribution of vectors), Reduction of habitat and increased contact with wild animals and Introduction of naïve wild and domestic animals to new geographic areas where the disease is endemic and immunologically unknown for them (increases zoonotic pool within a geographic region).

Trans-boundary and emerging diseases are becoming ever more important since they can spread throughout an entire region, impact trading partners and commerce, tourism, consumer confidence and occur in distinct countries, with devastating economic and livelihood consequences. With the globalization of trade and the increasing movements of people, these major crises will continue to menace the global animal and human which is most of the time in developing countries.

Trans-boundary animal disease has a significant impact on national economies due to its high rates of morbidity and mortality in the susceptible animal populations, costs of control or eradication programmes and restrictions on international trade. These diseases negatively affect livestock or poultry production in the country because they impose heavy economic losses in the form of morbidity and mortality and are thus considered as vital threat to livestock production and livelihood of poor farmers, especially in those areas of the country where such diseases have assumed an endemic role. Livestock enterprises and animal production contribute significantly to the world economy, provide household source of income, food security, source of energy, draft power for crop cultivation, high quality animal proteins and vitamins (meat, milk), manure, raw materials (hides and skins) and bride price and generate a livelihood for 1.0 billion poor people in the world.

Transboundary Animal Disease

Trans-boundary animaldisease are transmissible diseases which have the potential for very serious and rapid spread, irrespective of national borders, which are of serious socio-economic or public health consequence and which are of major importance in the international trade of animals and animal products. Food and Agriculture Organization of the United Nations, defines Trans-Boundary Animal Diseases (TADs) as, Those diseases with an essential impact on the economy, trade and/or food security of a group of countries, which can be easily spread to other countries, reaching epidemic proportions and that require control and eradication cooperation between different nations.

Trans- boundary animal diseases (TADs), includes those that are zoonotic (infective to humans) and continue to cause widespread negative socio-economic impacts in an increasingly globalized world with a huge and increasing volume of regional and international trade in livestock and livestock products and the rapid movement of large numbers of people across continents through air travel. Several infectious zoonotic diseases have recently emerged, causing devastating economic losses in the countries affected.

Ways of Transmission of Trans-Boundary Animal Disease

Many cases of the Trans -Boundary AnimalDiseases (TADs) have been reported in various areas of the world during last decades. Such TADs are easily transmitted from one country to another, due to the rapid globalization including the increase of international trade in domestic and wild animals and animal products, to the expansion of human population, global climate changes, changes of agricultural production systems and to microbiological adaptation. The common ways of introduction of animal diseases to a new geographical location are through entry of live diseased animals and contaminated animal products. Other introductions result from the importation of contaminated biological products such as vaccines or germplasm or via entry of infected people (in case of zoonotic diseases). Even migration of animals and birds, or natural spreading by insect vectors or wind currents, could also spread diseases across geographical borders. Traditionally, trade, traffic and travel have been instruments of disease spread. Now, changing climate across the globe is adding to the misery.

International trade in live animals and animal products offers opportunities for pathogens and vectors to be transported across oceans and continents. However, with the exception of a few documented examples, such as, the multiplicity of routes of introduction, including active and passive dispersal of vectors, infected human hosts, animal movements and migration, transportation of goods and biological invasions such as, introduction, initial dispersal, establishment and spread, the specific contribution of globalization to disease emergence is inherently difficult to quantify.

Control of Trans-Boundary Animal Disease

Combatingdiseases is a necessity for farmers. Though a farmer's decision to control the diseases or not is a private one, the presence of an infectious disease in a farm poses a threat too adjacent and even distant farms and can even affect other animal species and develop into an epidemic. This situation where high stakes are involved demand the intervention and action from public agencies or governments.

As TADs are a concern globally, cumulative effort is needed at international level to minimize the spread of infectious diseases across the borders. Reducing man-made disasters that have adverse implications on climate, global warming and climate change either due to natural or anthropogenic influences are likely to predispose the animal population to newer infections. Therefore, collective efforts are needed to minimize adverse climatic changes.

Interrupting the human-livestock, wildlife transmission of infections

Diseases at the wildlife–livestock interface must become the focus for surveillance of emerging infectious diseases. Breaking the cycle of disease transmission would help control the spread of infections. Preventing incidence of trans-boundary animal disease can also be practiced by controlling disease transmitting vectors, minimizing the movement of animals across the borders and prompt practice of quarantine protocol. Geographic information system (GIS) and remote sensing could be utilized as early warning systems and in the surveillance and control of infectious diseases.

Surveillance and Control of Infectious Diseases

- Establishing regional biosecurity arrangement with capacity for early disease warning system for surveillance, monitoring and diagnosis of emerging disease threats.
- Undertaking animal breeding strategies to create disease resistant gene pools, enhancing host genetic resistance to disease by selective breeding of resistant animals is a smart strategy to improve natural immunity of animals to counter invading infections.
- Strengthening government policies to enhance agricultural/animal research and training and technology development. More funds need to be allocated for this purpose to build goal oriented research programs in combating TADs.
- Ensuring appropriate preparedness and response capacity to any emerging disease. Keeping in view that emerging infectious diseases

are a constant threat, detection capacity and then implement a timely response.

Challenges in Dealing with Trans-Boundary Animal Disease

- Requirement of novel systems having capacityof real-time surveillance of emerging diseases; for this, need driven research and service oriented scientific technology are a necessary at regional levels. Research emphasis has to be on specific detection and identification of the infectious agents.
- Need for epidemiological methods to assess the dynamics of infections in the self and neighbouring countries/regions. These methods should be of real-time utility. Research and development of disease diagnostic reagents those do not need refrigeration is also important.
- More importantly, they should be readily available as well as affordable, for use in pen-side test format.
- There are many diseases for which there is inadequate supply of vaccines or there are no vaccines available. Insufficient or lack of vaccine hampers the disease control programmes.
- Even if a technology is available, it has to be cheaper to adopt at the point of use. Need for ensuring public awareness of epidemic animal diseases. Many farmers are unaware of the emerging diseases.
- As such, unless reported to concerned regional authority, an emerging disease may go unnoticed. Shortage of government and private funding for research on emerging animal disease problems, Government as well as industries dealing with animal health should take initiative and appropriate sponsorship in these regard.
- Inadequate regulatory standards for safe international trade of livestock and livestock products. Otherwise, there would be a compromised situation in disease control strategies.

Impact of Transboundary Animal Disease on International Trade

Trans-boundary Animal Diseases (TADs) are highly contagious diseases of livestock in the world. Moreover, their economic importance is a major constraint in international trade losses themselves. Therefore, the countries which are free from major diseases will tend to protect their local agriculture by totally excluding the importation of livestock products from areas affected by specific animal diseases. Conversely, benefits of elimination of trans-boundary animal diseases can be very large as it offers the opportunity of gaining access to high-value export markets. Trans-boundary animal disease also has implications on domestic trade as the veterinary authority imposes restriction on animal movement as a part of disease control measure. In evaluating TAD impacts, it is now recognized that poverty implications, technical feasibility and political desirability play a role in disease response program selections in LDCs Furthermore, the cost benefit offset of eradication versus management programs will vary by disease, where high impact diseases do not necessarily carry high benefits from eradication.

Trans-boundary animal diseases including those that are zoonotic (infective to humans), continue to cause widespread negative socio-economic impacts in an increasingly globalized world. With a huge and increasing volume of regional and international trade in livestock and livestock products and the rapid movement of large numbers of people across continents through air travel. Several infectious zoonotic diseases have recently emerged, causing devastating economic losses in the countries affected. These have a wide-ranging impact on the livelihoods of farmers and on regional and international trade, food safety, public health and international travel and tourism. Disease pathogens continue to evolve and adapt themselves to animals and humans alike. Disease investigation indicates that many of these new diseases emerge in response to number critical factors, such as changes in climate, ecosystems, animal production systems and land use, all of which alter the interactions between pathogens and various hosts.

Loss of production and productivity is likely to influence the market price of the commodity, determined by the supply and demand effects induced by TADs. Limited supply can result in increase in market price but on the contrary public health concern associated with certain TADs may also decrease the demand. Moreover, relative effects on producers and consumers of the production shortfall will depend on the relative elasticities of demand and supply. Another dimension of the price and market effects is that it is not only the producers but also the traders who are directly affected international trade in livestock and livestock products continues to be seriously hindered by epizootic animal diseases, in particular those categorized as'transboundary animal diseases' (TADs). There is the danger that informal and illegal trans-border trade in livestock which abound in some countries could derail the revolution.

Economic Impacts

Trans-boundary animal diseases(TADs) cause significant economic losses throughout the world. But producers in less developed countries (LDCs) are at particular risk because livestock provide not only income and an asset base, but also food, draught power and various social functions. Trans-boundary diseases threaten food security, affect livelihoods of rural communities and disrupt local and international trade.

In general, livestock´s products have increased their international trade in 4% only in the last two decades. However, as a result of the emergence and re-emergence of various animal diseases, such as bovine spongiform encephalopathy (BSE), the annual growth of meat products decreased 2% in the late 1990s. Therefore, the cost of trans-boundary animal diseases relates to agricultural products, to the country's economy and international markets are massive. Thus, it is very important to create public policies focused to assure countries' food security (especially in developing nations) to avoid negative economic impacts caused by TADs, especially on the more susceptible social stratus. The World Bank has estimated that zoonotic disease outbreaks in the past 10 years have cost worldwide more than $US200 billion due to loss of trade, tourism and tax revenues.

Pests and animal diseases cause the loss of more than 40% in the global food supply, being a clear threat to the residual economies of developing countries and food security of its inhabitants. Many economic impacts are difficult to quantify and valuation also may be problematic. Such factors as animal welfare, human health and the environment are of obvious importance, but do not have market values and different people have different perceptions of their value. It is therefore impossible to provide objective assessments of the total cost of most animal diseases, especially the most serious ones that have wide-ranging effects. But the cost of animal disease can be enormous.

Specimens to be Collected and Tests Conducted in Diagnosing Certain Important Microbes

Viral Diseases of Animals and Birds

Viral Diseases of Cattle

Viral diseases	Etiology and species affected	Important symptoms	Specimens to be collected	Host system	Diagnosis
Rinderpest	Paramyxovirus Cattle, buffalo, sheep, goat, deer and pigs	Sharp rise in temperature, frothy salivation, mucopurulent occulonasal discharge, grayish white ulcers in the mouth, lower lips, gums, cheeks, tongue, foul smelling, shooting diarrhoea. In atypical forms the above may be mild and the diarrhoea may be absent	1. Defibrinated blood at the viraemic stage 2. Debris from the ulcers of gum and tongue 3. Saliva , eye and nasal secretions 4. Biopsy of prescapular lymphnode and spleen 5. Pieces of spleen and mesenteric lymphnodes from dead animals 6. Paired sera samples from the affected flock **Do not add glycerol saline**	In calf kidney cell culture, VERO cell culture	1. From symptoms, post mortem lesions – erosions in the GI tract, Zebra markings of rectum, 2. AGPT, CIE, ELISA and PHA tests 3. CFT 4. Neutralisation tests both in eggs as well as tissue culture 5.MHI (Measles haemagglutination inhibition test) 6. Biological tests using healthy young calves or sheep
Foot and Mouth Disease	Picorna virus Cattle, buffalo, pigs, sheep, goats, wild game animals and elephants	High fever, stringy salivation, smacking of lips, vesicles of the tongue, gums, dental pads, cheeks, around the muzzle, coronary band interdigital cleft and udder. In pigs sheep and goats foot lesions are common	1. Vesicular fluid 2. Epithelium from fresh lesions of the mouth (about 1gm) in 50% glycerine or phosphate buffered glycerol saline, pH7.4	Intra peritoneal inoculation of suckling mice, foot pad inoculation of guinea pig, BHK	From symptoms, 1. Neutralisation test in tissue cell culture 2. observation of CPE in tissue culture 3.CFT and .ELISA

Viral diseases	Etiology and species affected	Important symptoms	Specimens to be collected	Host system	Diagnosis
			3. Blood in anticoagulant 4. Serum/esophageal / pharyngeal fluids 5. Lymphnodes, thyroid and heart from dead animals	21 cell line, calf kidney cell culture and bovine thyroid cells	5. Guinea pig – lesions appear in the form of vesicles 6. mice –paralysis and death
Infectious bovine rhinotracheitis	Alpha herpes virus Cattle	Respiratory symptoms, abortions, sometimes genital infections, vulvo vaginitis in cows and balnoposthitis in bulls	1. Blood at the height of temperature 2. Nasal and vulval swabs 3. Tracheal and occular swabs 4. Aborted fetal liver, spleen and stomach contents 5. Paired sera samples 6. Milk sample 7. Foetal liver in 10% formal saline	In calf kidney and testes cell culture.	Symptoms, SNT, CFT, ELISA, FAT and PHA tests
Bovine viral diarrhoea and Mucosal disease	Flavi virus Cattle	Sharp rise in temperature, frothy salivation, mucopurulent occulo nasal discharge, grayish white ulcers in the mouth, lower lips, gums, cheeks, tongue, foul smelling diarrhoea. In atypical forms the above symptoms may be mild and diarrhoea may be absent	1. Defibrinated blood 2. Paired sera samples 3. Faecal samples with antibiotics 4. Nasal exudates from dead animals 5. Spleen, mesenteric lymphnode and abortus fetuses from dead animals	In bovine and ovine kidney, testes and spleen cell culture	Virus isolation in cell culture, FAT, AGID, SNT, In situ hybridization, ELISA, and PCR.

Viral diseases	Etiology and species affected	Important symptoms	Specimens to be collected	Host system	Diagnosis
Viral enteritis in young animals	corona, rota and astro viruses	Diarrhoea in young animals, faeces may be watery, brown, grey of light green; sometimes curds, blood and mucus present	Faecal swabs, intestinal segments on ice and lymphglands.	In calf kidney and testes cell culture	From symptoms, IEM, FAT, ELISA tests, detection of RNA fragments by PAGE
					analysis in rota virus infections
Ephemeral fever	Rhabdo virus Cattle	Sudden appearance of high fever, muscular shivering, lameness, stiffness of joints and enlargement of peripheral lymphnodes, reduction in milk yield, recumbency seen in some cases as a later manifestation	1. Defibrinated blood at the hyperthermic stage 2. Lymphnodes	Suckling mouse and susceptible cattle	Intracerebral inoculation into suckling mouse and by symptoms
Malignant catarrhal fever	Gamma herpes virus Catlle, buffalo, farmed deer	Long incubation period. High fever, degenerative changes in the URT and alimentary tract, diffuse necrosis and ulcers in oral and nasal mucousa, lips, hard and soft palate, cheeks, erosions in the abomasum, fundic folds and pyloric region, sweeling of eyelids, opacity of cornea, enlargement of lymphnodes, encephalitis	1. Thyroid 2. Defibrinated blood (buffy coats for leucocytes) 3. Lymphnode, thymus, spleen, bone marrow, ocular and nasal secretions. 4. Pieces of tissues in 10% formal saline for HP	1. Calf thyroid or Calf testis cells 2. rabbit inoculation	1. clinical symptoms 2. virus isolation

Viral diseases	Etiology and species affected	Important symptoms	Specimens to be collected	Host system	Diagnosis
Cow pox and buffalo pox	Pox virus	Small red papules on teat and udder. In buffaloes the lesions may be seen around the ears, ear flap, eyes, neck region and sometimes throughout the body	1. The lesion scrapings and dried scabs in 50% glycerol saline	1. in calf Kidney cell culture 2. CE inoculation – CAM route	1. symptoms 2. virus isolation 3. Pock lesions on CAM 4. AGPT
Papillomas (Warts)	Papilloma virus Various species of animals	Thickened rough nodules, sometimes large peduncle cauliflower like growth usually on head, neck, shoulder, udder, etc.	1. Biopsy or the lesions on ice or in buffered glycerine	-	1. lesions 2. direct EM

Viral Diseases Of Sheep and Goat

Viral diseases	Etiology and species affected	Important symptoms	Specimens to be collected	Host system	Diagnosis
Peste des Petitis Ruminants	Paramyxo virus Sheep and Goat	Subacute form: nasal catarrh Mucosal erosion, intermittent diarrhoea Acute form: fever, oculonasal discharge, erosions in respiratory and alimentary tract	1. Blood in anticoagulant during febrile period 2. Lymphnodes and tonsils 3. Lungs, affected intestinal mucous membrane 4. Serum	Primary lamb kidney cells and VERO cells	From symptoms, detection of viral antigen by AGID, CIE, Cross – neutralisation test and MHI
Blue tongue	Orbi virus Sheep and cattle	High rise of temperature, copious salivation, occulonasal discharge, swelling and hyperemia of the mucousa of the mouth, ulceration of the tongue, dental pad and lips. Tongue swollen, cyanotic, difficult to be retracted. Swelling and tenderness of the coronary band	1. Blood in anticoagulants from febrile animals 2. Spleen and Lymphnodes from dead animals 3. Lesion scrappings 4. Swabs from eye and nasal discharges and faecal swabs 5. Paired sera samples from live animals	In CE by Yolk sac route or i/v route In bovine and ovine kidney and testes cell culture BHK21, HeLa cell lines In new born mice	From symptoms, Cheery red embryo in yolk sac/ i/v route of inoculation Isolation of the virus in tissue culture IFT & CFT AGPT ELISA

Viral diseases	Etiology and species affected	Important symptoms	Specimens to be collected	Host system	Diagnosis
Sheep Pox	Pox virus	High rise of temp., increased respiratory rate, swollen eye lids, dermal odema with marked raised circular thickened plaques with congested borders, lesions in wool free areas. Lesions in the oral, intestinal and respiratory tracts. Nodules in internal organs	1. Skin nodules and scab material preserved in buffered glycerol 2. Pieces of affected tissues in 10% formal saline for HP 3. Portions of lung showing lesions from dead sheep	In goat kidney and sheep testis fibroblast cell cultures	Symptoms, virus isolation in cell culture, EM, demonstration of cytoplasmic inclusions, AGPT, CIE, ELISA and VNT
Goat pox	Pox virus	Pock on mucous membranes and skin	1. Scrapings from cutaneous lesions in 50% glycerol saline	In Goat/ sheep kidney cells	Virus isolation Lesions , AGPT
Contagious pustular dermatitis (Orf)	Pox virus Lambs and kids	Pox lesions on the lips – proliferative type of lesions in sheep (lambs) and goats(kids)	1. Skin lesions and scab in 50% glycerol saline	In lamb kidneys and testes cell culture	Lesions, direct EM examination, virus isolation, AGPT
Rift valley fever	***Bunya virus Sheep, Goat, cattle and man	High rise of temp. severe icterus, abortion in pregnant animals. Off feed, pass bloody urine, diarrhoea, dyspnoea, buccal erosion, necrosis of skin over udder and scrotum. Highly fatal in young animals	1.Blood in febrile stage 2. Liver, placenta, foetus, brain, kidney, heart and spleen 3. Pieces of the above tissues in 10 % formal saline in HP	Yolk sac or CAM route of CE inoculation In monkey kidney cells/ VERO/ Mouse brain cell cultures	Symptoms, virus isolation, demonstration of acidophilic intranuclear inclusion bodies

Viral diseases	Etiology and species affected	Important symptoms	Specimens to be collected	Host system	Diagnosis
Visna/Maedi	Lenti virus Sheep and Goats	**Maedi**: Respiratory distress but no coughing or nasal discharge. Clinical courses lasts for months or more than a year. Not usually seen in animals below 3-4 years of age. **Visna:** Lameness of one or both hind legs progress to front legs. Slight tremor of head and occasional blindness	1. Lungs, spleen, lymphnode, and salivary glands 2. Buffy coat cells, mammary glands 3. Pooled sera samples	1. BHK –21 cells / choroid plexes cells	Symptoms, virus isolation and histopathology. Antibody detection by AGPT and ELISA

Viral Diseases of Swine

Viral diseases	Etiology and species affected	Important symptoms	Specimens to be collected	Host system	Diagnosis
Swine Fever (Hog Cholera)	Flavi virus Pigs	High rise of temperature, severe leukopenia, conjunctivitis, dyspnoea, vomition, diarrhoea, purple discoloration of the abdomen, ears and snout. Weakness of hind quarters, staggering gait. Sometimes death without any symptoms	1. Whole blood at the febrile phase 2. Nasal and eye swabs 3. Mesenteric lymphnodes, pancreas, spleen and tonsil from dead animals 4. Pieces of tissues in 10% formal saline for HP 5. Paired sera samples	In pig kidney, spleen, testes, lymphnode cell cultures and PK 15 cell lines	Symptoms, PM lesions: button ulcers in the intestine, turkey egg kidney Viral isolation IFT, ELISA and SNT

African swine fever	**Asfi virus Pigs	Similar to swine fever, severe leukopenia, oculonasal discharge, dyspnoea, abortion, vomiting and diarrhoea. Skin lesions and joint lesions in the form of swelling may be seen in chronic cases.	1. Whole blood, buffy coat cells 2. Lymphoid tissue 3. Nasal and eye swabs 4. Paired sera samples	In yolk sac of CE In pig, bovine and chicken kidney cell culture	Symptoms, PM lesions: wide spread haemorrhage in lymphnodes has to be differentiated from European swine fever. Haemadsorption of pig RBCs on to infected animals macrophages. CFT and ELISA
Pseudo rabies	***Herpes virus Pigs, cattle, dogs, foxes and rabbits	Coughing, pneumonia, in coordination, paralysis and convulsions. Pregnant animals may show SMEDI syndrome. In cattle – intense pruritis leading to self mutilation	1. Lung, tonsils, kidney, spleen or brain of pigs 2. Nasopharynx, lungs and vagina of affected cattle.	In RK cells, PK 15 cell line In CE – Cam route In rabbit- s/c route	Symptoms, virus isolation, FAT, VNT and ELISA test.

Viral Diseases of Horse

Viral diseases	Etiology and species affected	Important symptoms	Specimens to be collected	Host system	Diagnosis
Equine rhino pneumonitis (Equine abortion)	Alpha herpes virus Horses	High fever, respiratory catarrh and abortions in pregnant mares. Aborted fetus autolysed or fetus with petechia of mucous membranes, excessive pleural fluid, spleenic enlargement and necrosis of liver	1. Deep nasal swabs and vaginal swabs 2. Lungs, liver, spleen and thymus of aborted fetus 3. Portions of the above tissues in 10% formal saline for HP 4. Paired sera samples	In equine kidney cell culture and PK 15 cell line	Symptoms with history of abortions. Demonstration of intranuclear inclusions in respiratory epithelium, hepatic cells of aborted foetus CIE, FAT, VNT etc.
African horse sickness	Orbi virus Horses, mules and zebras	High fever, pulmonary form, cardia form, mixed form, atypical form. Mortality ranges between 25-90%. Odematous swelling of eye lids, lips, cheeks, tongue and chest	1. Blood during febrile stage 2. Lymphnodes, spleen and lungs 3. Paired sera samples	In vero cell line and In CEs In intracerebral inoculation of suckling mouse	Symptoms, virus isolation, VNT, CFT, HI and AGPT tests.
Equine infectious anemia	Lenti virus Horses, mules and asses	Prevalent during late summer and autumn. Sudden rise in temperature. Temp. rise intermittent. Progressive weakness, anaemia, petechial haemorrhage in visible mucous membrane. Odema in the subcutis of ventral wall; enlargement of lymphnodes	1. Blood during febrile stage 2. Paired serum samples	In primary horse leucocytes cell culture	History and symptoms. Virus isolation from the whole blood or buffy coat, transmission experiments, AGID (Coggin's test)

Viral Diseases of Dog

Viral diseases	Etiology and species affected	Important symptoms	Specimens to be collected	Host system	Diagnosis
Rabies	Rhabdo virus All warm blooded animals including birds	Rapidly fatal encephalitis following a somewhat lengthy incubation period. Excitement followed by paralysis; death in 3-6 days following the onset of clinical symptoms	1. If possible send the whole carcass 2. Impression smears prepared from the Hippocampus major, cerebellum and ammons horn fixed in absolute ethanol or 10% acetone 3. Portions of Hippocampus major, cerebrum and cerebellum in 2% potassium dichromate or 50% glycerol saline in separate containers or a portion of the brain preserved over ice. 4. Corneal impression smears and skin biopsy	In suckling mice-i/cranial In CE-yolk sac route Many cell lines including BHK-21	Based on symptoms Demonstration of Negri bodies (Seller's stain) in the impression smear taken from hippocampus The mice will die within 2-3 weeks exhibiting paralytic symptoms (I/ cranial inoculation) FAT, RREID, PCR
Canine distemper	Paramyxo virus Dogs	Biphasic temperature. Bronchopneumonia, vomiting and diarrhoea, vesicular and pustular dermatitis particularly at the under aspect of the abdomen	1. Blood at febrile phase 2. Nasal or conjunctival secretions/ epithelium 3. Pustules or fluid from affected animals 4. Corneal impression smears and corneal swabs	In chicken embryo fibroblast cell culture, dog and ferret kidney cell culture In ferrets and pups	Symptoms, isolation of the virus, demonstration of inclusion bodies, neutralization test in chicken embryos, FAT, CFT and ELISA

Viral diseases	Etiology and species affected	Important symptoms	Specimens to be collected	Host system	Diagnosis
			5. Spleen, lung, stomach, intestine and urinary bladder from dead animals 6. Paired sera samples 7. Pieces of brain and lung in 10% formal saline		
Parvo viral diarrhoea (Canine parvo virus)	Dogs Parvo virus	Vomiting, Haemorrhagic diarrhoea/ myocarditis. Mild or subclinical in adults	1. Fresh faecal samples, 2. Mesenteric lymphnodes, intestinal scrapings, spleen, and liver 3. Paired serum samples	In canine foetal kidney, testes and lymphnode primary cell culture	Symptoms, HA (swine RBC), HI, IEM, FAT and ELISA tests.
Infectious canine hepatitis (Rubarth's Disease)	Dogs and Foxes Adeno virus	High fever, vomition, acute abdominal pain, mucopurulent oculonasal discharge, corneal opacity, blue eye in dogs. In foxes acute encephalitis- convulsions followed by paralysis coma and death.	1. Urine preferably from bladder without preservative. 2. Liver, spleen and lymphnodes 3. Pieces of the above tissues in 10% formal saline 4. Paired serum samples	In canine kidney cell culture	Symptoms, demonstration of intranuclear inclusions in hepatic cells, virus isolation, FAT and CFT

Viral Diseases of Poultry

Viral diseases	Etiology and species affected	Important symptoms	Specimens to be collected	Host system	Diagnosis
New castle disease (Ranikhet disease)			1. Spleen, brain, lungs, liver, trachea, proventriculus preserved over 50% glycerol saline or over ice 2. Paired sera samples	Cultivation in chicken embryos by allantoic route	Inoculation into 9-10 days old embryos by i/allantoic route. Embryos die after 24 hours depending on virulence with occipital haemorrhages and haemorrhages over the extremities HA, HI, SNT and ELISA
Avian infectious bronchitis	Corona virus Chicken	Rapid spread of coughing, gasping and rales among birds of all ages, particularly younger groups. There may be watery discharge, lacrymation and facial swelling, severe drop in egg production, deterioration in the quality of eggs- small size, misshape, haemorrhages in the albumen or yolk.	1. Lungs and tracheal scrappings 2. Tracheal swabs, trachea, bronchi, lungs, caecal tonsils, kidney 3. Pieces of above tissues in 10% formal saline for HP 4. Paired serum samples	In chicken embryos by allantoic route In primary chicken cell culture and tracheal organ cultures	Symptoms, virus isolation by CE inoculation –dwarfing of the embryo, VNT, FAT and HI

Viral diseases	Etiology and species affected	Important symptoms	Specimens to be collected	Host system	Diagnosis
Avian leucosis (Big liver disease)	Retro virus chicken	Develop only in chicks above 16-20 weeks of age, absence of nervous, skin, eye or muscular form of the disease. Uniform enlargement of the bursa with nodular lesions due to intra follicular infiltration of tumour cells- helpful to differentiate from MD	1. Affected portion of eyes, liver, spleen, kidney, gonads, bursa of fabricius, thymus, bone marrow and any other tumor tissue over ice and also portions of these tissues in buffered formalin for HP 2. Paired sera samples	In chicken embryo fibroblast cell culture	Virus isolation, HP, COFAL, Indirect FAT and ELISA
Chicken anemia	Circo virus chicken	Anemia, aplasia of bone marrow, lymphoid depletion, muscle haemorrhage and atrophy of thymus and bursa, enlarged liver.	1. Buffy coat cells 2. Whole blood for PCV 3. Thymus, bone marrow, liver and spleen 4. Above tissues in 10% formalin for HP	In susceptible day old chicks In MDCC-MSB1 cells	Virus isolation, blood picture: severe anemia and haematocrit value below 27%, lesions on thymus and bone marrow, FAT, ELISA and PCR.

Specimens to be Collected and Tests Conducted in Diagnosing Certain Important Microbes Bacterial Diseases of Animals and Birds

Name of the Disease	Brief description	Material to be collected and preservative if any	Brief description of Methods of examination.
Actinobacillosis	*Actinobacillosis lignieresii.* A disease caused by gram negative bacteria extremely pleomorphic (coccoid to filamentous) forms. Affects soft tissues causing abscess in cattle. Mammary abscesses in cow.	Pus from the abscesses. Portion of tongue preserved in ice. Sterile swabs from the lesions.	1. Microscopical: Pus smears made and crushed between slides, stained by Gram's stain; Gram negative bacilli surrounded by long radially disposed clubs. 2. Straus reaction: Inoculation into male Guinea pig produces orchitis. 3. Cultural examination from affected tissues and lymph nodes-serum or blood agar.
Actinomycosis	*Actinomyces bovis*. Gram positive, non acid-fast branching filamentous organism in artificial media. In infected tissues forms typical ray fungus granules. Affect hard tissues. Lesions seen commonly in mandible and tongue. Affects mostly cattle and pig. Infection in horse, dog and cat seen occasionally.	Fresh portion of affected tissues or scraping from abscesses preserved in ice.	1. Microscopical: Pus granules crushed between slides. Stained by Gram stain, Gram-positive filaments arranged like mycelia surrounded by Gram-negative club shaped bodies. 2. Cultural examination: Affected portion (pus) in serum or blood agar.
Anthrax	All domestic animals are susceptible to this highly septicemic disease seen mostly among cattle, sheep and pig. In man it causes malignant pustule or wool sorter's disease, caused by a Gram-positive *Bacillus anthracis,* a capsulated rod, non motile; sporulates outside the body. Sudden death bleeding from natural orifices. Animals are affected in lots. Postmortem examination should not be conducted. because Anthrax spores will be	1. Peripheral blood smear. 2. Swabs from fluid exuded from natural orifices. 3. Ear piece, muzzle - two or three pieces.	I. Microscopical: Blood smear stained by Leishman's method will reveal characteristic rod shaped organisms with truncated ends, possessing a distinct capsule and arranged singly or in very short chains. Blood smear stained by Methylene Blue will evince the Mc Fadyean methylene blue reaction.

Name of the Disease	Brief description	Material to be collected and preservative if any	Brief description of Methods of examination.
	distributed all overt the areas and Survive for long time. Carcasses should be incinerated or buried six feet deep with lime covering on all slide to a depth of one foot.		II. Cultural: 1. Cotton wool growth in broth clear supernatant. 2. Medusa head colonies on agar. 3. Inverted fir free growth in gelatin. III. Biological: Guinea pigs and mice are use. On i/p inoculation animal dies in 48 to 72 hours showing characteristic lesions. Blood smear reveals characteristic organism. IV. Serological: Ascolis thermoprecipitation test
Black quarter (B.Q.)	Seen commonly occurring as an endemic disease among cattle and buffaloes. Prevalent widely in Tamil Nadu. Caused by Gram-positive spore bearing anaerobe *Clostridium septicum.* Characterized by prevalence in animals over six months age. Affected animals evince swelling of hindquarter; sometimes forequarters also are affected resulting in lameness, which culminates in death of the animals. Few animals survive the infection.	1. Muscle impression smear. 2. Muscle piece -air dried.	1. Microscopical: Muscle impression smear stained by Claudius method reveals rods with oval sub terminal spores. 2. Biological: Intramuscular inoculation into a Guinea pig produces characteristic gangrenous necrosis and death. Extensive hemorrhagic oedema and emphysema with s/c hemorrhage is seen. Injection of a drop of 5% calcium chloride before inoculation accelerates the reaction. Clostridium chauvoei and Clostridium septicum can be differentiated by cultural characters and biological test.

Name of the Disease	Brief description	Material to be collected and preservative if any	Brief description of Methods of examination.
Brucellosis	Three organisms are of importance. *Brucellaabortus occurs* commonly in cattle, *Br. melitensis* occurs commonly in goats, and *Br. suis* occurs commonly in pigs. They cause abortion in pregnant animals by producing suppurative placentitis. In addition it may cause mastitis.	1. Serum preserved in 5% Chloroform. 2. Milk from affected quarter. 3. Smears from vaginal exudate and fetal stomach contents. 4. Swabs from vaginal exudate or fetal stomach contents, fetal cotyledons and uterine discharge.	1.Microscopical: Smears stained by Gram's staining reveal Gram negative coccobacilli. 2. Cultural: Tryptose agar, Dextrose agar or Liver infusion agar can be used. Increased tension of carbon dioxide favours the growth of Brucella abortus. 3.Biological: Intraperitoneal inoculation into a male guinea pig produces orchitis-Straus' test. 4. Serological: (a) Rapid plate test for screening (b) Tube agglutination test for confirmation (c) ABR test or MRT test with milk.
Bovine Lymphangitis (B.L.)	Caused by *Pasteurella pseudotuberculosis redentium type III.*	1. Pus smear 2. Pus swab.	1.Pus smear or Gram's staining reveal Gram negative rods. 2.Isolation and identification of the organism. 3.Straus's test.
Caseous Lymphadenits	Caused by *Corynebacterium pseudotuberculosis* and occurs in sheep characterized by the presence of causation necrosis in the lymph glands. The lesions are frequently confined to superficial lymph glands but generalized cases of the disease are not uncommon. In horse ulcerative lymphangitis is produced.	Swabs from abscesses.	Isolation and identification of the organism on sheep blood agar- haemolysis after 72 hours incubation.

Name of the Disease	Brief description	Material to be collected and preservative if any	Brief description of Methods of examination.
Calf Diphtheria.	Necrosis in the mouth, larynx and trachea of calves produced by *Fusobacterium necroporum* in liver abscesses are seen in cattle of all ages. In sheep 'Foot rot' is produced by this organism.	Scab from the lesions.	Mere isolation and identification of this non-sporulating Grams negative anaerobe do not confirm the etiological aspect. Biological test in rabbits is always necessary as this organism appears as secondary invader in most of the cases.
Calf Dysentery or White scours or Colibacillosis	Prevalent more commonly on colostrum deprived calves. Produced by certain pathogenic strains of *Escherichia coli.* This organism has been found to be pathogenic to most of the animals including man.	Rectal swabs.	Isolation of the organism in MacConkey agar- pink colour colonies are produced.
Calf Pneumonia	Typical pleuropneumonia is seen in calves, caused by *Pasteurella multocida.* The role of *Pasteurella hemolytica* in sheep, pneumonia had been pointed out. Its role in calves is yet to be assessed.	Swabs from pleura and lungs.	Isolation and identification of the organism on ox or sheep blood agar and Macconkey agar
Enterotoxaemia.	A toxic disease due to *Clostridium welchi Type* D, affecting lambs and sheep characterized by convulsion and sudden death.	Intestinal contents or a loop of intestine tied at both ends and preserved in 0.5% Chloroform.	Biological test-in mouse by i/v inoculation results in death with in few minutes Actual toxin can be determined by neutralization with specific toxins.
Glanders	Primarily seen in horses and mules. The disease in commonly classified into three types Pulmonary glanders, Nasal glanders and cutaneous glanders (Farey) caused by *Pseudomonas mallei.*		Biological test: Mallein test in living animals. Straus test positive. Cultural methods do not yield good results.

Name of the Disease	Brief description	Material to be collected and preservative if any	Brief description of Methods of examination.
Haemorrhagic Septicemia. (H.S.)	A disease characterized by septicemia affecting in various animals including cattle, buffaloes, sheep and goats caused by *Pasteurella multocida*. In fowls the disease is known as Fowl cholera.	1.Auricular vein smear. 2.Oedema fluid smear. 3.Heart blood smear. 4.Heart blood swabs. 5.Swabs from internal organs. 6.Long bone preserved in charcoal in case of extreme putrefaction.	1.Microscopical: Presence of bipolar organisms in blood smears stained by leishmans stain. 2.Cultural: Isolation and identification of the organisms from heart blood, internal organs and bone marrow. 3.Biological: in mouse and rabbits by i/p route.
Johne's disease (Paratuberculosis)	Chronic wasting disease with diarrhea caused by an acid-fast organism – *Mycobacterium paratuberculosis.*	Rectal pinch smear and Rectal washings.	1.Smears stained by Ziehl Nelson's method reveals acid-fast bacteria. 2.Cultural examination in Dorset's test in living animals. 3. Johnin test in living animal.
Leptospirosis	*Leptospiraicterohaemorrhagiae* and *L. canicola* occur in dogs and produce jaundice. *L.pomona* has been incriminated in bovines. Horses & swine are also affected.	Serum, urine	1.Microscopical: Dark field microscopical examination of serum and urine. Microscopical Agglutination test (MAT), slide or plate agglutination.
Listeriosis	*Listeria monocytogenes.* Encountered in sheep, cattle chickens, swine, wild animals horses and man. In sheep, ox and pig lesions are confined to the CNS. In bovines abortions also result. In chickens myocardial necrosis.	Brain, nasal content and heart blood of the fetus preserved on ice.	Isolation and identification of the organisms- small pleomorphic Gram positive rod, motile, beta hemolytic. Isolation augmented by prior chilling in ice.
Mastitis	*Str. agalactiae, Str. dysgalactiae, Str. uberis, Staph. aureus C. pyogenes.*	1. Smears from milk sediment. 2. Milk preserved in boric acid.	Microscopical examination of the smears stained by grams stain Cultural examination and isolation and identification of the organism.

Name of the Disease	Brief description	Material to be collected and preservative if any	Brief description of Methods of examination.
Pullorum disease & fowl typhoid	Pullorum disease is seen most commonly in baby chicks and called Bacillary white diarrhoea caused by *Salmonella enteritidis* Pullorum Fowl typhoid is seen more commonly in adult birds and caused by *Salmonella enteritidis* Gallinarum.	1. Cloacal swab. 2. Ailing chicks. 3. Serum from birds. 1. Piece of intestine tied at both. 2. Intestinal swab.	Isolation and identification of the organism. Agglutination test with Salmonella pullorum Colored Antigen (SPCA). Isolation and identification of the organism.
Pyaemia	Pyaemia means the presence of pus producing (pyogenic) organisms in the blood. This results in abscess formation. Several microorganisms caused this condition.	Pus smears, Pus swab.	Microscopical examination after staining with Grams and another with acid-fast method of staining. Isolation and identification of the organisms for confirmation.
Salmonellosis	Salmonellosis affects animals and man and cause variety of diseases. In horses abortion in lab, animals enteritis; in man typhoid; in sheep abortion etc., Depending upon the disease conditions swab should be collected.	Blood, milk, faeces, aborted fetuses, foetal membranes uterine discharges, heart blood and visceral organs.	Isolation and identification of the organism. Plate agglutination test and tube agglutination test
Strangles	Occurs in young horses. Characterized by catarrhal discharge with inflammation of the nasal mucous membranes followed quickly by a swelling of the adjacent lymphatic glands. Infection spreads through lymph channels. Fatal termination is rare but sometimes occurs due to septicemia, pyaemia and pneumonia. One of the causes is *Streptococcus equi*.	Swabs from abscess.	The isolation and identification of Streptococcus equi.
Streptococcal Infections	*Streptococcal agalactiae, Streptococcus dysgalactiae* and *Streptococcus uberis*. These three are called as mastitis streptococci. *Streptococcus zooepidermicus* produces severe infection often ending in septicemia and death.	Swabs and smears from lesions, milk smears and milk in case of mastitis.	Isolation and identification of the organism on Edward's medium

Name of the Disease	Brief description	Material to be collected and preservative if any	Brief description of Methods of examination.
Staphylococcal infections	*Staphylococcus aureus* is associated with suppurative infections and wound infections in man and animals. It is also responsible for mastitis.	Swabs and smears from lesions, milk smears and milk in case of mastitis	Isolation and identification of the organism in mannitol salt agar.
Swine Erysipelas	Characterized by chronic arthritis Acute form is characterized by sudden onset, rapid and high mortality. Urticarial form - Diamond skin lesions on the skin.	1. Liver, spleen, kidney, heart and synovial fluid 2. Serum	Isolation and identification of the organisms on ox or sheep blood agar.
Tetanus	Produced by the toxins of *Clostridium tetani.* (Tetanospasmin and tetanolysin). Horse, Pig, mouse, monkey, rabbit, sheep, cattle and goats are susceptible in that order Cat, dogs, birds and cold blooded animals are usually resistant.	1. Smar from the wound 2. Serum.	Necrotic tissue from a wound or wound exudate can be heated to 80oCfor 20 minutes and used to inoculate a blood agar plate. A tube of thioglycollate medium or cooked meat broth could also be inoculated and subcultured onto blood agar after 2-3 days of incubation. The blood agar plates are incubated at 37oC for 3-4 days under an atmosphere of H2 and CO2.
Tuberculosis	Produced by *Mycobacterium tuberculosis* (human type) *Mycobacterium bovis* (bovine type) and Mycobacterium avium (avian type) Practically all mammals and birds are affected. The disease is essentially chronic and the attack is faced by the lymph nodes. The lesions range from abscess formation to caeseation, necrosis and calcification. A highly wasting disease resulting in gradual loss of weight.	1. Smear from lesions. 2. Swabs from lesions. 3. Lesion preserved in ice.	1.Microscopical examination after the acid fast staining. 2.Cultural examination, Culture develop in 2 to 3 weeks- bundle of faggots. 3.Biological test in guinea pigs, rabbits and fowl depending upon the type. Disease is reproduced in 4 to 8 weeks. 4.In living animals, allergic “Tuberculin” Test.